PURE MATHE

A First Course

By J. K. Backhouse, S. P. T. Houldsworth and B. E. D. Cooper
PURE MATHEMATICS: A Second Course

PURE
MATHEMATICS
A First Course
SI Edition

BY

J. K. BACKHOUSE, M.A.

*Tutor at the Department of Educational Studies
in the University of Oxford
Formerly Head of the Mathematics Department
at Hampton Grammar School*

AND

S. P. T. HOULDSWORTH, M.A.

*Lately Headmaster of Sydney Grammar School
Formerly Assistant Master at Harrow School*

LONGMAN

LONGMAN GROUP LIMITED
Longman House
Burnt Mill, Harlow, Essex, U.K.

First Published 1957
Second Edition 1965
SI Edition   1971
Tenth impression 1981

ISBN 0 582 31797 5

Printed in Singapore by
Kyodo Shing Loong Printing Industries Pte Ltd

PREFACE

We have aimed at providing a year's course in pure mathematics for those who have completed the usual syllabuses in elementary mathematics. The reader should find himself well equipped to sit for the various papers in Additional Mathematics and Alternative Ordinary Mathematics set by the examining boards in this country, and may carry his reading beyond these requirements.

The reader is warned that the work should not be taken in the order in which the chapters are arranged, but that the branches should be developed simultaneously. We have tried to present the contents in such a way that it should be easy to find any particular topic. Chapter 1 introduces co-ordinates and the straight line. Thereafter, the arrangement is:

Chapters 2–7—Calculus;
Chapters 8–11—Algebra;
Chapters 12–16—Trigonometry, including the derivatives of trigonometrical functions;
Chapters 17–19—Coordinate Geometry.

Thus it is hoped that each teacher will be able to select without difficulty the course of his own choice from this book. Some of the chapters are too long to be worked straight through and should be broken up into two or more parts.

The individual reader has been kept in mind and he is advised to work through the questions in the text marked "Qu." as he reads, and the class teacher will find that many of these are suitable for oral work.

We are indebted to the Mathematical Association whose Reports have been frequently consulted during the writing of the book, and our thanks are due to the Northern Universities Joint Matriculation Board, the Oxford and Cambridge Schools Examination Board and the University of Cambridge Local Examinations Syndicate (whose questions are marked "C.") for permission to include questions from examination papers.

Finally, we wish to express our thanks to Mr. J. B. Morgan, Mr. B. E. D. Cooper and others for their advice and assistance, and to all those at Longman Group Ltd. who have helped in the production of this book.

J. K. B.

Harrow-on-the-Hill.　　　　　　　S. P. T. H.
January 1956.

PREFACE TO THE THIRD EDITION

Pure Mathematics—A First Course and A Second Course have both been modified to employ SI units (Système International d'Unités), the internationally agreed system of units.

For the most part this has involved simply a change of units but on occasions data has had to be altered. In a very few cases completely new questions have been substituted, and there have also been minor amendments to the text. However, apart from the units, the changes are of a minor nature and this edition is compatible with the previous ones.

J. K. B.

Oxford.　　　　　　　S. P. T. H.
October 1970.

Note.—The printing of this book has followed the recommendations of British Standard 1991: 1954, Letter Symbols, Signs and Abbreviations, modified to conform with the Système International d'Unités.

CONTENTS

CHAPTER 5

INTEGRATION

CHAPTER 6

FURTHER DIFFERENTIATION

NOTE: the derivatives of trigonometrical functions will be found in Chapter 16.

CHAPTER 7

INTEGRATION BY SUMMATION

CHAPTER 8

SOME USEFUL TOPICS IN ALGEBRA

CHAPTER 9

PERMUTATIONS AND COMBINATIONS

CHAPTER 10

SERIES

CHAPTER 11

THE BINOMIAL THEOREM

CHAPTER 12

THE GENERAL ANGLE AND PYTHAGORAS' THEOREM

CHAPTER 13

COMPOUND ANGLES

CHAPTER 19

FURTHER TOPICS IN COORDINATE GEOMETRY

CHAPTER 20

VARIATION AND EXPERIMENTAL LAWS

CHAPTER 1

COORDINATES AND THE STRAIGHT LINE

Coordinates

1.1 The first thing that a reader new to this stage of mathematics will discover is that number, and the methods of algebra, may be brought to bear upon geometrical ideas to a much greater extent than before, and with great clarity and economy. To do this we must have a way of describing exactly and briefly the position of a point in a plane (i.e. a flat surface).

We may think for a moment of the pirate of old, who buried his treasure chest on a large flat featureless island, but was able to locate it when he returned. Starting at the most westerly point, he measured 100 paces due East, and then from there 100 paces due North. There, he knew, was the exact spot at which to dig.

This illustrates the method we shall use to fix the position of a point on a plane. Two straight lines cutting at right angles fix our directions, and we start our measurement from their point of intersection O. (Fig. 1.1.)

The point O is called the **origin**. The **x-axis** is drawn across the page, and the **y-axis** is drawn up the page; units of distance are marked off on them, positive in one direction, negative in the other.

Consider the point A. To reach A from O we travel 4 units in the direction of Ox, and then 1 unit in the direction of Oy.

The **x-coordinate** (or *abscissa*) of A is $+ 4$.
The **y-coordinate** (or *ordinate*) of A is $+ 1$.

We say that the **coordinates of** A are (4, 1), or that A is the point (4, 1). The x-coordinate is always given first, thus we distinguish between the points A(4, 1) and B(1, 4). By use of the sign of the coordinates we distinguish between the points A(4, 1) and C($-$ 4, $-$ 1).

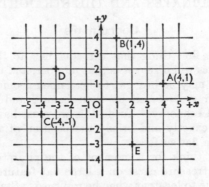

Fig. 1.1

Qu. 1. Write down the coordinates of the points D, E, O in Fig. 1.1. $D = (-3, 2)$ $E = (2, -3)$ $O = (0,0)$

Qu. 2. Sketch your own axes and plot the points P(2, 4), Q($-$ 5, 7), R(4, $-$ 2), S(0, 3), T(2, 0).

The length of a straight line

1.2. Example 1. *Find the length of the straight line joining* A(2, 1) *and* B(5, 5).

AC and CB are drawn parallel to the x-axis and y-axis respectively (Fig. 1.2). Applying Pythagoras' theorem to the right-angled triangle ABC,

$$AB^2 = AC^2 + CB^2$$
$$= (5 - 2)^2 + (5 - 1)^2,$$
$$= 9 + 16.$$

$$\therefore \ AB = \sqrt{25},$$
$$\therefore \ AB = 5.$$

Notice that, if A had been the point $(-2, 1)$ in the above example, the length of AC would still be the *difference* between the x-coordinates of A and B, since it would be $5 - (-2) = 5 + 2 = 7$.

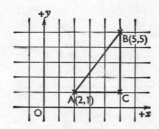

Fig. 1.2

Qu. 3. Find the lengths of the straight lines joining the following pairs of points:

(i) A(3, 2) and B(8, 14),

(ii) C(−1, 3) and D(4, 7),

(iii) E(p, q) and F(r, s).

The mid-point of a straight line

1.3. Example 2. *Find the mid-point of the straight line joining* A(2, 1) *and* D(6, 5).

Let M, the mid-point of AD, have coordinates (p, q). FM and ED are drawn parallel to Oy; AFE is drawn parallel to Ox. (Fig. 1.3.)

In the $\triangle$ADE, applying the mid-point theorem, since M is the mid-point of AD, and MF is parallel to DE, F is the mid-point of AE.

Thus $\hspace{4em}$ AF = FE.

$$\therefore p - 2 = 6 - p,$$

$$\therefore p = \frac{6 + 2}{2};$$

$$\therefore p = 4.$$

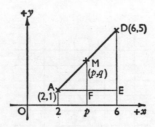

Fig. 1.3

The x-coordinate of M is seen to be the average of those of A and D. The y-coordinate of M may be found similarly.

$$q = \frac{5 + 1}{2},$$

$$\therefore q = 3.$$

$\therefore$ the mid-point of AD is (4, 3).

In practice, of course, the working would be presented in shortened form thus:

the mid-point of AD is $\left(\dfrac{6 + 2}{2}, \dfrac{5 + 1}{2}\right)$, i.e. (4, 3).

Qu. 4. Find the coordinates of the mid-points of the straight lines joining the following pairs of points:

(i) A(4, 2) and B(6, 10), $\hspace{2em}$ (ii) C(− 5, 6) and D(3, 2),
(iii) E(− 6, − 1) and F(3, − 4), $\hspace{1em}$ (iv) G(p, q) and H(r, s).

Exercise 1a

1. Find the lengths of the straight lines joining the following pairs of points:

   (i) A(1, 2) and B(5, 2),   (ii) C(3, 4) and D(7, 1),

  (iii) E(− 2, 3) and F(4, 3),   (iv) G(− 6, 1) and H(6, 6),

   (v) J(− 4, − 2) and K(3, − 7),

  (vi) L(− 2, − 4) and M(− 10, − 10).

2. Find the coordinates of the mid-points of the lines AB, CD, etc., in No. 1.

3. Find the distance of the point (− 15, 8) from the origin.

4. P, Q, R are the points (5, − 3), (− 6, 1), (1, 8) respectively. Show that △PQR is isosceles, and find the coordinates of the mid-point of the base.

5. Repeat No. 4 for the points L(4, 4), M(− 4, 1), N(1, − 4).

6. A and B are the points (− 1, − 6) and (5, − 8) respectively. Which of the following points lie on the perpendicular bisector of AB?

   (i) P(3, − 4),   (ii) Q(4, 0),   (iii) R(5, 2),   (iv) S(6, 5).

7. Three of the following four points lie on a circle centre the origin. Which are they, and what is the radius of the circle?

       A(− 1,7), B(5, − 5), C(− 7, 5), D(7, − 1).

8. A and B are the points (12, 0) and (0, − 5) respectively. Find the length of AB, and the length of the median, through the origin O, of the triangle OAB.

The gradient of a straight line

1.4. Consider the straight line passing through A(1, 1) and B(7, 2). (Fig. 1.4.) If we think of the x-axis as horizontal, and the line through A and B as a road, then someone walking from A to B would rise a vertical distance CB whilst at the same time he moves a horizontal distance AC.

The gradient of the road is $\dfrac{CB}{AC} = \dfrac{2 - 1}{7 - 1} = \dfrac{1}{6}$. Instead of the two points A and B we might just as well have taken any other two points on the line, D and E; the gradient would then be expressed as $\frac{FE}{DF}$, which is the same as $\frac{CB}{AC}$, since the triangles ABC and DEF are similar.

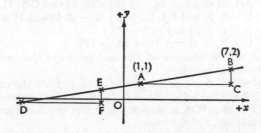

Fig. 1.4

DEFINITION

The gradient of a straight line is $\dfrac{the\ increase\ in\ y}{the\ increase\ in\ x}$ in moving from one point on the line to another. In moving from A to B, since both x and y increase by positive amounts, the gradient is positive.

But now consider the gradient of PQ. (Fig. 1.5.) In moving from P to Q, the increase in x is $+2$, but since y

Fig. 1.5

decreases, we may say the increase in y is -3. Thus the gradient of PQ is $-\frac{3}{2}$.

Until the reader is accustomed to the idea of positive and negative gradient it may help to think of it this way. In travelling along a line with x increasing (i.e. moving from left to right across the page) if going UPhill the gradient is POSITIVE: whereas if going DOWNhill the gradient is NEGATIVE. In calculating gradients a figure should not be necessary, but one similar to Fig. 1.5 will help in the first few examples.

Example 3. *Find the gradient of the line joining* R(4, 8) *and* S(5, − 2).

$$\text{The gradient of RS} = \frac{y \text{ coord. of R} - y \text{ coord. of S}}{x \text{ coord. of R} - x \text{ coord. of S}},$$

$$= \frac{8 - (-2)}{4 - 5},$$

$$= \frac{10}{-1},$$

$$= -10.$$

[Remember that the coordinates of R must appear first in the denominator *and* numerator (or second in both). In this case $\dfrac{8 - (-2)}{4 - 5}$ and $\dfrac{-2 - 8}{5 - 4}$ both give the correct gradient.]

Qu. 5. Find the gradients of the lines joining the following pairs of points:

(i) (4, 3) and (8, 12),   (ii) (− 2, − 3) and (4, 6),

(iii) (5, 6) and (10, 2),   (iv) (− 3, 4) and (8, − 6),

(v) (− 5, 3) and (2, 3),   (vi) (2, 1) and (2, 9),

(vii) (p, q) and (r, s),   (viii) (0, a) and (a, 0),

(ix) (0, 0) and (a, b).

Qu. 6. A and B are the points (3, 4) and (7, 1) respectively. Use Pythagoras' theorem to prove that OA is perpendicular to AB. Calculate the gradients of OA and AB, and find their product.

Qu. 7. Repeat Qu. 6 for the points A(5, 12) and B(17, 7).

Parallel and perpendicular lines

1.5. The gradient of a straight line was defined in §1.4; it may be proved that it is also the tangent of the angle between the line and the positive direction of the x-axis.

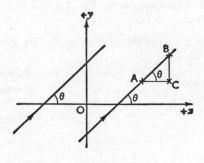

Fig. 1.6

In Fig. 1.6 the gradient of AB is $\dfrac{CB}{AC}$, which is tan θ. The reader familiar with the tangent of an obtuse angle will appreciate that this covers negative gradient as well.

Since parallel lines make equal corresponding angles with the x-axis, *parallel lines have equal gradients*.

Qu. 6 and 7 of §1.4 will have led the reader to discover a useful property of the gradients of perpendicular lines. This we will now prove.

Consider the two straight lines AB and CD which cut at right angles at E.   EF is drawn perpendicular to the x-axis. (Fig. 1.7.)

$$a + \theta = 90°.$$
$$a + \beta = 90°.$$
$$\therefore \theta = \beta.$$

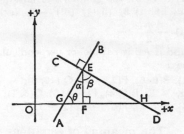

Fig. 1.7

Let the gradient of AB be m, then

$$m = \frac{FE}{GF} = \tan \theta.$$

The gradient of CD $= -\dfrac{FE}{FH}$,

$$= -\frac{1}{\tan \beta},$$

$$= -\frac{1}{\tan \theta},$$

$$= -\frac{1}{m}.$$

∴ the gradient of AB × the gradient of CD

$$= m \times \left(-\frac{1}{m}\right) = -1.$$

In general, *if two lines are perpendicular, the product of their gradients is* -1. Or in other words, if the gradient of one is m, the gradient of the other is $-\dfrac{1}{m}$.

Qu. 8. Write down the gradients of lines perpendicular to lines of gradient (i) 3, (ii) $\frac{1}{4}$, (iii) -6, (iv) $-\frac{2}{3}$, (v) $2m$, (vi) $-\frac{b}{a}$, (vii) $-\frac{m}{2}$.

Qu. 9. Find if AB is parallel or perpendicular to PQ in the following cases:

(i) A(1, 4), B(6, 6), P(2, $-$ 1), Q (12, 3);
(ii) A($-$ 1, $-$ 1), B(0, 4), P($-$ 4, 3), Q(6, 1);
(iii) A(0, 3), B(7, 2), P(6, $-$ 1), Q($-$ 1, $-$ 2).

The meaning of equations

1.6. The bare statement "P is the point (x, y)" means that P can be anywhere in the plane. Previously, if we have been asked to find P, we have been given some data which enabled us to find one pair of numerical values for x and y, and so to fix the position of P.

Suppose however that the data is in the form of the equation $y = x^2 - 2x$. This does not give one pair of values for x and y, it gives as many as we like to find. But P is not now free to be anywhere in the plane, since for any chosen value of x there is only one corresponding value of y; P is now restricted to positions whose coordinates (x, y) satisfy this relationship $y = x^2 - 2x$.

The reader will be familiar with the process of making a table of values as shown below, in which certain suitable values of x are chosen, and the corresponding values of y calculated.

Table of values for $y = x^2 - 2x$

x	-1	$-\frac{1}{2}$	0	$\frac{1}{2}$	1	$\frac{3}{2}$	2	$\frac{5}{2}$	3
x^2	1	$\frac{1}{4}$	0	$\frac{1}{4}$	1	$2\frac{1}{4}$	4	$6\frac{1}{4}$	9
$-2x$	2	1	0	-1	-2	-3	-4	-5	-6
y	3	$1\frac{1}{4}$	0	$-\frac{3}{4}$	-1	$-\frac{3}{4}$	0	$1\frac{1}{4}$	3

From this we find that the points $(-1, 3)$, $(-\frac{1}{2}, 1\frac{1}{4})$, $(0, 0)$, etc., have coordinates which satisfy the relationship

$y = x^2 - 2x$, and by plotting these points and drawing a smooth curve through them, we obtain all the possible positions of P within the range of values we have chosen for x. (Fig. 1.8).

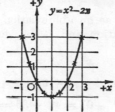

Fig. 1.8

Just as coordinates are used to name a point, so an equation is used to name a curve, and we refer to "the curve $y = x^2 - 2x$".

It must be stressed that the equation is the condition that the point (x, y) should lie on the curve.

Thus, only if $b = a^2 - 2a$ does the point (a, b) lie on the curve $y = x^2 - 2x$, and in that case we say that the co-ordinates of the point *satisfy* the equation.

If $q \neq p^2 - 2p$, the point (p, q) does *not* lie on the curve $y = x^2 - 2x$.

Example 4. *Do the points* $(-3, 9)$ *and* $(14, 186)$ *lie on the curve* $y = x^2$?

(i) The point $(-3, 9)$:

When $x = -3$, $y = x^2 = (-3)^2 = +9$,

$\therefore$ $(-3, 9)$ does lie on the curve $y = x^2$.

(ii) The point $(14, 186)$:

When $x = 14$, $y = x^2 = 14^2 = 196$.

$\therefore$ $(14, 186)$ does *not* lie on the curve $y = x^2$.

The next example illustrates another way of presenting this idea.

Example 5. *Does the point* $(-7, 6)$ *lie on the curve* $x^2 - y^2 = 14$?

[We use L.H.S. as an abbreviation for "The left-hand side" of the equation and R.H.S. for "The right-hand side".]

$$x^2 - y^2 = 14.$$

When $x = -7$ and $y = +6$,

L.H.S. $= (-7)^2 - 6^2 = 49 - 36 = 13.$

R.H.S. $= 14.$

The coordinates of the point do not satisfy the equation
$\therefore$ $(-7, 6)$ does *not* lie on the curve $x^2 - y^2 = 14.$

Qu. 10. Find the y-coordinates of the points on the curve $y = 2x^2 - x - 1$ for which $x = 2, -3, 0.$

Qu. 11. Find the x-coordinates of the points on the curve $y = 2x + 3$ for which the y-coordinates are $7, 3, -2.$

Qu. 12. Find the points at which the curve in Qu. 10 cuts (i) the x-axis, and (ii) the y-axis.

Qu. 13. Determine whether the following points lie on the given curves:

  (i) $y = 6x + 7, (1, 13),$ (ii) $y = 2x + 2, (13, 30),$

(iii) $3x + 4y = 1, (-1, \frac{1}{2}),$ (iv) $y = x^3 - 6, (2, -2),$

  (v) $xy = 36, (-9, -4),$ (vi) $x^2 + y^2 = 25, (3, -4).$

The relationship between a curve and its equation gives rise to two main groups of problems.

Firstly there are those problems in which we are given the equation, and from it we are required to find the curve. With this type the reader will already be familiar, in such work as the graphical solution of quadratic and other equations.

Secondly there are those problems in which we are given

some purely geometrical facts about the curve, and from these we are required to discover the equation. It is this second type of problem with which we are now mainly concerned, but first we shall discuss a few more simple equations, to see what they represent.

$y = x$. This equation is satisfied by the coordinates of the points $(0, 0)$, $(1, 1)$, $(2, 2)$, $(3, 3)$, etc., and it is readily seen to represent a straight line through the origin. Its gradient is 1.

$x = 2$. Whatever the value of its y-coordinate, provided that its x-coordinate is 2, a point will lie on this curve. The points $(2, 0)$, $(2, 1)$, $(2, 2)$, $(2, 3)$, etc., lie on the curve, which is a straight line parallel to the y axis, 2 units from it, on the side on which x is positive.

Qu. 14. Make a rough sketch of the lines represented by the following equations. Write down the gradient of each:

  (i) $y = 3$,   (ii) $y = 2x$,   (iii) $y = 3x$,
  (iv) $y = \frac{1}{2}x$,   (v) $y = -x$.

The equation $y = mx + c$

1.7. We come now to the second type of problem mentioned above, in which from some geometrical facts about a curve we discover its equation. And the examples we do will, in turn, help us to interpret straight line equations more skilfully.

Example 6. *Find the equation of the straight line of gradient 4 which passes through the origin.*

If $P(x, y)$ is any point on the line, other than O, the gradient of the line may be written $\dfrac{y}{x}$. (Fig. 1.9.)

$$\therefore \frac{y}{x} = 4.$$

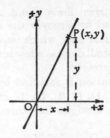

Fig. 1.9

Hence $y = 4x$ is the required equation.

Qu. 15. Write down the equations of the straight lines through the origin having gradients (i) $\frac{1}{3}$, (ii) -2, (iii) m.

Qu. 16. Rearrange the following equations in the form $y = mx$, and hence write down the gradients of the lines they represent:

(i) $4y = x$,   (ii) $5x + 4y = 0$,   (iii) $3x = 2y$,

(iv) $\dfrac{x}{4} = \dfrac{y}{7}$,   (v) $\dfrac{x}{p} - \dfrac{y}{q} = 0$.

Example 7. *Find the equation of the straight line of gradient 3 which cuts the y-axis at $(0, 1)$.*

Let $P(x, y)$ be any point on the line other than $(0, 1)$.

The gradient of the line may be written $\dfrac{y - 1}{x}$. (Fig. 1.10.)

$$\therefore \frac{y - 1}{x} = 3.$$

Hence $y = 3x + 1$ is the required equation.

Qu. 17. By the method of Example 7, find the equations of the straight lines of given gradients cutting the y-axis at the named points:

(i) gradient 3, $(0, 2)$,   (ii) gradient 3, $(0, 4)$,

(iii) gradient 3, (0, − 1),   (iv) gradient $\frac{1}{5}$, (0, 2),
  (v) gradient $\frac{1}{5}$, (0, 4).

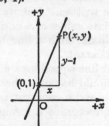

Fig. 1.10

If a straight line cuts the y-axis at the point $(0, c)$, the distance of this point from the origin is called the **intercept** on the y-axis.

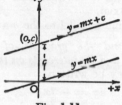

Fig. 1.11

Then the equation of a straight line of gradient m, making an intercept c on the y-axis (Fig. 1.11) is

$$\frac{y - c}{x} = m,$$

i.e. $$y = mx + c.$$

This line is parallel to $y = mx$, which passes through the origin, and it is m, the coefficient of x, which in each case determines the gradient. The effect of altering the value of the number c (c being the intercept on the y-axis) is to raise or lower the line, without altering its gradient; the sign of c determines whether the line cuts the y-axis above or below the origin.

We might be tempted to think at this stage that in $y = mx + c$ we have found the form in which all straight line equations may be written. But we get into trouble if we try to find appropriate values of m and c to give a straight line parallel to the y-axis; such a line has as its equation $x = k$, where k is a constant.

The various straight line equations we have met are summarized below. It should be noted that only terms of the first degree in x and y and a constant term occur; this, in fact, is how we may recognize a straight line, or *linear*, equation.

$y = mx + c$ is a line of gradient m, passing through $(0, c)$.

$y = mx$ is a line of gradient m, passing through the origin.

$y = c$ is a line of zero gradient (i.e. parallel to the x-axis).

$x = k$ is a line parallel to the y-axis.

Example 8. *Find the gradient of the straight line* $7x + 4y + 2 = 0$, *and its intercepts on the axes.*

The equation may be written

$$4y = -7x - 2,$$
or $$y = -\tfrac{7}{4}x - \tfrac{1}{2}.$$

This is now in the form $y = mx + c$, where $m = -\tfrac{7}{4}$, and $c = -\tfrac{1}{2}$, and we see that the gradient is $-\tfrac{7}{4}$, and that the intercept on the y-axis is $-\tfrac{1}{2}$. In fact, to find the intercepts on each axis it is better to go back to the original equation $7x + 4y + 2 = 0$.

To find the intercept on the y-axis:
putting $x = 0$, $4y + 2 = 0$, $\therefore\ y = -\tfrac{1}{2}$.

To find the intercept on the x-axis:
putting $y = 0$, $7x + 2 = 0$, $\therefore\ x = -\tfrac{2}{7}$.

The intercepts on the x-axis and y-axis are $-\tfrac{2}{7}$ and $-\tfrac{1}{2}$ respectively.

Qu. 18. Arrange the following equations in the form $y = mx + c$, hence write down the gradient of each line; also find the intercepts on the y-axis:

(i) $3y = 2x + 6$, (ii) $x - 4y + 2 = 0$,

(iii) $3x + y + 6 = 0$, (iv) $7x = 3y + 5$,

(v) $y + 4 = 0$, (vi) $lx + my + n = 0$.

Qu. 19. Write down the equations of (i) the x-axis, (ii) the y-axis, (iii) a straight line parallel to the y-axis through $(4, 0)$, (iv) a straight line parallel to the x-axis making an intercept of -7 on the y-axis.

Exercise 1b

1. Find the gradients of the straight lines joining the following pairs of points:

   (i) $(4, 6)$ and $(9, 15)$, (ii) $(5, -11)$ and $(-1, 3)$,
   (iii) $(-2\frac{1}{2}, -\frac{1}{2})$ and $(4\frac{1}{2}, -1)$, (iv) $(7, 0)$ and $(-3, -2)$.

2. Show that the three given points are in each case collinear, i.e. they lie on the same straight line:

   (i) $(0, 0)$, $(3, 5)$, $(21, 35)$, (ii) $(-3, 1)$, $(1, 2)$, $(9, 4)$,
   (iii) $(-3, 4)$, $(1, 2)$, $(7, -1)$,
   (iv) $(1, 2)$, $(0, -1)$, $(-2, -7)$.

3. Find the gradients of the straight lines which make the following angles with the x-axis, the angle in each case being measured anti-clockwise from the positive direction of the x-axis:

   (i) $45°$, (ii) $135°$, (iii) $60°$, (iv) $150°$.

4. Find if AB is parallel or perpendicular to PQ in the following cases:

(i) $A(4, 3)$,	$B(8, 4)$,	$P(7, 1)$,	$Q(6, 5)$;
(ii) $A(-2, 0)$,	$B(1, 9)$,	$P(2, 5)$,	$Q(6, 17)$;
(iii) $A(8, -5)$,	$B(11, -3)$,	$P(1, 1)$	$Q(-3, 7)$;
(iv) $A(-6, -1)$,	$B(-6, 3)$,	$P(2, 0)$,	$Q(2, -5)$;
(v) $A(4, 3)$,	$B(-7, 3)$,	$P(5, 2)$,	$Q(5, -1)$;
(vi) $A(3, 1)$,	$B(7, 3)$,	$P(-3, 2)$,	$Q(1, 0)$.

5. Show that $A(-3, 1)$, $B(1, 2)$, $C(0, -1)$, $D(-4, -2)$ are the vertices of a parallelogram.

6. Show that P(1, 7), Q(7, 5), R(6, 2), S(0, 4) are the vertices of a rectangle.   Calculate the lengths of the diagonals, and find their point of intersection.

7. Show that D(-2, 0), E($\frac{1}{2}$, $1\frac{1}{2}$), F($3\frac{1}{2}$, $-3\frac{1}{2}$) are the vertices of a right-angled triangle, and find the length of the shortest side, and the mid-point of the hypotenuse.

8. Find the y-coordinates of the points on the curve $y = x^2 + 1$ for which the x-coordinates are -3, 0, 1, 5.   Find the coordinates of points on the curve whose y-coordinates are 5, and 17.   Sketch the curve.

9. Find the coordinates of the points on the curve $y = x^3$ for which $x = -3$, -1, 1, 3; and also of the points for which $y = -8$, $0 + 8$.   Sketch the curve.

10. Determine whether the following points lie on the given curve:

   (i) $y = 3x - 5$, $(-1, -8)$,
   (ii) $5x - 2y + 7 = 0$, $(1, -1)$,
   (iii) $y = x^3$, $(-4, 64)$,   (iv) $x^2y = 1$, $(-2, \frac{1}{4})$.

11. Find the intercepts on the axes made by the straight line $3x - 2y + 10 = 0$.   Hence find the area of the triangle enclosed by the axes and this line.

12. Find the coordinates of the points at which the following curves cut the axes:

   (i) $y = x^2 - x - 12$,   (ii) $y = 6x^2 - 7x + 2$,
   (iii) $y = x^2 - 6x + 9$,   (iv) $y = x^3 - 9x^2$,
   (v) $y = (x + 1)(x - 5)^2$,   (vi) $y = (x^2 - 1)(x^2 - 9)$.

13. Plot the following points on squared paper, and write down the equations of the straight lines passing through them, in the form $y = mx + c$:

   (i) $(-1, -1)$, $(0, 0)$, $(4, 4)$,
   (ii) $(-1, 1)$, $(0, 0)$, $(1, -1)$,
   (iii) $(-4, -2)$, $(0, 0)$, $(8, 4)$,
   (iv) $(0, -4)$, $(4, -2)$, $(6, -1)$,
   (v) $(-5, 2)$, $(-5, 0)$, $(-5, -2)$,
   (vi) $(-3, 7)$, $(3, 3)$, $(6, 1)$.

14. Write down the equation of the straight line
   (i) through (5, 11) parallel to the x-axis,

(ii) which is the perpendicular bisector of the line joining (2, 0) and (6, 0),

(iii) through $(0, -10)$ parallel to $y = 6x + 3$,

(iv) through $(0, 2)$ parallel to $y + 8x = 0$,

(v) through $(0, -1)$ perpendicular to $3x - 2y + 5 = 0$.

15. Find the equation of the straight line joining the origin to the mid-point of the line joining A(3, 2) and B(5, -1).

16. P(-2, -4), Q(-5, -2), R(2, 1), S are the vertices of a parallelogram. Find the coordinates of M, the point of intersection of the diagonals, and of S.

17. (i) Write down the gradient of the straight line joining (a, b) and (p, q). Write down the two conditions that these points should lie on the line $y = 7x - 3$. From these deduce the gradient of the line.

(ii) Repeat for the line $3x + 2y - 1 = 0$, and check your result by writing the equation in the form $y = mx + c$.

The use of suffixes

1.8. When we wish to refer to points whose coordinates are not given, it is convenient to write them as

$$(x_1, y_1) \quad \text{read as} \quad \text{``}x \text{ one, } y \text{ one''}$$
$$(x_2, y_2) \quad \text{read as} \quad \text{``}x \text{ two, } y \text{ two'', etc.}$$

It is important to write the number (the suffix) at the bottom of the letter, so as to avoid confusion between x_2 and x^2, x_3 and x^3, and so on. This is a suitable point at which to summarize some of the early results of this chapter, using this notation.

If A and B are the points (x_1, y_1) and (x_2, y_2) respectively,

The length of AB is $\sqrt{\{(x_1 - x_2)^2 + (y_1 - y_2)^2\}}$.

The mid-point of AB is $\left(\dfrac{x_1 + x_2}{2}, \dfrac{y_1 + y_2}{2}\right)$.

The gradient of AB is $\dfrac{y_1 - y_2}{x_1 - x_2}$ or $\dfrac{y_2 - y_1}{x_2 - x_1}$,

The condition for A to lie on $ax + by + c = 0$ is $ax_1 + by_1 + c = 0$.

Finding the equation of a straight line

1.9. The method of Example 7 in §1.7 can of course be used to find the equation of any straight line provided (i) that we know one point through which the line passes, and (ii) that we know, or can calculate, the gradient. Two examples will illustrate this.

Example 9. *Find the equation of the straight line of gradient* $-\frac{2}{3}$, *which passes through* $(-4, 1)$.

Let $P(x, y)$ be any point on the line other than $(-4, 1)$. (Fig. 1.12.)

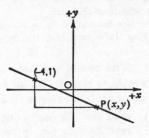

Fig. 1.12

The gradient of the line may be written

$$\frac{y - 1}{x - (-4)} = \frac{y - 1}{x + 4}.$$

But the gradient is given as $-\frac{2}{3}$,

$$\therefore \frac{y - 1}{x + 4} = -\frac{2}{3},$$

$$\therefore 3(y - 1) = -2(x + 4),$$

$$\therefore 3y - 3 = -2x - 8.$$

Hence the required equation is

$$2x + 3y + 5 = 0.$$

Example 10. *Find the equation of the straight line joining the points* $(-5, 2)$ *and* $(3, -4)$.

The gradient of the line $= \dfrac{2 - (-4)}{-5 - 3} = \dfrac{6}{-8} = -\tfrac{3}{4}.$

If $\mathsf{P}(x, y)$ is any point on the line other than $(3, -4)$, the gradient may be written

$$\frac{y - (-4)}{x - 3} = \frac{y + 4}{x - 3}$$

$$\therefore \frac{y + 4}{x - 3} = -\tfrac{3}{4},$$

$$\therefore 4(y + 4) = -3(x - 3),$$

$$\therefore 4y + 16 = -3x + 9.$$

Hence the required equation is

$$3x + 4y + 7 = 0.$$

Examples 9 and 10 illustrate the most direct approach. The equation as first written is the direct statement of the condition that the point (x, y) should lie on the given line.

Another method is given below as an alternative solution to Example 9. We know that the equation $y = mx + c$ represents a straight line of gradient m; so the equation $y = -\tfrac{2}{3}x + c$ represents any line of gradient $-\tfrac{2}{3}$, according to the value of the constant c, and our problem is to find the appropriate value of c for the given line. To do this we use the fact that if the point (x_1, y_1) lies on the straight line $y = mx + c$, its coordinates satisfy the equation of the line i.e. $y_1 = mx_1 + c$.

Example 9. (*Alternative solution.*)
The equation is of the form $y = -\tfrac{2}{3}x + c$.
Since $(-4, 1)$ lies on this line,

$$1 = -\tfrac{2}{3}(-4) + c,$$

$$\therefore \; c = 1 - \tfrac{8}{3} = -\tfrac{5}{3}.$$

Hence the required equation is $y = -\tfrac{2}{3}x - \tfrac{5}{3}$, or

$$2x + 3y + 5 = 0.$$

Qu. 20. Use the methods of Examples 9 (first solution) and 10 to find the equations of the straight lines

(i) through $(4, -3)$, of gradient $\tfrac{5}{2}$,

(ii) joining $(-3, 8)$ and $(1, -2)$.

Qu. 21. Using the method of Example 9 (alternative solution) find the equations of the straight lines

(i) through $(5, -2)$, of gradient $\tfrac{3}{4}$,

(ii) joining $(-2, 5)$ and $(3, -7)$.

Qu. 22. Write down the equation of the straight line through (x_1, y_1) of gradient m.

Points of intersection

1.10. If the two straight lines $x + y - 1 = 0$ and $2x - y - 8 = 0$ cut at the point $P(a, b)$ then the coordinates of P satisfy the equation of each line, and we may write

$$a + b - 1 = 0,$$
$$2a - b - 8 = 0.$$

The solution of these equations is $a = 3$, $b = -2$, which tells us that the given lines cut at $(3, -2)$. In practice we obtain the result by solving the equations simultaneously for x and y.

Qu. 23. Find the points of intersection of the following pairs of straight lines:

(i) $2x - 3y = 6$ and $4x + y = 19$,

(ii) $y = 3x + 2$ and $2x + 3y = 17$,

(iii) $y = c$ and $y = mx + c$,

(iv) $x = -a$ and $y = mx + c$.

Qu. 24. Can you find the point of intersection of

$$3x - 2y - 10 = 0 \text{ and } 4y = 6x - 7?$$

Qu. 25. Find the points of intersection of the curve $y = 12x^2 + x - 6$ and the x-axis.

Exercise 1c

1. Find the equations of the straight lines of given gradients, passing through the points named:

   (i) 4, (1, 3), (ii) 3, (− 2, 5),
   (iii) $\frac{1}{3}$, (2, − 5), (iv) − $\frac{3}{4}$, (7, 5),
   (v) $\frac{1}{2}$, ($\frac{1}{3}$, − $\frac{1}{2}$), (vi) − $\frac{5}{3}$, ($\frac{1}{4}$, − 3).

2. Find the equations of the straight lines joining the following pairs of points:

   (i) (1, 6) and (5, 9), (ii) (3, 2) and (7, − 3),
   (iii) (− 3, 4) and (8, 1), (iv) (− 1, − 4) and (4, − 3),
   (v) ($\frac{1}{2}$, 2) and (3, $\frac{1}{3}$), (vi) (− $\frac{1}{2}$, 0) and (5, 11).

3. Find the points of intersection of the following pairs of straight lines:

   (i) $x + y = 0$, $y = − 7$,
   (ii) $y = 5x + 2$, $y = 3x − 1$,
   (iii) $3x + 2y − 1 = 0$, $4x + 5y + 3 = 0$,
   (iv) $5x + 7y + 29 = 0$, $11x − 3y - 65 = 0$.

4. Find the equation of the straight line

   (i) through (5, 4), parallel to $3x − 4y + 7 = 0$,
   (ii) through (− 2, 3), parallel to $5x − 2y − 1 = 0$,
   (iii) through (4, 0), perpendicular to $x + 7y + 4 = 0$,
   (iv) through (− 2, − 3), perpendicular to $4x + 3y − 5 = 0$.

5. Find the equation of the perpendicular bisector of AB, where A and B are the points (− 4, 8) and (0, − 2) respectively.

6. Repeat No. 5 for the points A(7, 3) and B(− 6, 1).

7. Find the equation of the straight line joining A(10, 0) and B(0, − 7). Also find the equation of the median through the origin, O, of the triangle OAB.

8. P, Q, R are the points (3, 4), (7, − 2), (− 2, − 1) respectively. Find the equation of the median through R of the triangle PQR.

9. Calculate the area of the triangle formed by the line $3x − 7y + 4 = 0$ and the axes.

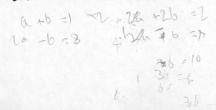

10. Find the circumcentre of the triangle with vertices $(-3, 0)$, $(7, 0)$, $(9, -6)$. Show that the point $(1, 2)$ lies on the circumcircle.

11. Find the equation of the straight line through $P(7, 5)$ perpendicular to the straight line AB whose equation is $3x + 4y - 16 = 0$. Calculate the length of the perpendicular from P to AB.

12. ABCD is a rhombus. A is the point $(2, -1)$, and C is the point $(4, 7)$. Find the equation of the diagonal BD.

13. $L(-1, 0)$, $M(3, 7)$, $N(5, -2)$ are the mid-points of the sides BC, CA, AB respectively of the triangle ABC. Find the equation of AB.

14. Find the points of intersection of $x^2 = 4y$ and $y = 4x$.

15. The straight line $x - y - 6 = 0$ cuts the curve $y^2 = 8x$ at P and Q. Calculate the length of PQ.

Exercise 1d (Miscellaneous)

1. Find the equation of the line joining the points $(6, 3)$ and $(5, 8)$. Show also that these two points are equidistant from the point $(-2, 4)$.

2. What is the equation of the straight line joining the points $A(7, 0)$ and $B(0, 2)$? Obtain the equation of the straight line AC such that the x-axis bisects the angle BAC.

3. Find the equations of the following straight lines, giving each in the form $ax + by + c = 0$:

    (i) the line joining the points $(2, 4)$ and $(-3, 1)$,

    (ii) the line through $(3, 1)$ parallel to the line $3x + 5y = 6$,

    (iii) the line through $(3, -4)$ perpendicular to the line

$$5x - 2y = 3.$$

4. Write down the condition that the straight lines $y = m_1x + c_1$ and $y = m_2x + c_2$ should be at right angles. Find the equations of the straight lines through the point $(3, -2)$ which are (i) parallel, and (ii) perpendicular to the line $2y + 5x = 17$.

5. The points A, B, C have coordinates $(7, 0)$, $(3, -3)$, $(-3, 3)$ respectively. Find the coordinates of D, E, F, the mid-points of BC, CA, AB respectively. Find the equations of the lines AD, BE, and the coordinates of K, their

point of intersection.   Prove that C, K, F are in a straight
line

6. Find the equation of the straight line

   (i) joining the points $(-3, 2)$ and $(1, -4)$,
   (ii) through $(-1, 3)$ parallel to the line $2x + 7y - 8 = 0$,
   (iii) through $(2, -3)$ perpendicular to the line

$$5x - 2y - 11 = 0.$$

7. Find the equations of the lines passing through the point
   $(4, -2)$ and respectively (i) parallel, (ii) perpendicular
   to the line $2x - 3y - 4 = 0$.   Find also the coordinates
   of the foot of the perpendicular from $(4, -2)$ to
   $2x - 3y - 4 = 0$.

8. A line is drawn through the point $(2, 3)$ making an angle
   of $45°$ with the positive direction of the x-axis, and it meets
   the line $x = 6$ at P.   Find the distance of P from the
   origin O, and the equation of the line through P perpendicular to OP.

9. Prove that the points $(-5, 4)$, $(-1, -2)$, $(5, 2)$ lie at
   three of the corners of a square.   Find the coordinates of
   the fourth corner, and the area of the square.

10. The vertices of a quadrilateral ABCD are A$(4, 0)$, B$(14, 11)$,
    C$(0, 6)$, D$(-10, -5)$.   Prove that the diagonals AC,
    BD bisect each other at right angles, and that the length
    of BD is four times that of AC.

11. The coordinates of the vertices A, B, C of the triangle
    ABC are $(-3, 7)$, $(2, 19)$, $(10, 7)$ respectively.

    (i) Prove that the triangle is isosceles.
    (ii) Calculate the length of the perpendicular from B to
         AC, and use it to find the area of the triangle.

12. A triangle ABC has A at the point $(7, 9)$, B at $(3, 5)$, C at
    $(5, 1)$.   Find the equation of the line joining the mid-
    points of AB and AC; and find also the area of the triangle
    enclosed by the line and the axes.

13. One side of a rhombus is the line $y = 2x$, and two opposite
    vertices are the points $(0, 0)$ and $(4\frac{1}{2}, 4\frac{1}{2})$.   Find the
    equations of the diagonals, the coordinates of the other two
    vertices, and the length of the side.

14. Prove that the four points $(4, 0)$, $(7, -3)$, $(-2, -2)$,

$(-5, 1)$ are the vertices of a parallelogram and find the equations of its diagonals.

15. Find the equation of the line which is parallel to the line $x + 4y - 1 = 0$, and which passes through the point of intersection of the lines $y = 2x$ and $x + y - 3 = 0$.

16. Find the equations of the lines which pass through the point of intersection of the lines $x - 3y = 4$ and $3x + y = 2$, and are respectively parallel and perpendicular to the line $3x + 4y = 0$.

17. The three straight lines $y = x$, $2y = 7x$, and $x + 4y - 60 = 0$ form a triangle. Find the equations of the three medians, and calculate the coordinates of their point of intersection.

18. The points $D(2, -3)$, $E(-1, 7)$, $F(3, 5)$ are the mid-points of the sides BC, CA, AB respectively of a triangle. Find the equations of its sides.

19. Prove that the points $(1, -1)$, $(-1, 1)$, $(\sqrt{3}, \sqrt{3})$ are the vertices of an equilateral triangle. Find the coordinates of the point of intersection of the medians of this triangle.

20. The points $A(-7, -7)$, $B(8, -1)$, $C(4, 9)$, D are the vertices of the parallelogram ABCD. Find the coordinates of D. Prove that ABCD is a rectangle and find its area.

21. Find the equation of the line which is parallel to the line $3x + 4y = 12$ and which makes an intercept of 5 units on the x-axis. Find also the equation of the line which is perpendicular to the given line and which passes through the point $(4, 5)$.

22. A, B, C are the points $(1, 6)$, $(-5, 2)$, $(3, 4)$ respectively. Find the equations of the perpendicular bisectors of AB and BC. Hence find the coordinates of the circumcentre of the triangle ABC.

23. Find the equation of the straight line joining the feet of the perpendiculars drawn from the point $(1, 1)$ to the lines $3x - 3y - 4 = 0$ and $3x + y - 6 = 0$.

24. Through the point $A(1, 5)$ is drawn a line parallel to the x-axis to meet at B the line PQ whose equation is $3y = 2x - 5$. Find the length of AB and the sine of the angle between PQ and AB; hence show that the length of the perpendicular from A to PQ is $18 \div \sqrt{13}$. Calculate the area of the triangle formed by PQ and the axes.

CHAPTER 2

THE GRADIENT OF A CURVE*

The gradient of a curve

2.1 So far we have only discussed the gradient of a straight line. A man walking up the ramp AB (Fig. 2.1) is climbing a gradient of $\frac{2}{7}$.

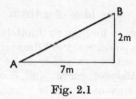

Fig. 2.1

Let us now consider a man walking up the slope represented by the curve CPD. (Fig. 2.2.) Between C and D

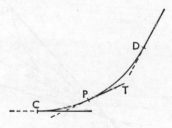

Fig. 2.2

* NOTE.—Most of the questions in the text in this chapter should be worked by the pupils themselves.

27

the gradient is steadily increasing.   If, when he had reached
the point P, the gradient had stopped increasing, and had
remained constant from then on, he would have climbed up
the slope represented by the straight line PT, the tangent
to the curve at P.   Thus in walking up the slope CD, when
the man is at the point P (and only at that instant) he is
climbing a gradient represented by the gradient of PT.

DEFINITION

*The gradient of a curve at any point is the gradient of the
tangent to the curve at that point.*

The idea of a limit

2.2.   If we wish to find approximately the gradient of a
curve at a certain point, we could draw the curve, draw the
tangent at that point by eye, and measure its gradient.   But
to develop our study of curves and their equations, it is
important that we should discover a method of calculating
exactly the gradient of a curve at any point; to do this we
shall think of a tangent to a curve in the following way.

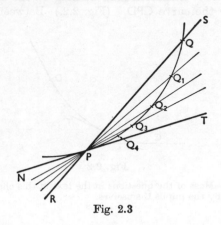

Fig. 2.3

First we start with two distinct points on a curve, P and Q (Fig. 2.3), and the chord PQ is drawn and produced in both directions. Now consider RPQS as a straight rod hinged at P, which is rotated clockwise about P to take up successive positions shown by PQ_1, PQ_2, PQ_3, etc. Notice that the points at which it cuts the curve, Q_1, Q_2, Q_3, are successively nearer the fixed point P. The nearer this second point of intersection approaches P, the nearer does the gradient of the chord approach the gradient of the tangent NPT. *By taking Q sufficiently close to P, we can make the gradient of the chord PQ as near as we please to the gradient of the tangent at P.*

To see precisely how this happens, place the edge of a ruler along RPQS and then rotate it clockwise about P. You will see the second point of intersection approach P along the curve, until it actually coincides with P when the ruler lies along the tangent NPT. Using an arrow to denote "tends to" or "approaches" we may write:

as $Q \rightarrow P$ along the curve

the gradient of the chord $PQ \rightarrow$ the gradient of the tangent at P,

the tangent at P is called the **limit** of the chord PQ (or more exactly of the secant RPQS), and

the gradient of the curve at P is the limit of the gradient of the chord PQ.

Qu. 1. A regular polygon of n sides is inscribed in a circle. What is the limit of the polygon as n increases infinitely (as $n \rightarrow \infty$)?

Qu. 2. OP is a radius of a circle centre O. A straight line PQR cuts the circumference at Q. What is the limit of the angle OPR as Q approaches P along the circumference?

Qu. 3. P is a point on the straight line $y = \frac{1}{3}x$. Q is the foot of the perpendicular from P to the x-axis. As P approaches O, the origin, what happens to PQ and QO? What can you say about the value of $\frac{PQ}{QO}$?

The gradient of $y = x^2$ at $(2, 4)$

2.3. We shall now use this idea of a tangent being the limit of a chord, to find the gradient of the curve $y = x^2$ at a particular point, namely $(2, 4)$.

P is the point $(2, 4)$ on the curve $y = x^2$. (Fig. 2.4.)

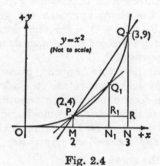

Fig. 2.4

Q is another point on the curve, which we take first as $(3, 9)$. Then, as the chord PQ rotates clockwise about P, Q moves along the curve to Q_1, and then nearer and nearer to P. By studying the behaviour of the gradient of PQ as this is happening we hope to be able to deduce the gradient of the tangent at P.

$$\text{The gradient of PQ} = \frac{RQ}{PR},$$

$$= \frac{RQ}{MN},$$

$$= \frac{NQ - NR}{ON - OM},$$

$$= \frac{9 - 4}{3 - 2},$$

$$= 5.$$

If Q now moves to the position Q_1, whose coordinates are $(2\frac{1}{2}, 6\frac{1}{4})$,

$$\text{the gradient of } PQ_1 = \frac{N_1Q_1 - N_1R_1}{ON_1 - OM},$$

$$= \frac{6\frac{1}{4} - 4}{2\frac{1}{2} - 2},$$

$$= \frac{2\frac{1}{4}}{\frac{1}{2}},$$

$$= 4\frac{1}{2}.$$

We now let Q approach yet closer to P along the curve, and the following table gives the gradient of the chord PQ as it approaches the gradient of the tangent at P.

ON (x coord. of Q)	NQ (y coord. of Q)	PR (ON — 2)	RQ (NQ — 4)	$\dfrac{RQ}{PR}$ Gradient of PQ	
3	9	1	5	5	
$2\frac{1}{2}$	$6\frac{1}{4}$	$\frac{1}{2}$	$2\frac{1}{4}$	$\dfrac{2\frac{1}{4}}{\frac{1}{2}}$	$= 4\frac{1}{2}$
2·1	4·41	0·1	0·41	$\dfrac{0·41}{0·1}$	$= 4·1$
2·01	4·0401	0·01	0·0401	$\dfrac{0·0401}{0·01}$	$= 4·01$
2·001	4·004 001	0·001	0·004 001	$\dfrac{0·004\ 001}{0·001}$	$= 4·001$

Comparing the first and last columns of this table, we see that for each position of Q, the gradient of PQ exceeds 4 by the same amount as the x-coordinate of Q exceeds 2. The actual equality is not important; what is important is that these values we have taken so far suggest that by taking Q sufficiently near P (i.e. by taking the x-coordinate of Q sufficiently near 2) we can make the gradient of PQ as near 4 as we please. This suggests that the limit of the gradient of PQ is 4, and that the gradient of the tangent at P is 4.

Qu. 4. Draw a figure similar to Fig. 2.4, taking P as the point (1, 1). Taking the x-coordinate of Q successively as 2, $1\frac{1}{2}$, 1·1, 1·01, make out a table similar to the one above. What appears to be the limit of the gradient of PQ in this case?

Qu. 5. Add a last line to your table for Qu. 4 by taking the x-coordinate of Q to be $1 + h$. What happens to Q as $h \to 0$? What happens to the gradient of PQ as $h \to 0$? Deduce the gradient of $y = x^2$ at (1, 1).

Qu. 6. Add a last line to the table in the book, taking the x-coordinate of Q as $(2 + h)$. Deduce the gradient of $y = x^2$ at (2, 4).

The gradient function of $y = x^2$

2.4. We now use the method suggested in Qu. 5 to find the gradient of $y = x^2$ at *any* point.

P is the point (a, a^2), and Q is another point on the curve whose x-coordinate is $a + h$. (Fig. 2.5.)

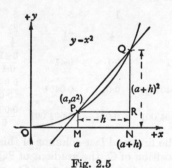

Fig. 2.5

$$RQ = NQ - NR,$$
$$= (a + h)^2 - a^2,$$
$$= 2ah + h^2,$$

and $$PR = h.$$

The gradient of the chord PQ is $\dfrac{RQ}{PR} = \dfrac{2ah + h^2}{h}$,

$$= 2a + h.$$

As we let the chord rotate clockwise about P, Q approaches P along the curve, and the gradient of the chord PQ → the gradient of the tangent at P, and $h \to 0$.

But as $h \to 0$, the gradient of the chord PQ, $(2a + h) \to 2a$. It follows that the gradient of the tangent at P is $2a$.

Thus the gradient of $y = x^2$ at (a, a^2) is $2a$, and since a is the x-coordinate of the point (a, a^2), the gradient of $y = x^2$ at (x, x^2) is $2x$.

Just as x^2 is the expression in which we substitute a value of x to find the corresponding y coordinate and plot a point on the curve $y = x^2$, so we now have another expression, $2x$, in which we can substitute the value of x to find the gradient at that point.

$2x$ is called the **gradient function** of the curve $y = x^2$.

Example 1. *Find the coordinates of the points on the curve $y = x^2$, given by $x = 4$ and -10, and find the gradient of the curve at these points.*

(i) $y = x^2.$

When $x = 4$, $y = 16.$

             The gradient function $= 2x.$

When $x = 4$,   the gradient $= 8.$

      $\therefore$ the point is $(4, 16)$, and the gradient is 8.

(ii) When $x = -10$, $y = x^2 = +100.$

       The gradient function $= 2x = -20.$

$\therefore$ the point is $(-10, 100)$, and the gradient is -20.

Qu. 7. Calculate the gradients of the tangents to $y = x^2$ at the points given by $x = -1\frac{1}{2}, -1, +\frac{1}{2}, +2$.

Qu. 8. Use the method of §2.4 to find the gradient functions of the following curves, making a sketch in each case, and compare each result with the gradient function of $y = x^2$: (i) $y = 3x^2$, (ii) $y = 5x^2$, (iii) $y = \frac{1}{2}x^2$, (iv) $y = cx^2$, where c is a constant, (v) $y = x^2 + 3$, (vi) $y = x^2 + k$, where k is a constant.

Clearly we need an abbreviation for the statement "the gradient function of $y = x^2$ is $2x$". A convenient way of writing this is,

$$\text{"if } y = x^2$$
$$\operatorname{grad} y = 2x.\text{"}$$

The process of finding the gradient function of a curve is known as **differentiation**, and it is useful if we understand "grad" also to be an instruction to differentiate. Thus,

$$\operatorname{grad}(x^2) = 2x.$$

The differentiation of x^3

2.5. P is any point (a, a^3) on the curve $y = x^3$. Q is another point on the curve with x-coordinate $(a + h)$. (Fig. 2.6.)

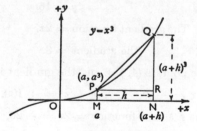

Fig. 2.6

$$RQ = NQ - NR,$$
$$= (a + h)^3 - a^3,$$
$$= a^3 + 3a^2h + 3ah^2 + h^3 - a^3,$$
$$= 3a^2h + 3ah^2 + h^3.$$
$$PR = h.$$

The gradient of $PQ = \dfrac{RQ}{PR}$,

$$= \frac{3a^2h + 3ah^2 + h^3}{h},$$

$$= 3a^2 + 3ah + h^2.$$

As Q approaches P along the curve, $h \to 0$, and the terms $3ah$ and h^2 each tend to zero; therefore the gradient of $PQ \to 3a^2$.

It follows that the gradient of $y = x^3$ at (a, a^3) is $3a^2$, or

$$\text{grad } x^3 = 3x^2.$$

Qu. 9. Use the method of §2·5 to find grad x^4. [Hint: $(a + h)^4 = a^4 + 4a^3h + 6a^2h^2 + 4ah^3 + h^4$.]

Qu. 10. Differentiate $2x^3$ by the same method.

Summary of results

2.6. We have now confirmed the following:

$$\text{grad } x^2 = 2x.$$
$$\text{grad } x^3 = 3x^2.$$
$$\text{grad } x^4 = 4x^3.$$

The form of these results suggests that the rule for differentiating a power of x is *multiply by the index, and reduce the index by* 1; this means that grad x^5 would be $5x^4$, grad x^6 would be $6x^5$, and so on.

At this stage we must dispense with a formal proof of the validity of this process in general, and we shall assume that

$$\textbf{grad } x^n = nx^{n-1}$$

when n is any positive whole number.

It is now time to link up these ideas with our earlier work on a straight line, and to extend them further.

$$y = c$$

Straight lines of this form, such as $y = 4$ and $y = -2$, are parallel to the x-axis, and have zero gradient.

It follows that   grad $4 = 0$,
and   grad $-2 = 0$.

Thus, *if we differentiate a constant we get* 0.

[Note that this does agree with the general result, grad $x^n = nx^{n-1}$. Since $x^0 = 1$ (see §8.3), we may write grad $4 = $ grad $4x^0 = 0 \times 4x^{-1} = 0$.]

$$y = mx$$

We know that a straight line of this form has gradient m, e.g. $y = x$ has gradient 1, and $y = 3x$ has gradient 3.

Thus   grad $x = 1$.

[Again, this agrees with the general result, since grad $x^1 = 1 . x^0 = 1$.]

Also   grad $3x = 3$ grad $x = 3.1 = 3$,
and as Qu. 8 showed,

grad $3x^2 = 3$ grad $x^2 = 3 . 2x = 6x$.

Thus, *if a function has a constant factor, that constant remains unchanged as a factor of the gradient function.* (Fig. 2.7.)

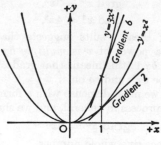

Fig. 2.7

Qu. 11. Differentiate:

(i) $4x^3$, (ii) $5x^4$, (iii) ax^2, (iv) $4x^n$, (v) Kx^{n+1}.

The differentiation of a polynomial

So far we have differentiated functions of one term only. What happens if there are two or more terms?

$$y = mx + c$$

The straight lines $y = 3x$, $y = 3x + 4$, and $y = 3x - 2$ all have gradient 3.

Thus

$$\text{grad } 3x = 3,$$
$$\text{grad } (3x + 4) = 3,$$
$$\text{grad } (3x - 2) = 3.$$

The above lines are parallel, and as we discovered in §1.7, the effect of giving the different values $c = 0$, $+ 4$ and $- 2$, is to raise or lower the line, but not to alter its gradient.

Clearly the same applies to the curves $y = x^2$, $y = x^2 + 4$ and $y = x^2 - 2$. (Fig. 2.8.) At the point on each curve

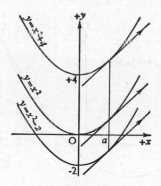

Fig. 2.8

for which $x = a$, the tangents are parallel, each having gradient $2a$.

$$\operatorname{grad} x^2 = 2x,$$
$$\operatorname{grad} (x^2 + 4) = 2x,$$
$$\operatorname{grad} (x^2 - 2) = 2x.$$

In the above cases where the function consists of two terms, we should get the same result by differentiating each term separately. Thus,

$$\operatorname{grad} (x^2 + 4) = \operatorname{grad} x^2 + \operatorname{grad} 4$$
$$= 2x + 0$$
$$= 2x.$$

This leads us to investigate whether this method is valid in general.

$$y = x^2 + 3x - 2$$

To find the gradient function of this curve, let P be any point $(a, a^2 + 3a - 2)$ on it. Q is another point on the curve with x-coordinate $(a + h)$. (Fig. 2.9.)

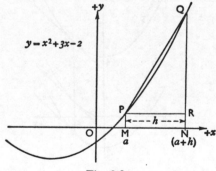

Fig. 2.9

$$RQ = NQ - NR,$$
$$= \{(a + h)^2 + 3(a + h) - 2\} - \{a^2 + 3a - 2\},$$
$$= a^2 + 2ah + h^2 + 3a + 3h - 2 - a^2 - 3a + 2,$$
$$= 2ah + h^2 + 3h.$$
$$PR = h.$$

The gradient of $PQ = \dfrac{RQ}{PR} = \dfrac{2ah + h^2 + 3h}{h},$

$$= 2a + h + 3.$$

As Q approaches P along the curve, $h \to 0$ and the gradient of $PQ \to 2a + 3$.

It follows that the gradient of $y = x^2 + 3x - 2$ at $(a, a^2 + 3a - 2)$ is $2a + 3$, or

$$\text{grad } (x^2 + 3x - 2) = 2x + 3.$$

Now, if we try differentiating each term separately,

$$\text{grad } (x^2 + 3x - 2) = \text{grad } x^2 + \text{grad } 3x + \text{grad} - 2$$
$$= 2x + 3 + 0$$
$$= 2x + 3.$$

Thus the gradient function of the sum of a number of terms is obtained by differentiating each term separately.

Qu. 12. Differentiate:

(i) $x^3 + 2x^2 + 3x,$ (ii) $4x^4 - 3x^2 + 5,$

(iii) $ax^2 + bx + c.$

Products and quotients

A special method of dealing with products and quotients will be met later, but for the present we must reduce a function in this form to the sum of a number of terms before differentiating. (The reader may check that to differentiate each factor separately in the following examples does *not* lead to the correct result.)

$$\text{grad } \{x^2(2x + 3)\} = \text{grad } (2x^3 + 3x^2) = 6x^2 + 6x.$$

$$\text{grad } \left\{\frac{x^3 + 4x^2}{x}\right\} = \text{grad } (x^2 + 4x) = 2x + 4.$$

Qu. 13. Differentiate:

(i) $x^2(4x - 2)$,   (ii) $(x + 3)(x - 4)$,   (iii) $\dfrac{5x^3 + 3x^2}{x^2}$.

Exercise 2a

Write down the gradient functions of the following curves:

1. $y = x^{12}$.   **2.** $y = 3x^7$.   **3.** $y = 5x$.

4. $y = 5x + 3$.   **5.** $y = 3$.   **6.** $y = 5x^2 - 3x$.

7. $y = 3x^4 - 2x^3 + x^2 - x + 10$.

8. $y = 2x^4 + \frac{1}{3}x^3 - \frac{1}{4}x^2 + 2$.   **9.** $y = ax^3 + bx^2 + cx$.

10. $y = 2x(3x^2 - 4)$.   **11.** $y = \dfrac{10x^5 + 3x^4}{2x^2}$.

Differentiate the following functions:

12. $- x$.   **13.** $+ 10$.   **14.** $4x^3 - 3x + 2$.

15. $\frac{1}{2}ax^2 - 2bx + c$.   **16.** $2(x^2 + x)$.   **17.** $3x(x - 1)$.

18. $\frac{1}{3}(x^3 - 3x + 6)$.   **19.** $(x + 1)(x - 2)$.

20. $3(x + 1)(x - 1)$.   **21.** $\dfrac{(x + 3)(2x + 1)}{4}$.

22. $\dfrac{2x^3 - x^2}{3x}$.   **23.** $\dfrac{x^4 + 3x^2}{2x^2}$.

Find the y-coordinate, and the gradient, at the points on the following curves corresponding to the given values of x:

24. $y = x^2 - 2x + 1$, $x = 2$.   **25.** $y = x^2 + x + 1$, $x = 0$.

26. $y = x^2 - 2x$, $x = -1$.

27. $y = (x + 2)(x - 4)$, $x = 3$.

28. $y = 3x^2 - 2x^3$, $x = -2$.   **29.** $y = (4x - 5)^2$, $x = \frac{1}{2}$.

Find the coordinates of the points on the following curves at which the gradient has the given values:

30. $y = x^2$; 8.   **31.** $y = x^3$; 12.   **32.** $y = x(2 - x)$; 2.

33. $y = x^2 - 3x + 1$; 0.   **34.** $y = x^3 - 2x + 7$; 1.
35. $y = x^3 - 6x^2 + 4$; -12.   **36.** $y = x^4 - 2x^3 + 1$; 0.
37. $y = x^2 - x^3$; -1.   **38.** $y = x(x - 3)^2$; 0.

Tangents and normals

2.7. DEFINITION. *A normal to a curve at a point is the straight line through the point at right angles to the tangent at the point.* (Fig. 2.10.)

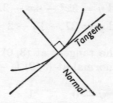

Fig. 2.10

We are now able to find the equations of tangents and normals.

Example 1. *Find the equation of the tangent to the curve $y = x^3$ at the point* (2, 8).

$$y = x^3.$$
$$\therefore \operatorname{grad} y = 3x^2.$$
When $x = 2$, $\operatorname{grad} y = 12.$

Thus the gradient of the tangent at (2, 8) is $+12$. Its equation is

$$\frac{y - 8}{x - 2} = 12,$$

$$\therefore y - 8 = 12x - 24,$$

$\therefore$ the equation of the tangent is $12x - y - 16 = 0$.

Example 3. *Find the equation of the normal to the curve* $y = (x^2 + x + 1)(x - 3)$ *at the point where it cuts the x-axis.*

$$y = (x^2 + x + 1)(x - 3).$$

When $y = 0$, $(x^2 + x + 1)(x - 3) = 0$,

But $x^2 + x + 1 = 0$ has no real roots,

$$\therefore x = +3.$$

$\therefore$ the curve cuts the x-axis at $(3, 0)$.

$$y = x^3 - 2x^2 - 2x - 3.$$

$$\therefore \operatorname{grad} y = 3x^2 - 4x - 2.$$

When $x = 3$, $\operatorname{grad} y = 27 - 12 - 2 = 13$.

The gradient of the tangent at $(3, 0)$ is $+13$, therefore the gradient of the normal at $(3, 0)$ is $-\frac{1}{13}$ (see §1.5) and its equation is

$$\frac{y - 0}{x - 3} = -\frac{1}{13},$$

$$\therefore 13y = -x + 3,$$

$\therefore$ the equation of the normal is $x + 13y - 3 = 0$.

Exercise 2b

1. Find the equations of the tangents to the following curves at the points corresponding to the given values of x:

   (i) $y = x^2$, $x = 2$;   (ii) $y = 3x^2 + 2$, $x = 4$;

   (iii) $y = 3x^2 - x + 1$, $x = 0$;

   (iv) $y = 3 - 4x - 2x^2$, $x = 1$;

   (v) $y = 9x - x^3$, $x = -3$.

2. Find the equations of the normals to the curves in No. 1 at the given points.

3. Find the equation of the tangent and the normal to the curve $y = x^2(x - 3)$ at the point where it cuts the x-axis. Sketch the curve.

4. Repeat No. 3 for the curve $y = x(x - 4)^2$.

5. Find the equation of the tangent to the curve $y = 3x^3 - 4x^2 + 2x - 10$ at the point of intersection with the y-axis.

6. Repeat No. 5 for the curve $y = x^2 - 4x + 3$.

7. Find the values of x for which the gradient function of the curve $y = 2x^3 + 3x^2 - 12x + 3$ is zero. Hence find the equations of the tangents to the curve which are parallel to the x-axis.

8. Repeat No. 7 for the curve $y = 2x^3 - 9x^2 + 10$.

Exercise 2c (Miscellaneous)

1. Find the gradient of the curve $y = 6x - x^3$ at the point where $x = 1$. Find the equation of the tangent to the curve at this point. Where does this tangent meet the line $y = x$?

2. Find the equation of the tangent at the point $(2, 4)$ to the curve $y = x^3 - 2x$. Also find the coordinates of the point where the tangent meets the curve again.

3. Find the equation of the tangent to the curve $y = x^3 - 9x^2 + 20x - 8$ at the point $(1, 4)$. At what points of the curve is the tangent parallel to the line $4x + y - 3 = 0$?

4. Find the equation of the tangent to the curve $y = x^3 + \frac{1}{2}x^2 + 1$ at the point $(-1, \frac{1}{2})$. Find the coordinates of another point on the curve where the tangent is parallel to that at the point $(-1, \frac{1}{2})$.

5. Find the points of intersection with the x-axis of the curve $y = x^3 - 3x^2 + 2x$, and find the equation of the tangent to the curve at each of these points.

6. Find the equations of the normals to the parabola $4y = x^2$ at the points $(-2, 1)$ and $(-4, 4)$. Show that the point of intersection of these two normals lies on the parabola.

7. Find the equation of the tangent at the point $(1, -1)$ to the curve $y = 2 - 4x^2 + x^3$. What are the coordinates of the point where the tangent meets the curve again? Find the equation of the tangent at this point.

8. Find the coordinates of the point P on the curve $8y = 4 - x^2$ at which the gradient is $\frac{1}{2}$. Write down the equation of the tangent to the curve at P. Find also the equation of the tangent to the curve whose gradient is $-\frac{1}{2}$, and the coordinates of its point of intersection with the tangent at P.

9. Find the equations of the tangents to the curve $y = x^3 - 6x^2 + 12x + 2$ which are parallel to the line $y = 3x$.

10. Find the coordinates of the points of intersection of the line $x - 3y = 0$ with the curve $y = x(1 - x^2)$. If these points are in order P, O, Q, prove that the tangents to the curve at P and Q are parallel, and that the tangent at O is perpendicular to them.

11. Find the equations of the tangent and normal to the parabola $x^2 = 4y$ at the point $(6, 9)$. Also find the distance between the points where the tangent and normal meet the y-axis.

12. The curve $y = (x - 2)(x - 3)(x - 4)$ cuts the x-axis at the points $P(2, 0)$, $Q(3, 0)$, $R(4, 0)$. Prove that the tangents at P and R are parallel. At what point does the normal to the curve at Q cut the y-axis?

13. Find the equation of the tangent at the point $P(3, 9)$ to the curve $y = x^3 - 6x^2 + 15x - 9$. If O is the origin, and N is the foot of the perpendicular from P to the x-axis, prove that the tangent at P passes through the mid-point of ON. Find the coordinates of another point on the curve the tangent at which is parallel to the tangent at the point $(3, 9)$.

14. A tangent to the parabola $x^2 = 16y$ is perpendicular to the line $x - 2y - 3 = 0$. Find the equation of this tangent and the coordinates of its point of contact.

15. Find the equation of the tangent to $y = x^2$ at the point $(1, 1)$ and of the tangent to $y = \frac{1}{6}x^3$ at the point $(2, \frac{4}{3})$. Show that these tangents are parallel, and find the distance between them.

16. The point (h, k) lies on the curve $y = 2x^2 + 18$.   Find the gradient at this point and the equation of the tangent there.   Hence find the equations of the two tangents to the curve which pass through the origin.

17. For the curve $y = x^2 + 3$ show that $y = 2ax - a^2 + 3$ is the equation of the tangent at the point whose x-coordinate is a.   Hence find the coordinates of the two points on the curve, the tangents at which pass through the point $(2, 6)$.

CHAPTER 3

VELOCITY AND ACCELERATION

Gradient and velocity

3.1 The reader will have met "travel graphs" in his study of mathematics. One such graph is shown in Fig. 3.1,

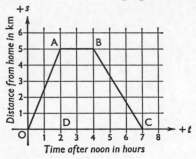

Fig. 3.1

representing a man walking to see a friend who lives five km away, staying two hours, and then returning home. On his outward journey represented by **OA**, he travels 5 km in 2 hours, and his velocity, $\frac{5}{2}$ km/h, is represented by $\dfrac{\text{DA}}{\text{OD}}$, the gradient of **OA**.

Whilst with his friend his velocity is zero; this is represented by the gradient of **AB**.

On his return journey, the gradient of **BC** gives his velocity as $-\frac{5}{3}$ km/h. The negative sign denotes that he is now travelling in the opposite direction; he is *decreasing* the distance from home.

This type of graph in which the distance, *s*, is plotted against the time, *t*, is called a *space-time graph*.

Variable velocity

When velocity is variable, as in a car journey, we may be concerned with the average velocity, which we need to define.

DEFINITION

$$Average\ velocity\ is\ \frac{total\ distance\ travelled}{total\ time\ taken}\ or\ \frac{increase\ in\ s}{increase\ in\ t}.$$

When a car is increasing its speed or slowing down, the moving speedometer needle nevertheless gives a reading *at any instant*, and we must now deal with the idea of *the velocity at an instant*.

Suppose that a car, starting from rest, increases its velocity steadily up to 80 km/h. Then the space-time graph is similar to the curve OPQ in Fig. 3.2. The point P we shall

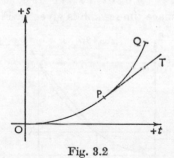

Fig. 3.2

take to correspond to the instant at which the speedometer needle reaches the 60-km/h mark. If from that instant onward the velocity had instead been kept constant at

60 km/h, then the space-time graph would have consisted of the curve OP and the straight line PT of gradient 60.

It would appear that PT is the tangent at P to the original space-time curve OPQ (like cotton under tension leading off a reel), and in that case its gradient would be the same as the gradient of the curve OPQ at P. This suggests that, when the velocity is variable, we mean, by the velocity at an instant, the velocity represented by the gradient of the space-time curve at the corresponding point. However, we must proceed to find a precise definition.

Velocity at an instant

3.2 We consider a stone falling from rest, its velocity steadily increasing. It can be verified by experiment that under certain conditions, it will be s m below its starting point t seconds after the start, where s is given by the formula $s = 4 \cdot 9t^2$. From this we may make a table of values giving the position of the stone at different times.*

Value of t	0	0·5	1·0	1·5	2·0	2·5	3·0
Value of s	0	1·2	4·9	11·0	19·6	30·6	44·1

Part of the space-time graph is given in Fig. 3.3.

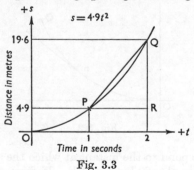

Fig. 3.3

* Throughout §3.2, including Qu. 1 to 5, we work to one decimal place.

From $t = 1$ to $t = 2$, the *average velocity* is represented by the gradient of the chord PQ.

$$\frac{RQ}{PR} = \frac{19{\cdot}6 - 4{\cdot}9}{2 - 1} = 14{\cdot}7.$$

∴ the average velocity is $14{\cdot}7$ m/s.

Qu. 1. How far does the stone move in the interval $t = 1$ to $t = 1{\cdot}5$? What is the average velocity during this interval?

Qu. 2. Repeat Qu. 1 for the intervals (i) $t = 1$ to $t = 1{\cdot}1$, and (ii) $t = 1$ to $t = 1 + h$.

The smaller we make the time interval (letting $Q \rightarrow P$ along the curve), the nearer the average velocity (the gradient of PQ) approaches the velocity given by the gradient of the curve at P.

Now we have seen that the gradient of the curve at P is the limit of the gradient of PQ as $Q \rightarrow P$ (§2.2); this leads to the following definition.

DEFINITION

The velocity at an instant is the limit of the average velocity for an interval following that instant, as the interval tends to zero.

Qu. 3. From your answer to Qu. 2 (ii) determine the actual velocity at the instant when $t = 1$.

Qu. 4. Calculate the distance moved, and the average velocity during the following intervals:

  (i) $t = 2$ to $t = 3$,   (ii) $t = 2$ to $t = 2{\cdot}5$,
 (iii) $t = 2$ to $t = 2{\cdot}1$,   (iv) $t = 2$ to $t = 2 + h$.

Deduce the velocity when $t = 2$.

The definition given above identifies the velocity at an instant with the gradient of the space-time graph for the corresponding value of t. If we are given s in terms of t we can therefore find an expression for the velocity of the stone at any instant by differentiation.

When $s = 4{\cdot}9t^2$, the velocity, v m/s, is given by

$$v = \text{grad } s = 9{\cdot}8t.$$

Thus

$$\text{when } t = 0, \quad v = 0,$$
$$\text{when } t = 1, \quad v = 9\cdot8,$$
$$\text{when } t = 2, \quad v = 19\cdot6, \quad \text{etc.}$$

Qu. 5. A stone is thrown vertically downwards from the top of a cliff, and the depth below the top, s m, after t s, is given by the formula $s = 2t + 4\cdot9t^2$.

  (i) Where is the stone after 1, 2, 3, 4 s?
  (ii) What is its velocity at these times?
  (iii) What is its average velocity during the 3rd second (from $t = 2$ to $t = 3$)?

The symbols Δs and Δt

3.3. The idea of gradient helped us to arrive at the definition of velocity at an instant. It is instructive to take the definition as our starting point; and now, without reference to graphical ideas, we shall again demonstrate that velocity is found by differentiating the expression for s in terms of t. To do this it is convenient to introduce some new symbols, which will be of great use from now onwards.

Again we deal with the stone which falls s metres from rest in t seconds. Suppose that it falls a further small distance Δs metres in the additional small interval of time Δt seconds.

[The symbol Δt, read as "delta t", is used to denote a small increase, or *increment*, in time. Note that Δt is a single symbol; it does not mean Δ multiplied by t. Δs is the corresponding *increment* in distance.]

The average velocity for the time interval Δt (i.e. from t to $t + \Delta t$) is $\dfrac{\Delta s}{\Delta t}$ m/s, and we now obtain an expression for this in terms of t.

Since the stone falls $(s + \Delta s)$ metres in $(t + \Delta t)$ seconds

$$s + \Delta s = 4\cdot9(t + \Delta t)^2,$$

i.e. $\qquad s + \Delta s = 4\cdot9t^2 + 9\cdot8t\,.\,\Delta t + 4\cdot9\,.\,(\Delta t)^2.$

But $\qquad s \qquad = 4\cdot9t^2,$

and subtracting,

$$\Delta s = \qquad 9{\cdot}8t \,.\, \Delta t + 4{\cdot}9 \,.\, (\Delta t)^2.$$

To find the average velocity between time t and time $(t + \Delta t)$ we divide each side by Δt, giving

$$\frac{\Delta s}{\Delta t} = 9{\cdot}8t + 4{\cdot}9 \,.\, \Delta t.$$

As $\Delta t \to 0$ the R.H.S. $\to 9{\cdot}8t$.

By the definition of velocity at an instant, the velocity, v m/s, at time t, is the limit of $\dfrac{\Delta s}{\Delta t}$ as $\Delta t \to 0$, hence

$$v = 9{\cdot}8t.$$

The fact that this process is identical with that of finding the gradient function of $s = 4{\cdot}9t^2$ is readily seen from Fig. 3.4.

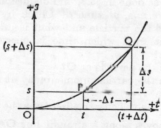

Fig. 3.4

Exercise 3a

1. A stone is thrown vertically upwards at 35 m/s. It is s m above the point of projection t s later, where $s = 35t - 4{\cdot}9t^2$.

   (i) What is the distance moved, and the average velocity during the 3rd second (from $t = 2$ to $t = 3$)?

   (ii) Find the average velocities for the intervals $t = 2$ to
        $t = 2\cdot5$, $t = 2$ to $t = 2\cdot1$, $t = 2$ to $t = 2 + h$.
  (iii) Deduce the actual velocity at the end of the 2nd second.

2. A stone is thrown vertically upwards at $24\cdot5$ m/s from a
   point on the level with but just beyond a cliff ledge. Its
   height above the ledge t s later is $4\cdot9t(5 - t)$ m. If its
   velocity is v m/s, differentiate to find v in terms of t.
   (i) When is the stone at the ledge level?
   (ii) Find its height and velocity after 1, 2, 3, and 6 s.
   (iii) What meaning is attached to a negative value of s?
        A negative value of v?
   (iv) When is the stone momentarily at rest? What is the
        greatest height reached?
   (v) Find the total distance moved during the 3rd second.

3. A particle moves along a straight line so that it is s m from
   a fixed point O on the line t s after a given instant, where
   $s = 3t + t^2$. After $(t + \Delta t)$ s it is $(s + \Delta s)$ m from O.
   Find the average velocity during the time interval t to
   $(t + \Delta t)$ as was done in §3.3, and deduce an expression for
   the velocity v m/s, at time t. Check by differentiation.
   (ii) Where is the particle and what is its velocity at the
        instant from which time is measured (i.e. when
        $t = 0$)?
   (ii) When is the particle at O?
   (iii) When is the particle momentarily at rest? Where is
        it then?
   (iv) What is the velocity the first time the particle is at O?

4. A particle moves along a straight line OA in such a way
   that it is s m from O t s after the instant from which time
   is measured, where $s = 6t - t^2$. A is to be taken as being
   on the positive side of O.
   (i) Where is the particle when $t = 0, 2, 3, 4, 6, 7$? What
       is the meaning of a negative value of s?
   (ii) Differentiate the given expression to find the velocity,
        v m/s, in terms of t. Find the value of v when
        $t = 0, 2, 4, 6$. What is the meaning of a negative
        value of v?

(iii) When and where does the particle change its direction of motion?

5. A slow train which stops at every station passes a certain signal box at noon. Its motion between the two adjacent stations is such that it is s km past the signal box t min past noon, where $s = \frac{1}{6}t + \frac{1}{6}t^2 - \frac{1}{27}t^3$. Find
    (i) the time of departure from the first station, and the time of arrival at the second,
    (ii) the distance of each station from the signal box,
    (iii) the average velocity between the stations,
    (iv) the velocity with which the train passes the signal box.
6. Repeat No. 5 in the case where $s = \frac{1}{72}t(36 - 3t - 2t^2)$.
7. A stone is thrown vertically downwards at 19·6 m/s from the top of a cliff 24·5 m high. It is s m below the top after t s, where $s = 19·6t + 4·9t^2$. Calculate the velocity with which it strikes the beach below.

Constant acceleration

3.4. Earlier in this chapter we used the formula $s = 4·9t^2$ for a stone falling from rest. On differentiation $v = \text{grad } s = 9·8t$. The stone's velocity is 9·8, 19·6, 29·4, 39·2 . . . m/s at the end of successive seconds, and it is steadily increasing by 9·8 m/s in each second. This *rate* at which the stone's velocity increases is called its *acceleration*. This particular formula is based on the assumption that gravity is producing a *constant* acceleration of 9·8 m per second per second, written usually as 9·8 m/s² or 9·8 ms⁻².

Fig. 3.5 shows the corresponding *velocity-time graph*. The

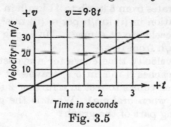

Fig. 3.5

equation $v = 9 \cdot 8t$ (being of the form $y = mx$) represents graphically a straight line through the origin of gradient $9 \cdot 8$. In this case then, the acceleration is represented by the gradient of the velocity-time graph.

Qu. 6. A stone is thrown vertically downwards with a velocity of 10 m/s, and gravity produces on it an acceleration of $9 \cdot 8$ m/s².
  (i) What is the velocity after 1, 2, 3, t s?
  (ii) Sketch the velocity-time graph.

If a particle has an initial velocity u m/s and a constant acceleration a m/s², then its velocity after t s is $(u + at)$ m/s and the equation $v = u + at$ (being of the form $y = mx + c$) represents a straight line of gradient a.

Thus when acceleration is constant, it is represented by *the gradient* of the straight line *velocity-time graph*.

Exercise 3b

(Acceleration is constant.)

1. At the start and end of a two-second interval, a particle's velocity is observed to be 5, 10 m/s. What is its acceleration?
2. A body starts with velocity 15 m/s, and at the end of the 11th second its velocity is 48 m/s. What is its acceleration?
3. Express an acceleration of 5 m/s² in (i) km/h per s, (ii) km/h².
4. A car accelerates from 5 km/h to 41 km/h in 10 s. Express this acceleration in (i) km/h per s, (ii) m/s², (iii) km/h².
5. A car can accelerate at 4 m/s². How long will it take to reach 90 km/h from rest?
6. Sketch the velocity time curve for a cyclist who, starting from rest, reaches 3 m/s in 5 s, travels at that speed for 20 s, and then comes to rest in a further 2 s. What is his acceleration when braking? What is the gradient of the corresponding part of the graph?

7. An express train reducing its velocity to 40 km/h, has to apply the brakes for 50 s. If the retardation produced is 0·5 m/s², find its initial velocity in km/h.

Variable acceleration

3.5. A car starts from rest and moves a distance s m in t seconds, where $s = \frac{1}{6}t^3 + \frac{1}{4}t^2$. If its velocity after t s is v m/s, then $v = \text{grad } s = \frac{1}{2}t^2 + \frac{1}{2}t$. The following table gives some corresponding values of v and t:

t	0	1	2	3	4
v	0	1	3	6	10

The increases in velocity during the first four seconds are 1 m/s, 2 m/s, 3 m/s, 4 m/s respectively. Since the rate of increase of the velocity is not constant in this case, we shall first investigate the average rate of increase over a given time interval.

DEFINITION

Average acceleration is $\dfrac{increase\ in\ v}{increase\ in\ t}$.

Thus from $t = 0$ to $t = 2$,

the average acceleration $= \dfrac{3 - 0}{2} = 1\frac{1}{2}$ m/s²,

and from $t = 2$ to $t = 4$,

the average acceleration $= \dfrac{10 - 3}{2} = 3\frac{1}{2}$ m/s².

Clearly the acceleration itself is increasing with the time, and the next step is to define what is meant by the acceleration at an instant.

DEFINITION

The acceleration at an instant is the limit of the average acceleration for an interval following that instant, as the interval tends to zero.

Using the notation of §3.3, if Δv is the small increase in velocity which occurs in time Δt, then the average acceleration for that interval is $\dfrac{\Delta v}{\Delta t}$, and the acceleration at time t is the limit of this as $\Delta t \to 0$.

Reference to the velocity-time graph given in Fig. 3.6

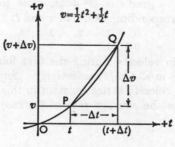

Fig. 3.6

shows that the average acceleration $\dfrac{\Delta v}{\Delta t}$ is the gradient of the chord PQ, and the limit is the gradient of the graph at P.

Thus an expression for the acceleration at time t may be found by differentiating the expression for v.

Example 1. *A car starts from rest and moves a distance s m in t s, where $s = \frac{1}{6}t^3 + \frac{1}{4}t^2$. What is the initial acceleration, and the acceleration at the end of the 2nd second?*

$$s = \tfrac{1}{6}t^3 + \tfrac{1}{4}t^2.$$
$$v = \operatorname{grad} s = \tfrac{1}{2}t^2 + \tfrac{1}{2}t.$$
$$a = \operatorname{grad} v = t + \tfrac{1}{2}.$$

When $t = 0$, $\qquad a = \tfrac{1}{2}$,
and when $t = 2$, $\qquad a = 2\tfrac{1}{2}$.

Hence the required accelerations are $\frac{1}{2}$ m/s², and $2\frac{1}{2}$ m/s².

Before reading Example 2 the reader should refer once again to the definitions of *average velocity* and *average acceleration*. In particular it should be noted that (i) average velocity is not the same as the average of the initial and final velocities (unless the acceleration is constant); and (ii) average acceleration is not necessarily the same as the average of the initial and final accelerations.

Example 2. *A particle moves along a straight line in such a way that its distance from a fixed point O on the line after t s is s m, where $s = \frac{1}{6}t^4$. Find* (i) *its velocity after 3 s, and after 4 s,* (ii) *its average velocity during the 4th second,* (iii) *its acceleration after 2 s, and after 4 s, and* (iv) *its average acceleration from $t = 2$ to $t = 4$.*

$$s = \tfrac{1}{6}t^4.$$
$$v = \text{grad } s = \tfrac{2}{3}t^3.$$
$$a = \text{grad } v = 2t^2.$$

(i) When $t = 3$,
$$v = \tfrac{2}{3}\,.\,3^3 = 18 \text{ m/s,}$$
and when $t = 4$,
$$v = \tfrac{2}{3}\,.\,4^3 = 42\tfrac{2}{3} \text{ m/s.}$$

Hence after 3 s and 4 s, the velocity is 18 m/s and $42\frac{2}{3}$ m/s respectively.

(ii) When $t = 3$,
$$s = \tfrac{81}{6} = 13\tfrac{1}{2} \text{ m,}$$
and when $t = 4$,
$$s = \tfrac{256}{6} = 42\tfrac{2}{3} \text{ m,}$$

∴ the average velocity during the 4th second
$$= \frac{42\tfrac{2}{3} - 13\tfrac{1}{2}}{1} = 29\tfrac{1}{6} \text{ m/s.}$$

(iii) When $t = 2$,
$$a = 2 \cdot 2^2 = 8 \text{ m/s}^2,$$
and when $t = 4$,
$$a = 2 \cdot 4^2 = 32 \text{ m/s}^2.$$

(iv) When $t = 2$,
$$v = \tfrac{2}{3} \cdot 2^3 = 5\tfrac{1}{3} \text{ m/s},$$
and when $t = 4$,
$$v = \tfrac{2}{3} \cdot 4^3 = 42\tfrac{2}{3} \text{ m/s}$$

The change in velocity $= 37\tfrac{1}{3}$ m/s,

∴ the average acceleration from $t = 2$ to $t = 4$ is

$$\frac{37\tfrac{1}{3}}{2} \text{ m/s}^2 = 18\tfrac{2}{3} \text{ m/s}^2.$$

Exercise 3c

1. A stone is thrown vertically upwards, and after t s its height is h m, where $h = 10 \cdot 5t - 4 \cdot 9t^2$. Determine, with particular attention to the signs, the height, velocity and acceleration of the stone (i) when $t = 1$, (ii) when $t = 2$, and (iii) when $t = 3$. Also state clearly in each case whether the stone is going up or down, and whether its speed is increasing or decreasing.

2. A stone is thrown downwards from the top of a cliff, and after t s it is s m below the top, where $s = 20t + 4 \cdot 9t^2$. Find how far it has fallen, its velocity, and its acceleration at the end of the first second.

3. A ball is thrown vertically upwards and its height after t s is s m where $s = 25 \cdot 2t - 4 \cdot 9t^2$. Find
   (i) its height and velocity after 3 s,
   (ii) when it is momentarily at rest,
   (iii) the greatest height reached,
   (iv) the distance moved in the 3rd second,
   (v) the acceleration when $t = 2\tfrac{4}{7}$.

4. A particle moves in a straight line so that after t s it is s m from a fixed point O on the line, where $s = t^4 + 3t^2$.

Find

(i) the acceleration when $t = 1$, $t = 2$, and $t = 3$,

(ii) the average acceleration between $t = 1$ and $t = 3$.

5. At the instant from which time is measured a particle is passing through O and travelling towards A, along the straight line OA. It is s m from O after t s where $s = t(t - 2)^2$.

  (i) When is it again at O?

  (ii) When and where is it momentarily at rest?

  (iii) What is the particle's greatest displacement from O, and how far does it move, during the first 2 s?

  (iv) What is the average velocity during the 3rd second?

  (v) At the end of the 1st second where is the particle, which way is it going, and is its speed increasing or decreasing?

6. Repeat Qu. 5 (v) for the instant when $t = -1$.

7. A particle moves along a straight line so that after t s, its distance from O a fixed point on the line is s m where $s = t^3 - 3t^2 + 2t$.

  (i) When is the particle at O?

  (ii) What is its velocity and acceleration at these times?

  (iii) What is its average velocity during the 1st second?

  (iv) What is its average acceleration between $t = 0$ and $t = 2$?

Exercise 3d (Miscellaneous)

1. The distance of a moving point from a fixed point in its straight line of motion is s m, at a time t s after the start. If $s = \frac{1}{10}t^2$, find the distances travelled from rest by the end of the 1st, 2nd, 3rd, 4th, and 5th seconds.

  Draw a graph plotting distance against time, taking 2 cm to represent both 1 m and 1 s. Draw a tangent to your graph at the point where $t = 3 \cdot 5$ and measure its slope; deduce the velocity of the moving point when $t = 3 \cdot 5$.

2. A point moves along a straight line so that, at the end of t s, its distance from a fixed point on the line is $t^3 - 2t^2 + t$ m. Find the velocity and acceleration at the end of 3 s.

3. A particle moves in a straight line and its distance (s m) from the point at which it is situated at zero time is given in terms of the time (t s) by the formula $s = 45t + 11t^2 - t^3$. Find the velocity and acceleration after 3 s, and prove that the particle will come to rest after 9 s.   (C)

4. A particle moves along the x-axis in such a way that its distance x cm from the origin after t s is given by the formula $x = 27t - 2t^2$. What are its velocity and acceleration after 6·75 s? How long does it take for the velocity to be reduced from 15 cm/s to 9 cm/s, and how far does the particle travel meanwhile?

5. A point moves along a straight line OX so that its distance x cm from the point O at time t s is given by the formula $x = t^3 - 6t^2 + 9t$. Find
   (i) at what times and in what positions the point will have zero velocity,
   (ii) its acceleration at those instants,
   (iii) its velocity when its acceleration is zero.

6. A particle moves in a straight line so that its distance x cm from a fixed point O on the line is given by $x = 9t^2 - 2t^3$ where t is the time in seconds measured from O. Find the speed of the particle when $t = 3$. Also find the distance from O of the particle when $t = 4$, and show that it is then moving towards O.

7. A particle moves along the x-axis in such a way that its distance x cm from the origin after t s is given by the formula $x = 7t + 12t^2$. What distance does it travel in the nth second? What are its velocity and acceleration at the end of the nth second?

CHAPTER 4

MAXIMA AND MINIMA

The symbols Δx, Δy and $\dfrac{dy}{dx}$

4.1. In Chapter 3 we met the symbols Δs and Δt, and to extend the scope of differentiation it is convenient to denote small increases in x and y as Δx and Δy in the same way. If P is the point (x, y) on a curve, and Q is another point, and if the increase in x in moving from P to Q is Δx, then the corresponding increase in y is Δy; thus Q is the point $(x + \Delta x, y + \Delta y)$. (Fig. 4.1.)

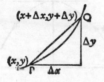

Fig. 4.1

The gradient of the chord PQ is $\dfrac{\Delta y}{\Delta x}$, and the gradient of the curve at P is the limit of $\dfrac{\Delta y}{\Delta x}$, as $\Delta x \to 0$. Up to now we have denoted this limit as "grad y" to keep in mind the fundamental idea of gradient in relation to differentiation. We will in future adopt the usual practice of writing this limit as $\dfrac{dy}{dx}$, the symbol $\dfrac{d}{dx}$ being an instruction to

61

differentiate.* Thus, if $y = x^2, \dfrac{dy}{dx} = 2x$; or we may write

$\dfrac{d}{dx}(x^2) = 2x$. The gradient function will also be referred

to in future as the *derived function*, or *derivative*.

Qu. 1. Find $\dfrac{dy}{dx}$ when

(i) $y = x^2 - 4x$,   (ii) $y = 3x^2 - 3$,
(iii) $y = 2x^3 - 5x^2 + 1$,   (iv) $y = x(x - 2)$,
(v) $y = x(x + 1)(x - 3)$.

Greatest and least values

4.2. Fig. 4.2 represents the path of a stone thrown from O, reaching its greatest height AB, and striking the ground at C. Between O and A, when the stone is climbing, the

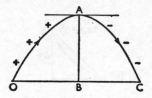

Fig. 4.2

gradient is positive but steadily decreases to zero at A. Past A the stone is descending, and the path has a negative gradient.

The curve $y = x^2$ of which we made much use earlier on, is called a parabola. A more general equation of this type

* NOTE.—This notation $\dfrac{d}{dx}$ serves to indicate that we are differentiating with respect to x. Thus $\dfrac{d}{dy}(y^3) = 3y^2$, and $\dfrac{d}{dt}(2t^2) = 4t$.

of curve is of the form $y = ax^2 + bx + c$; when a is positive, we get a curve like a valley, such as DEF in Fig. 4.3, on which y has a least value (GE); when a is negative, we get a curve like a mountain top, such as OAC in Fig. 4.3, on which y has a greatest value (BA).

We must now investigate the sign of the gradient on either side of the maximum (A) or the minimum (E); this is best seen in Fig. 4.3 where the *positive* gradient on approach

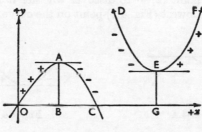

Fig. 4.3

If we allow our eye to travel along each curve in Fig 4.3 from left to right (the direction in which x increases), we notice that in passing through A, where y has a greatest value, the gradient is zero and is changing sign *from positive to negative*; on the other hand in passing through E, where y has a least value, the gradient is zero and is changing sign *from negative to positive*. This distinction enables us to investigate the highest or lowest point on a parabola without going to the length of plotting the curve in detail.

Example 1. *Find the greatest or least value of y on the curve $y = 4x - x^2$. Sketch the curve.*

$$y = 4x - x^2.$$

$$\frac{dy}{dx} = 4 - 2x,$$

$$= 2(2 - x).$$

The gradient is zero when

$$2(2 - x) = 0,$$
$$x = 2,$$
and $$y = 4.2 - 2^2 = 4$$

We must now investigate the sign of the gradient on either side of the point $(2, 4)$ to discover whether it is a highest (Fig. 4.4) or lowest (Fig. 4.5) point on the curve.

Fig. 4.4   Fig. 4.5

Just to the left of $(2, 4)$,

     x is just less than 2, and $\dfrac{\mathrm{d}y}{\mathrm{d}x}$ is positive.

Just to the right of $(2, 4)$,

     x is just greater than 2, and $\dfrac{\mathrm{d}y}{\mathrm{d}x}$ is negative.

Thus Fig. 4.4. gives the shape of the curve at $(2, 4)$, and the greatest value of y is $+ 4$.

To make a rough sketch of the curve, we find where it cuts the axes.

$$y = 4x - x^2.$$
When $x = 0$, $y = 0$;

     $\therefore$ the curve passes through $(0, 0)$.

When $y = 0$,

$$4x - x^2 = 0,$$
$$x(4 - x) = 0,$$
$$x = 0 \text{ or } 4,$$

     $\therefore$ the curve passes through $(0, 0)$ and $(4, 0)$.

From this information we can make the sketch (Fig. 4.6).

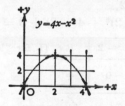

Fig. 4.6

Qu. 2. Find the coordinates of the points on the following curves where the gradient is zero.

(i) $y = 4x - 2x^2$, (ii) $y = 3x^2 + 2x - 5$,
(iii) $y = 4x^2 - 6x + 2$.

At this stage the reader must be clear about the meaning of "greater than" and "less than" in respect of negative numbers. For example, $-3 \cdot 1$ is *less* than -3, and $-2 \cdot 9$ is *greater* than -3.

Qu. 3. Find the values of x for which the following derived functions are zero, and determine whether the corresponding graphs have a highest or a lowest point for these values of x:

(i) $\dfrac{dy}{dx} = 5 - 3x$, (ii) $\dfrac{dy}{dx} = 6x - 7$,

(iii) $\dfrac{dy}{dx} = 2x + 3$, (iv) $\dfrac{dy}{dx} = -4 - 5x$.

The investigation of the sign of the gradient may be conveniently laid out in the way shown in the following example.

Example 2. *Find the greatest or least value of* $x^2 + 4x + 3$ *and the value of* x *for which it occurs.*

Let
$$y = x^2 + 4x + 3,$$
$$\frac{dy}{dx} = 2x + 4,$$
$$= 2(x + 2).$$

The gradient is zero when

$$x = -2,$$

and $\qquad y = (-2)^2 + 4(-2) + 3 = -1$

Value of x	L	-2	R
Sign of $\dfrac{dy}{dx}$	$-$	0	$+$

[L for "left", R for "right"]

When $x = -2$, $x^2 + 4x + 3$ has the least value -1.

This method can be used to solve some practical problems, as in the following example.

Example 3. 1000 m *of fencing is to be used to make a rectangular enclosure. Find the greatest possible area, and the corresponding dimensions.*

If the length is x m, the width will be $(500 - x)$ m, and the area, A m², is given by

$$A = x(500 - x),$$

or $\qquad A = 500x - x^2.$

[This problem could now be solved by drawing accurately the graph of area plotted against length (Fig. 4.7), and reading off the greatest area (N M) and the corresponding length (ON).

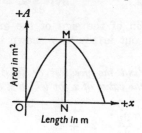

Fig. 4.7

In practice it is, of course, much quicker to continue, along the lines of Example 2, by finding the greatest value of $500x - x^2$, without plotting a graph.]

$$\frac{dA}{dx} = 500 - 2x,$$
$$= 2(250 - x),$$

which is zero when

$$x = 250,$$

and

$$A = 250(500 - 250) = 62\,500.$$

Value of x	L	250	R
Sign of $\dfrac{dA}{dx}$	+	0	−

The greatest area is 62 500 m², when the length is 250 m, and the width is 250 m.

Exercise 4a

1. Find $\dfrac{dy}{dx}$ when
   (i) $y = 3x^2 - 2x + 5$,   (ii) $y = 5x^2 + 4x - 6$,
   (iii) $y = 2x(1 - x)$,   (iv) $y = (x + 1)(3x - 2)$,
   (v) $y = 3(2u - 1)(4x + 3)$.

2. Find the coordinates of the points on the following curves where the gradient is zero:
   (i) $y = x^2 + 5x - 2$,   (ii) $y = 5 + 9x - 7x^2$,
   (iii) $y = x(3x - 2)$,   (iv) $y = (2 + x)(3 - 4x)$.

3. Find the values of x for which the following derived functions are zero, and determine whether the corresponding graphs have a highest or a lowest point for these values of x:
   (i) $\dfrac{dy}{dx} = 2x - 5$,   (ii) $\dfrac{dy}{dx} = \tfrac{1}{2}x + 3$,
   (iii) $\dfrac{dy}{dx} = \tfrac{1}{3} - \tfrac{1}{4}x$,   (iv) $\dfrac{dy}{dx} = -5 - \tfrac{1}{5}x$.

4. Find the greatest or least values of the following functions:
  (i) $x^2 - x - 2$,   (ii) $x(4 - x)$,
  (iii) $15 + 2x - x^2$,   (iv) $(2x + 3)(x - 2)$.

5. Sketch the graphs of the functions in No. 4.

6. A ball is thrown vertically upwards from ground level and its height after t s is $(15 \cdot 4t - 4 \cdot 9t^2)$ m. Find the greatest height it reaches, and the time it takes to get there.

7. A farmer has 100 m of metal railing with which to form two adjacent sides of a rectangular enclosure, the other two sides being two existing walls of the yard, meeting at right angles. What dimensions will give him the maximum possible area?

8. A stone is thrown into a mud bank and penetrates $(1200t - 36\,000t^2)$ cm in t s after impact. Calculate the maximum depth of penetration.

9. A rectangular sheep pen is to be made out of 1000 m of fencing, using an existing straight hedge for one of the sides. Find the maximum area possible, and the dimensions necessary to achieve this.

10. An aeroplane flying level at 250 m above the ground suddenly swoops down to drop supplies, and then regains its former altitude. It is h m above the ground t s after beginning its dive, where $h = 8t^2 - 80t + 250$. Find its least altitude during this operation, and the interval of time during which it was losing height.

11. Fig. 4.8 represents the end view of the outer cover of a match box, AB and EF being gummed together, and assumed to be the same length. If the total length of edge (ABCDEF) is 12 cm, calculate the lengths of AB and BC which will give the maximum possible cross-section area.

Fig. 4.8

To differentiate $\dfrac{1}{x}$

4.3. In §2.6 we reached the conclusion that if n was any positive whole number, grad $x^n = nx^{n-1}$, or using the new notation, $\dfrac{\mathrm{d}}{\mathrm{d}x}(x^n) = nx^{n-1}$. We shall now investigate whether this general result does, in fact, cover a case where n is a negative whole number, by finding the gradient function of the curve $y = \dfrac{1}{x}$, i.e. $y = x^{-1}$ (§8.3).

$P(x, y)$ and $Q(x + \Delta x, y + \Delta y)$ are two points on the curve $y = \dfrac{1}{x}$. Since Q lies on the curve, its coordinates must satisfy the equation (§1.6),

$$\therefore\ y + \Delta y = \frac{1}{x + \Delta x}.$$

But
$$y = \frac{1}{x}.$$

Subtracting,
$$\Delta y = \frac{1}{x + \Delta x} - \frac{1}{x},$$
$$= \frac{x - x - \Delta x}{x(x + \Delta x)},$$
$$= \frac{-\Delta x}{x^2 + x.\Delta x}.$$

$$\therefore\ \frac{\Delta y}{\Delta x} = \frac{-1}{x^2 + x.\Delta x}.$$

As $\Delta x \to 0$, $\dfrac{\Delta y}{\Delta x} \to \dfrac{-1}{x^2}$,

$$\therefore\ \frac{\mathrm{d}y}{\mathrm{d}x} = \frac{-1}{x^2}.$$

Thus the general statement $\dfrac{\mathrm{d}}{\mathrm{d}x}(x^n) = nx^{n-1}$ does apply when $n = -1$, since we have shown that $\dfrac{d}{dx}(x^{-1}) = -x^{-2}$;

and we shall now assume it to be valid when n is any positive or negative whole number, or zero.

Qu. 4.   Write down the derivative of

(i) x^{-4},   (ii) $\dfrac{3}{x^2}$,   (iii) $\dfrac{2}{x^3}$,   (iv) $\dfrac{1}{2x^3}$,

(v) $\dfrac{1}{x^m}$,   (vi) $2x^2 - 3x + 4 + \dfrac{5}{x}$,   (vii) $\dfrac{x^3 + 3x - 4}{x^2}$.

Maxima and minima

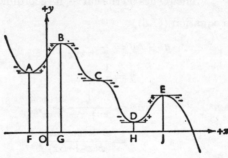

Fig. 4.9

4.4.   In §4.3 we were dealing with a type of curve whose gradient was zero only at one point.   With a more complicated curve (Fig. 4.9) the gradient may be zero at a number of points, and the possible shapes fall into three categories. In this case, moving along the curve from left to right, that is with x increasing,

(i) at A and D, the gradient is changing from negative to positive, and these are called **minimum points**; FA and HD are **minimum values** of y (or **minima**),

(ii) at B and E, the gradient is changing from positive to negative, and these are called **maximum points**; GB and JE are **maximum values** of y (or **maxima**).

The reader will note that the words maximum and minimum are used in the sense of greatest and least only in the immediate vicinity of the point; this local meaning is brought out clearly in this curve, since a maximum value, JE is in fact less than a minimum value, FA.

(iii) At C the gradient is zero, but is *not* changing sign; this is a **point of inflexion**, which may be likened to the point on an S-bend at which a road stops turning left and begins to turn right, or vice versa. The gradient of a curve at a point of inflexion need not be zero (the reader should be able to spot four more in Fig. 4.9); however at this stage we are concerned only with searching for maxima and minima, and we need to bear in mind points of inflexion only as a third possibility at points where the gradient is zero.

At any point where the gradient of a curve is zero, y is said to have a **stationary value**. Any maximum or minimum point is called a **turning point**, and y is said to have a **turning value** there.

Qu. 5. Copy the following figures, and on each draw the tangents at all points where the gradient is zero, and mark in

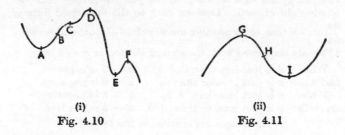

(i)

Fig. 4.10

(ii)

Fig. 4.11

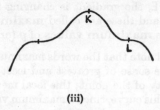

(iii)

Fig. 4.12

the sign of the gradient.   State whether the points marked are maxima, minima, or points of inflexion.

Example 4. *Investigate the stationary values of the function $x^4 - 4x^3$.*

Let
$$y = x^4 - 4x^3.$$
$$\frac{dy}{dx} = 4x^3 - 12x^2,$$
$$= 4x^2(x - 3),$$

which is zero when $x = 0$ or $+3$.

When $x = 0$, $y = 0$, and when $x = 3$, $y = -27$. Thus the stationary values of the function occur at $(0, 0)$ and at $(+3, -27)$.

[We now find the shape of the curve at these points by investigating the sign of the gradient just to the left and just to the right of each.   Looking back to the factorized form of $\frac{dy}{dx}$, we see that $4x^2$ is positive for all values of x other than zero, so we are concerned with the sign of the factor $x - 3$ only.

When x is just less than 0, $x - 3$ is negative,
and when x is just greater than 0, $x - 3$ is negative.
When x is just less than $+3$, $x - 3$ is negative,
and when x is just greater than $+3$, $x - 3$ is positive.

These signs are entered in the table.]

Value of x	L	0	R	L	+ 3	R
Sign of $\dfrac{dy}{dx}$	—	0	—	—	0	+

INFL. MIN.

The stationary values of $x^4 - 4x^3$ are 0 and $- 27$; $(0, 0)$ is a point of inflexion; $(3, -27)$ is a minimum point.

The following example further illustrates the advisability of arranging the gradient function in a convenient factorized form, and brings out an important point in the investigation of the sign of the gradient for negative values of x.

Example 5. *Find the turning values of y on the graph $y = 5 + 24x - 9x^2 - 2x^3$, and distinguish between them.*

$$y = 5 + 24x - 9x^2 - 2x^3.$$

$$\frac{dy}{dx} = 24 - 18x - 6x^2 = - 6(x^2 + 3x - 4)$$

$$= - 6(x + 4)(x - 1),$$

which is zero when $x = - 4$ or 1.

When $x = - 4$,

$$y = 5 + 24.(- 4) - 9.(- 4)^2 - 2.(- 4)^3 = - 107,$$

and when $x = 1$,

$$y = 5 + 24 - 9 - 2 = 18.$$

Thus the stationary values of y occur at $(- 4, - 107)$ and $(1, 18)$.

[In completing the gradient table we must remember the negative factor $- 6$, and find the sign of each factor $(x + 4)$ and $(x - 1)$; we shall then see if there are one, two or three negative factors, and so determine the sign of $\dfrac{dy}{dx}$.

Let us pay particular attention to the point $(-4, -107)$, and the sign of the factor $(x + 4)$. To the *left*, when x is just *less* than -4 (e.g. $-4\cdot1$), $(x + 4)$ is negative, $(x - 1)$ is also negative, thus $\dfrac{dy}{dx}$ has three negative factors and is negative. To the *right*, when x is just greater than -4 (e.g. $-3\cdot9$), $(x + 4)$ is now positive, $(x - 1)$ is still negative, thus $\dfrac{dy}{dx}$ has two negative factors, and is positive.]

Value of x	L	-4	R	L	1	R
Sign of $\dfrac{dy}{dx}$	$-$	0	$+$	$+$	0	$-$

MIN.   MAX.

The turning values of y are -107 and 18; -107 is a minimum value; 18 is a maximum value.

Exercise 4b

1. Write down the values of x for which the following derived functions are zero, and prepare in each case a gradient table as in the foregoing examples, showing whether the corresponding points on the graphs are maxima, minima or points of inflexion:

(i) $\dfrac{dy}{dx} = 3x^2$,

(ii) $\dfrac{dy}{dx} = -4x^3$,

(iii) $\dfrac{dy}{dx} = (x - 2)(x - 3)$,

(iv) $\dfrac{dy}{dx} = (x + 3)(x - 5)$,

(v) $\dfrac{dy}{dx} = (x + 1)(x + 6)$,

(vi) $\dfrac{dy}{dx} = -(x - 1)(x - 3)$,

(vii) $\dfrac{dy}{dx} = -x^2 + x + 12$,

(viii) $\dfrac{dy}{dx} = -x^2 - 5x + 6$,

(ix) $\dfrac{dy}{dx} = 15 - 2x - x^2$,

(x) $\dfrac{dy}{dx} = 5x^4 - 27x^2$,

(xi) $\dfrac{dy}{dx} = 1 - \dfrac{4}{x^2}$.

2. Find any maximum or minimum values of the following functions:

(i) $4x - 3x^3$,   (ii) $2x^3 - 3x^2 - 12x - 7$,

(iii) $x^2(x - 4)$,   (iv) $x + \dfrac{1}{x}$,

(v) $x(2x - 3)(x - 4)$.

3. Find the turning points on the following curves, and state whether y has a maximum or minimum value at each:

(i) $y = x(x^2 - 12)$,   (ii) $y = x^3 - 5x^2 + 3x + 2$,

(iii) $y = x^2(3 - x)$,   (iv) $y = 4x^2 + \dfrac{1}{x}$,

(v) $y = x(x - 8)(x - 15)$.

4. Investigate the stationary values of y on the following curves:

(i) $y = x^4$,   (ii) $y = 3 - x^3$,

(iii) $y = x^3(2 - x)$,   (iv) $y = 3x^4 + 16x^3 + 24x^2 + 3$.

Fig. 4.13

5. Fig. 4.13 represents a rectangular sheet of metal 8 cm by 5 cm. Equal squares of side x cm are removed from each corner, and the edges are then turned up to make an open box of volume V cm³. Show that $V = 40x - 26x^2 + 4x^3$, and find the maximum possible volume, and the corresponding value of x.

6. Repeat No. 5 when the dimensions of the sheet of metal are 8 cm by 3 cm, showing that in this case

$$V = 24x - 22x^2 + 4x^3.$$

7. The size of a parcel despatched through the post used to be limited by the fact that the sum of its length and girth (perimeter of cross-section) must not exceed 6 feet. What was

the volume of the largest parcel of square cross-section
which was acceptable for posting? (Let the cross-section
be a square of side x feet.)

8. Repeat No. 7 for a parcel of circular cross-section, leaving
π in your answer.

9. A chemical factory wishes to make a cylindrical container,
of thin metal, to hold 10 cm^3, using the least possible area
of metal. If the outside surface is S cm^2, and the radius is
r cm, show that $S = 2\pi r^2 + \dfrac{20}{r}$ and hence find the required
radius and height for the container. (Leave π in your
answer.)

10. Repeat No. 9 showing that whatever may be the given
volume, the area of metal will always be least when the
height is twice the radius.

11. 64 cm^3 of butter is to be made into a slab of square cross-
section. Calculate the required length if the total surface
area is to be as small as possible.

12. An open cardboard box with a square base is required to
hold 108 cm^3. What should be the dimensions if the area
of cardboard used is as small as possible?

Curve sketching

4.5. We have seen in §4.4 how maxima and minima
problems may be solved without direct use of the relevant
graph. Frequently however the determination of maximum and minimum points is a valuable aid in sketching a
curve, when all that is required is a general idea of the shape
and position, and a full table of values is not justified.

Example 6. *Sketch the curve* $y = 4x^3 - 3x^4$.

(i) To find where the curve meets the x-axis, put $y = 0$,
then
$$4x^3 - 3x^4 = 0.$$
$$\therefore x^3(4 - 3x) = 0.$$

Therefore the curve meets the x-axis at the points $(0, 0)$
and $(\tfrac{4}{3}, 0)$.

(ii) To find where the curve meets the y-axis, put $x = 0$. The curve meets the y-axis at the origin.

(iii) To find stationary points:

$$y = 4x^3 - 3x^4.$$
$$\therefore \frac{\mathrm{d}y}{\mathrm{d}x} = 12x^2 - 12x^3 = 12x^2(1 - x),$$

which is zero when $x = 0$ or 1.

Therefore $(0, 0)$ and $(1, 1)$ are stationary points.

Value of x	L	0	R	L	1	R
Sign of $\dfrac{\mathrm{d}y}{\mathrm{d}x}$	+	0	+	+	0	−

INFL.   MAX.

Hence $(0, 0)$ is a point of inflexion and $(1, 1)$ is a maximum.

These results may now be used to sketch the curve, as in Fig. 4.14.

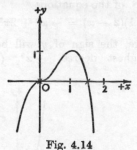

Fig. 4.14

Exercise 4c

Find where the following curves meet the axes. Find, also, the coordinates of their stationary points and use these results to sketch the curves.

1. $y = 3x^2 - x^3$.   2. $y = x^3 - 6x^2$.

3. $y = x^3 - 2x^2 + x$.   4. $y = (x + 1)^2(2 - x)$.

5. $y = x^2(x - 2)^2$.   6. $y = x^4 - 8x^3$.

7. $y = x^4 - 10x^2 + 9$.   8. $y = x^4 + 32x$.

9. $y = 4x^5 - 5x^4$.   10. $y = 3x^5 - 5x^3$.

11. $y = 2x^5 + 5x^2$.

4.6. Another useful approach to curve sketching is shown in the next example.

Example 7. *Sketch the curve* $y = (x + 1)(x - 1)(2 - x)$.

(i) To find where the curve meets the x-axis, put $y = 0$, then

$$(x + 1)(x - 1)(2 - x) = 0.$$

Therefore the curve meets the x-axis at $(-1, 0)$, $(1, 0)$, $(2, 0)$.

(ii) To find where the curve meets the y-axis, put $x = 0$. Thus the curve meets the y-axis at $(0, -2)$.

(iii) To examine the behaviour of the curve "at infinity", expand the R.H.S. of the equation:

$$y = (x^2 - 1)(2 - x) = -x^3 + 2x^2 + x - 2.$$

Now, if x is large, the sign of y will be determined by the term of highest degree, $-x^3$. (If $x = 100$, say,

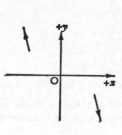

Fig. 4.15

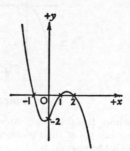

Fig. 4.16

$y = -1\,000\,000 + 20\,000 + 100 - 2$; or if $x = -100$, $y = 1\,000\,000 + 20\,000 - 100 - 2$. In either case the term in x^3 predominates.)

If x is large and positive, y is large and negative, and if x is large and negative, y is large and positive. Thus the behaviour of the curve as $x \to +\infty$ and $x \to -\infty$ is illustrated by Fig. 4.15.

The curve is then sketched, as in Fig. 4.16.

Distance, velocity and acceleration graphs

4.7. Useful physical interpretations of the graphical ideas discussed in §4.4 are obtained from the space-time, velocity-time, and acceleration-time graphs for the motion of a particle, if we plot one above the other as in the following example.

Example 8. O *is a point on a straight line. A particle moves along the line so that it is s m from O, t s after a certain instant, where $s = t(t - 2)^2$. Describe the motion before and after $t = 0$.*

The *space-time graph* has the equation $s = t(t - 2)^2$. By the methods of §4.5 we may determine that the graph has a MAX. point $(\frac{2}{3}, \frac{32}{27})$, a MIN. point $(2, 0)$, and passes through $(0, 0)$. We thus arrive at the upper sketch in Fig. 4.17. The equation may be written $s = t^3 - 4t^2 + 4t$.

$$\therefore \frac{ds}{dt} = 3t^2 - 8t + 4 = (3t - 2)(t - 2).$$

Hence the *velocity-time graph* has the equation $v = (3t - 2)(t - 2)$. This graph has a MIN. point $(1\frac{1}{3}, -1\frac{1}{3})$, and passes through $(\frac{2}{3}, 0)$, $(2, 0)$, and $(0, 4)$; it is the middle sketch in Fig. 4.17.

Differentiating once again, $\dfrac{dv}{dt} = 6t - 8$, and so the *acceleration-time graph* has the equation $a = 6t - 8$, and is the bottom sketch in Fig. 4.17.

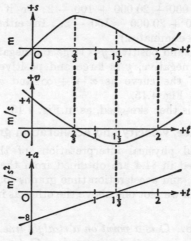

Fig. 4.17

Notice that the MAX. and MIN. values of s occur when $v\left(\text{i.e. } \dfrac{ds}{dt}\right)$ is zero, and that the MIN. value of v occurs when $a\left(\text{i.e. } \dfrac{dv}{dt}\right)$ is zero.

It is easy to visualize the motion of the particle as being along the Os axis of the space-time graph, its distance from O at any instant being given by the height of the graph for the corresponding value of t. Before $t = 0$, the particle is approaching O from the negative side; at $t = 0$, it is passing through O with velocity 4 m/s, and acceleration -8 m/s², hence its speed is decreasing. It comes momentarily to rest $\frac{32}{27}$ m from O (on the positive side) when $t = \frac{2}{3}$; it returns to O, where it is momentarily at rest when $t = 2$, and thereafter it moves away from O in the positive direction.

Some further points regarding the sign and direction of the velocity and acceleration deserve emphasis. Consider the three graphs between $t = 0$ and $t = 1\frac{1}{3}$; throughout this interval the acceleration is negative, and the velocity decreases from $+4$ m/s to $-1\frac{1}{3}$ m/s. The effect of the negative acceleration is to *decrease* the speed when the velocity is positive ($t = 0$ to $t = \frac{2}{3}$), and to *increase* the speed when the velocity is negative ($t = \frac{2}{3}$ to $t = 1\frac{1}{3}$). The reader should note the distinction between the *speed* and the *velocity*, the speed being the numerical value of the velocity, irrespective of sign.

Qu. 6. In Example 8, give the signs of the velocity, and acceleration, and state if the speed is increasing or decreasing, when (i) $t = 1\frac{1}{2}$, (ii) $t = 3$, (iii) $t = 1\frac{1}{4}$.

Exercise 4d

1. Make a rough sketch of each of the following curves by finding the points of intersection with the axes, and by investigating the behaviour of y as $x \to +\infty$ and as $x \to -\infty$. (Do *not* find maximum and minimum points):

   (i) $y = (x + 2)(x - 3)$,   (ii) $y = (5 + x)(1 - x)$,
   (iii) $y = x(x + 1)(x + 2)$,
   (iv) $y = (2 + x)(1 + x)(3 - x)$,
   (v) $y = (x - 1)(x - 3)^2$,   (vi) $y = (x + 4)^2(x - 3)$,
   (vii) $y = -x(x - 7)^2$,   (viii) $y = x^2(5 - x)$,
   (ix) $y = (x - 2)^3$,   (x) $y = (x - 3)^4$,
   (xi) $y = -x(x - 4)^3$.

2. A particle moves along a straight line OB so that t s after passing O it is s m from O, where $s = t(2t - 3)(t - 4)$. Deduce expressions for the velocity and acceleration in terms of t, and sketch the space-, velocity-, and acceleration-time graphs as in Fig. 4.17. Briefly describe the motion, and when $t = 2$ find

   (i) where the particle is,
   (ii) if it is going towards or away from B,
   (iii) its speed,

  (iv) if its speed is increasing or decreasing,
  (v) the rate of change of the speed.
3. Answer the questions in No. 2 for the instant when $t = 1$.
4. With the data of No. 2, when is the particle moving at its greatest speed away from B, and where is it then?
5. A particle is moving along a straight line OA in such a way that t s after passing through O for the first time it is s m from O where $s = - t(t - 8)(t - 15)$. A is taken to be on the positive side of O. Deduce expressions for the velocity and acceleration in terms of t, and sketch the three graphs as in Fig. 4.17. Briefly describe the motion.

  (i) Describe in detail the motion and position of the particle when $t = 10$;
  (ii) when is it moving towards A?
  (iii) when is it travelling at its greatest speed towards A?

6. A car in a traffic jam starts from rest with constant acceleration 2 m/s², and when its velocity reaches 6 m/s it remains constant at that figure for 4 s, and it is then reduced to zero in 6 s at a constant retardation. Sketch the space-, velocity-, and acceleration-time graphs for this motion.

Exercise 4e (Miscellaneous)

1. Find the coordinates of the points on the following curves at which y is a maximum or minimum:

  (i) $y = x^3 - 6x^2 + 9x + 2$,
  (ii) $y = 2x^3 - 3x^2 - 12x + 8$,
  (iii) $y = x^3 - 3x$,   (iv) $y = 4x^3 - 3x^2 - 6x + 4$,
  (v) $y = x^2(x^2 - 8)$,
  (vi) $y = 2(x + 1)(x - 1)^2 + 1$.

2. Find the turning points of the graph $y = 2x^3 + 3x^2 - 12x + 7$, distinguishing between maximum and minimum values. Show that the graph passes through $(1, 0)$ and one other point on the x-axis. Draw a rough sketch of the curve.

3. If $y = x^4 - 2x^2 + 1$, find the values of x for which y is a minimum and draw a rough sketch of the curve.

4. The equation of a curve is $y = x^3 - x^4 - 1$. Has y a maximum or a minimum value (i) when $x = \frac{3}{4}$, (ii) when $x = 0$?

5. Prove that there are two points on the curve $y = 2x^2 - x^4$ at which y has a maximum value, and one point at which y has a minimum value. Give the equations of the tangents to the curve at these three points.

6. A point P whose x-coordinate is a is taken on the line $y = 3x - 7$. If Q is the point $(4, 1)$ show that $PQ^2 = 10a^2 - 56a + 80$. Find the value of a which will make this expression a minimum. Hence show that the coordinates of N, the foot of the perpendicular from Q to the line, are $(2\frac{4}{5}, 1\frac{2}{5})$. Find the equation of QN.

7. The tangent to the curve of $y = ax^2 + bx + c$ at the point where $x = 2$ is parallel to the line $y = 4x$. Given that y has a minimum value of -3 where $x = 1$ find the values of a, b and c.

8. Find the equation of the tangent to the curve $xy = 4$ at the point P whose coordinates are $\left(2t, \dfrac{2}{t}\right)$. If O is the origin and the tangent at P meets the x-axis at A and the y-axis at B, prove
(i) that P is the mid-point of AB,
(ii) that the area of the triangle OAB is the same for all positions of P.

9. Find the equations of the normals to the curve $xy = 4$ which are parallel to the line $4x - y - 2 = 0$.

10. A solid rectangular block has a square base. Find its maximum volume if the sum of the height and any one side of the base is 12 cm.

11. A man wishes to fence in a rectangular enclosure of area 128 m^2 One side of the enclosure is formed by part of a brick wall already in position. What is the least possible length of fencing required for the other three sides?

12. The angle C of triangle ABC is always a right angle.
(i) If the sum of CA and CB is 6 cm, find the maximum area of the triangle.
(ii) If, on the other hand, the hypotenuse AB is kept equal to 4 cm, and the sides CA, CB allowed to vary, find the maximum area of the triangle.

13. A piece of wire of length l is cut into two parts of lengths x and $l - x$. The former is bent into the shape of a square, and the latter into a rectangle of which the base is double the height. Find an expression for the sum of the areas of these two figures. Prove that the only value of x for which this sum is a maximum or a minimum is $x = \dfrac{8l}{17}$; and find which it is.

14. A farmer has a certain length of fencing and uses it all to fence in two square sheep-folds. Prove that the sum of the areas of the two folds is least when their sides are equal.

15. Prove that, if the sum of the radii of two circles remains constant, the sum of the areas of the circles is least when the circles are equal.

16. An open tank is to be constructed with a horizontal square base and four vertical rectangular sides. It is to have a capacity of 32 m³. Find the least area of sheet metal of which it can be made.

17. A sealed cylindrical jam tin is of height h cm and radius r cm. The area of its total outer surface is A cm² and its volume is V cm³. Find an expression for A in terms of r and h. Taking $A = 24\pi$, find
   (i) an expression for h in terms of r, and hence an expression for V in terms of r;
   (ii) the value of r which will make V a maximum.

18. (i) A variable rectangle has a constant perimeter of 20 cm. Find the lengths of the sides when the area is a maximum.
   (ii) A variable rectangle has a constant area 36 cm². Find the lengths of the sides when the perimeter is a minimum.

19. A cylinder is such that the sum of its height and the circumference of its base is 5 m. Express the volume (V m³) in terms of the radius of the base (r m). What is the greatest volume of the cylinder?

20. An open tank is to be constructed with a square base and vertical sides so as to contain 500 m³ of water. What must be the dimensions if the area of sheet metal used in its construction is to be a minimum?

21. The length of a rectangular block is twice the width, and the total surface area is 108 cm². Show that, if the width

of the block is x cm, the volume is $\frac{4}{3}x(27 - x^2)$ cm³. Find
the dimensions of the block when its volume is a maximum.

22. A circular cylinder open at the top is to be made so as to
have a volume of 1 m³. If r m is the radius of the base,
prove that the total outside surface area is $\left(\pi r^2 + \dfrac{2}{r} \right)$ m².
Hence prove that this surface area is a minimum when
the height equals the radius of the base.

23. A match box consists of an outer cover, open at both ends,
into which slides a rectangular box without a top. The
length of the box is one and a half times its breadth, the
thickness of the material is negligible, and the volume of
the box is 25 cm³. If the breadth of the box is x cm, find,
in terms of x, the area of material used. Hence show that,
if the least area of material is to be used to make the box,
the length should be 3·7 cm approximately.

24. Two opposite ends of a closed rectangular tank are squares
of side x m and the total area of sheet metal forming the
tank is S m². Show that the volume of the tank is
$\frac{1}{4}x(S - 2x^2)$ m³. If the value of S is 2400, find the value
of x for which the volume is a maximum.

25. The point P(x, y) lies on the curve $y = x^2$; the point A has
coordinates $(0, 1)$. Express AP² in terms of x. Hence
find the positions of P for which AP² is least, and verify
that for each of these positions the line AP is perpendicular
to the tangent to the curve at P.

26. A window is in the shape of a rectangle surmounted by a
semi-circle whose diameter is the width of the window.
If the perimeter is 10 m, show that the area is a maximum
when the width of the window is 2·8 m. (Take π as 3·14.)

27. The base of a right circular cone is a section of a sphere
centre O and of unit radius. The vertex of the cone is
on the sphere, and the centre of the sphere is within the
cone. If the distance of O from the base of the cone is x,
show that the volume of the cone is $\dfrac{\pi}{3}(1 + x - x^2 - x^3)$.
Hence find x so as to get the maximum cone. What fraction of the volume of the sphere is occupied by the maximum cone?

CHAPTER 5

INTEGRATION

The reverse of differentiation—geometrical interpretation.

5.1. Suppose that instead of an equation of a curve, we take as our starting point a gradient function. For example, what is represented geometrically by the equation $\frac{dy}{dx} = \frac{1}{3}$?

The constant gradient $\frac{1}{3}$ indicates a straight line; $y = \frac{1}{3}x$ is the equation of the straight line of this gradient through the origin, and, on differentiation, it leads to $\frac{dy}{dx} = \frac{1}{3}$. But $y = \frac{1}{3}x$ is not the only possibility; any straight line of gradient $\frac{1}{3}$ may be written as $y = \frac{1}{3}x + c$, where c is a constant, and this is the most general equation which gives $\frac{dy}{dx} = \frac{1}{3}$.

Thus the equation $\frac{dy}{dx} = \frac{1}{3}$ represents the same as the equation $y = \frac{1}{3}x + c$, namely *all straight lines of gradient $\frac{1}{3}$*. (Fig. 5.1.)

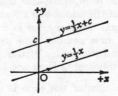

Fig. 5.1

Let us take another example, $\frac{dy}{dx} = 2x$. We know that $y = x^2$ is a curve with this gradient function; but the most general equation leading to $\frac{dy}{dx} = 2x$ on differentiation is $y = x^2 + c$, where c is a constant.

Thus the equation $\frac{dy}{dx} = 2x$ represents the same as the equation $y = x^2 + c$, namely the family of curves "parallel" to $y = x^2$ (see Fig. 2.8 on p. 37).

We have found that

$$\text{if} \quad \frac{dy}{dx} = \tfrac{1}{3}, \quad \text{then} \quad y = \tfrac{1}{3}x + c.$$

Also $\qquad$ if $\quad \frac{dy}{dx} = 2x, \quad$ then $\quad y = x^2 + c.$

This process of finding the expression for y in terms of x when given the gradient function—in other words, the reverse of differentiation—is called **integration**.

$x^2 + c$ is called the **integral** of $2x$ with respect to x.

The constant c, which, unless further data is given, cannot be determined, is called the **arbitrary constant** of integration.

We know that when we differentiate a power of x, the index is reduced by 1, since $\frac{d}{dx}(x^n) = nx^{n-1}$. In this reverse process of integration we must therefore increase the index by 1, thus

$$\text{if} \quad \frac{dy}{dx} = x, \quad y = \frac{x^2}{2} + c,$$

and $\qquad$ if $\quad \frac{dy}{dx} = 5x^2, \quad y = 5 \cdot \frac{x^3}{3} + c.$

The reader should check these by differentiating, and it will then be clear why the denominators 2 and 3 arise. The rule for integrating a power of x is seen to be "increase the index by 1, and divide by the new index". In other words,

if $\quad \dfrac{dy}{dx} = x^n, \quad y = \dfrac{x^{n+1}}{n+1} + c \quad$ (except when $n = -1$).

Qu. 1. Integrate with respect to x:

   (i) 2,   (ii) m,   (iii) $3x^2$,   (iv) $3x$,   (v) $3x^4$,
   (vi) $3 + 2x$,   (vii) $x - x^2$,   (viii) $ax + b$.

Just as the rule for differentiating powers of x applies to both positive and negative indices, so also does the rule for integration (except for x^{-1}). Thus if $\dfrac{dy}{dx} = \dfrac{1}{x^2} = x^{-2}$, then

$$y = \frac{x^{-2+1}}{-2+1} + c = \frac{x^{-1}}{-1} + c = -\frac{1}{x} + c.$$

The reader should check this last result by differentiating, and in fact should make a habit of doing this always. It is important to remember that the arbitrary constant is an essential part of each integral.

Qu. 2. Integrate with respect to x

   (i) $\dfrac{1}{x^3}$,   (ii) x^{-4},   (iii) $\dfrac{2}{x^2}$,   (iv) $\dfrac{1}{x^n}$.

Qu. 3. Why is the rule for integrating not valid when $n = -1$?

Reverting to our earlier examples, $\dfrac{dy}{dx} = \frac{1}{3}$ and $\dfrac{dy}{dx} = 2x$ are called **differential equations**, and $y = \frac{1}{3}x + c$ and $y = x^2 + c$ respectively are the **general solutions**. We saw that the differential equation $\dfrac{dy}{dx} = \frac{1}{3}$ represents all straight lines of gradient $\frac{1}{3}$; to be able to find the equation of a particular straight line of gradient $\frac{1}{3}$, we must find the appropriate value of c in the general solution $y = \frac{1}{3}x + c$, and to do this we need to know one point through which the line passes.   The reader should now read again the alternative solution of Example 9 in §1.9; it will be seen that the process of finding the equation of a straight line of given gradient passing through a given point may be thought of as finding a particular solution of a differential equation.

Qu. 4. $\dfrac{dy}{dx} = 4$.   Find y in terms of x, given that $y = 10$ when $x = -2$.   What does your solution represent graphically?

Exercise 5a

1. Integrate:
   (i) with respect to x: $\frac{1}{2}$, $\frac{1}{2}x^2$, $x^2 + 3x$, $(2x + 3)^2$, x^{-5}, $\dfrac{-2}{x^4}$;

   (ii) with respect to t: at, $\frac{1}{3}t^3$, $(t + 1)(t - 2)$, $\dfrac{1}{t^{n+1}}$, $\dfrac{1}{t^2} + 3 + 2t$;

   (iii) with respect to y: $-ay^{-2}$, $\dfrac{h}{y^2}$, $\dfrac{(y^2 + 2)(y^2 - 3)}{y^2}$.

2. Solve the following differential equations:
   (i) $\dfrac{dy}{dx} = 3ax^2$,   (ii) $\dfrac{ds}{dt} = 3t^3$,

   (iii) $\dfrac{ds}{dt} = u + at$,   (iv) $\dfrac{dx}{dt} = \left(1 + \dfrac{1}{t}\right)\left(1 - \dfrac{1}{t}\right)$,

   (v) $\dfrac{dy}{dt} = \dfrac{t^2 - 3t + 4}{t^3}$,   (vi) $\dfrac{dA}{dx} = \dfrac{(1 + x^2)(1 - 2x^2)}{x^2}$.

3. What is the gradient function of a straight line passing through $(-4, 5)$ and $(2, 6)$?   Find its equation.
4. A curve passes through the point $(3, -1)$ and its gradient function is $2x + 5$.   Find its equation.
5. A curve passes through the point $(2, 0)$ and its gradient function is $3x^2 - \dfrac{1}{x^2}$.   Find its equation.

6. The gradient of a curve at the point (x, y) is $3x^2 - 8x + 3$. If it passes through the origin, find the other points of intersection with the x-axis.

7. The gradient of a curve at the point (x, y) is $8x - 3x^2$, and it passes through the origin. Find where it cuts the x-axis, and find the equation of the tangent parallel to the x-axis.

8. Find s in terms of t if $\dfrac{ds}{dt} = 3t - \dfrac{8}{t^3}$, given that $s = 1\frac{1}{2}$ when $t = 1$.

9. Find A in terms of x if $\dfrac{dA}{dx} = \dfrac{(3x + 1)(x^2 - 1)}{x^5}$. What is the value of A when $x = 2$, if $A = 0$ when $x = 1$?

Velocity and acceleration

5.2. In Chapter 3 we used the formula $s = 4 \cdot 9t^2$ for a stone falling from rest, and it was explained that this is based on the assumption that the acceleration of the stone is 9·8 metres per second per second, or 9·8 m/s². We are now in a position to see how the formula is deduced from this assumption by the process of integration.

If the acceleration, $\dfrac{dv}{dt} = 9 \cdot 8$,

then $\qquad\qquad\qquad v = 9 \cdot 8t + c.$

Now if the stone falls from rest at the instant from which we measure the time, $v = 0$ when $t = 0$, and substituting these values in the last equation we get $c = 0$.

$$\therefore \quad v = 9 \cdot 8t.$$

This may be written $\dfrac{ds}{dt} = 9 \cdot 8t,$

from which $\qquad\qquad s = 4 \cdot 9t^2 + k.$

If s measures the distance below the initial position of the stone, $s = 0$ when $t = 0$, and substituting these values in the last equation, we get $k = 0$.

$$\therefore \quad s = 4 \cdot 9t^2.$$

Qu. 5. A stone is thrown vertically downwards from the top of a cliff at 15 m/s. Assuming that the acceleration due to gravity is 9·81 m/s², find expressions for its velocity and position t s later, by solving the differential equation $\dfrac{\mathrm{d}v}{\mathrm{d}t} = 9\cdot81$.

It again needs emphasising that displacement (s), velocity and acceleration in a straight line are positive in one direction, negative in the other, and it is important to decide at the outset which is to be taken as the positive direction. The reader should take upwards as positive in Qu. 6.

Qu. 6. A stone is thrown vertically upwards from the edge of a cliff at 19·6 m/s. Assuming that gravity produces a downwards acceleration of 9·8 m/s², deduce the velocity and position of the stone after 1, 3 and 5 s. Explain the sign of each answer, taking upwards as positive.

Example 1. *Fig. 5.2 represents part of a conveyor belt, the dots being small articles on it at 1 m spacing. Initially the belt is at rest with the article R 7 m short of O, a fixed mark on a wall. The belt is accelerated from rest so that its velocity is 0·1t m/s, t s after starting. Find (i) the position of R when $t = 10$, and (ii) the distance moved by R between $t = 3$ and $t = 5$.*

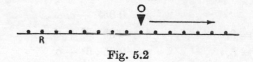

Fig. 5.2

(i) If the distance from O at time t s is s m (positive to the right of O, negative to the left), then it is true of each article that its velocity, $\dfrac{\mathrm{d}s}{\mathrm{d}t} = 0\cdot 1t$, and also, by integration, that

$$s = 0\cdot 05t^2 + c.$$

However, this last equation does not give us the distance of any particular article from O, until we have discovered the appropriate value of c. Since when $t = 0$, $s = c$, the arbitrary constant of integration in this case represents the initial position of an article.

In the case of R, when $t = 0$,

$$s = -7.$$

Substituting in the last equation, $-7 = 0 + c$,

$$\therefore c = -7.$$

Therefore the distance of R from O at time t s is s m where

$$s = 0\cdot 05t^2 - 7.$$

When $t = 10$, $s = 0\cdot 05 \,.\, 100 - 7 = -2$.

$\therefore$ R is 2 m short of O at this instant.

(ii) The distance moved by each article in any given interval is the same, therefore we are not concerned with any particular numerical value for the constant of integration, and we shall leave c in our working.

As before, since $\dfrac{\mathrm{d}s}{\mathrm{d}t} = 0\cdot 1t$,

$$s = 0\cdot 05t^2 + c.$$

When $t = 3$, $s = 0\cdot 05 \,.\, 3^2 + c.$

When $t = 5$, $s = 0\cdot 05 \,.\, 5^2 + c.$

The distance moved between $t = 3$ and $t = 5$ is

$$(0\cdot 05 \,.\, 5^2 + c) - (0\cdot 05 \,.\, 3^2 + c) \text{ m}$$

$$= 0\cdot 05 \,.\, 25 + c - 0\cdot 05 \,.\, 9 - c \text{ m} = 0\cdot 8 \text{ m}.$$

(ii) (*Alternative lay out.*) The following square bracket notation is an instruction to substitute and subtract, and shortens the working.

$$\frac{ds}{dt} = 0 \cdot 1\, t,$$

$$\therefore s = 0 \cdot 05\, t^2 + c.$$

The distance moved between $t = 3$ and $t = 5$ is

$$\left[0 \cdot 05 t^2 + c \right]_3^5 \text{m} = (0 \cdot 05 \, . \, 25 + c) - (0 \cdot 05 \, . \, 9 + c)\, \text{m},$$

$$= 1 \cdot 25 + c - 0 \cdot 45 - c\, \text{m},$$

$$= 0 \cdot 8\, \text{m}.$$

Qu. 7. Evaluate:

(i) $\left[3t + 8 \right]_2^5$, (ii) $\left[3t^2 - t + k \right]_1^4$,

(iii) $\left[t^2 - t \right]_{-2}^{+1}$, (iv) $\left[t^3 - 3t^2 + t \right]_{-3}^{-2}$.

Qu. 8. A particle moves in a straight line with velocity $2t^2$ m/s, t s after the start. Find the distance moved in the 3rd second.

Qu. 9. With the data of Example 1, answer the following questions.

(i) Find the position of R when $t = 20$.

(ii) Find the position when $t = 10$ of the article initially at O.

(iii) An article N is $2 \cdot 2$ m past O when $t = 2$; find its position when $t = 10$.

(iv) An article T is $99 \cdot 95$ m short of O when $t = 1$; find its initial position.

Exercise 5b

1. A stone is thrown vertically downwards at 20 m/s from the top of a cliff. Assuming that gravity produces on it an acceleration of $9 \cdot 81$ m/s^2, deduce, from the differential equation $\dfrac{dv}{dt} = 9 \cdot 81$, expressions for its velocity and position t s later.

2. A stone is thrown vertically upwards from ground level at 12 m/s, at a point immediately above a well. Taking The downwards direction as positive, deduce, from the differential equation $\dfrac{dv}{dt} = 9{\cdot}8$, expressions for the stone's velocity and position t s later. Find the velocity and position after 1, 2, 3 s, explaining the sign of each answer.

3. Find the displacement (s) in terms of time (t) from the following data:

   (i) $\dfrac{ds}{dt} = 3$, $s = 3$ when $t = 0$,

   (ii) $v = 4t - 1$, $s = 0$ when $t = 2$,

   (iii) $v = (3t - 1)(t + 2)$, $s = 1$ when $t = 2$,

   (iv) $v = t^2 + 5 - \dfrac{2}{t^2}$, $s = \frac{1}{3}$ when $t = 1$.

4. Evaluate:

   (i) $\Big[8t + c\Big]_1^5$,   (ii) $\Big[3t^2 + 5t\Big]_2^{10}$,

   (iii) $\Big[t^2 - 4t\Big]_{-3}^0$,   (iv) $\Big[2t^3 - t^2 - t\Big]_{-2}^{-1}$.

5. Find s in terms of t, and the distance moved in the stated interval (the units being metres and seconds), given that

   (i) $\dfrac{ds}{dt} = 4t + 3$, $t = 0$ to $t = 2$,

   (ii) $v = t^2 - 3$, $t = 2$ to $t = 3$,

   (iii) $v = (t - 1)(t - 2)$, $t = -1$ to $t = 0$,

   (iv) $v = t + 3 - \dfrac{1}{t^2}$, $t = 10$ to $t = 20$.

6. If a particle moves in a straight line so that its acceleration in terms of the time is At (A being a constant), deduce expressions for the velocity and displacement at time t.

7. Deduce expressions for v and s from the following data, determining the constants of integration whenever possible:

   (i) $a = 3t$, $s = 0$ and $v = 3$ when $t = 0$,

   (ii) $a = 2 + t$, $s = -3$ and $v = 0$ when $t = 0$,

   (iii) $a = 10 - t$, $v = 2$ when $t = 1$, $s = 0$ when $t = 0$,

    (iv) $a = \frac{1}{2}t$, $v = 5$ when $t = 0$,

    (v) $a = t^2$, $s = 10$ when $t = 1$.

8. A system of particles move along a straight line OA so that t s after a certain instant their velocity is v m/s where $v = 3t$.

    (i) One of the particles is at O when $t = 0$. Find its position when $t = 3$.

    (ii) A second particle is 4 m past O when $t = 1$. Find its position when $t = 0$.

    (iii) A third particle is 10 m short of O when $t = 2$. Find its position when $t = 4$.

    (iv) Find the distance moved by the particles during the 3rd second.

9. A particle moves along a straight line OA with velocity $(6 - 2t)$ m/s. When $t = 1$ the particle is at O.

    (i) Find an expression for its distance from O in terms of t, and deduce the net change in position which takes place between $t = 0$ and $t = 5$.

    (ii) By finding the time at which it is momentarily at rest, calculate the actual distance through which it moves during the same interval.

    (iii) Sketch the space-time and velocity-time graphs from $t = 0$ to $t = 6$.

10. A stone is thrown vertically upwards from ground level with a velocity of 12·6 m/s. If the acceleration due to gravity is 9·8 m/s², deduce, from the differential equation $\frac{dv}{dt} = -9\cdot8$, expressions for its velocity and its height t s later. Find

    (i) the time to the highest point,

    (ii) the greatest height reached,

    (iii) the distance moved through by the stone during each of the first two seconds of motion.

11. A train runs non-stop between two stations P and Q, and its velocity t hours after leaving P is $60t - 30t^2$ km/h. Find

    (i) the distance between P and Q,

    (ii) the average velocity for the journey,

    (iii) the maximum velocity attained.

12. A stopping train travels between two adjacent stations so that its velocity is v km/min, t min after leaving the first, where $v = \frac{4}{5}t(1 - t)$.   Find
   (i) the average velocity for the journey in km/h,
   (ii) the maximum velocity in km/h.

13. The formula connecting the velocity and time for the motion of a particle is $v = 1 + 4t + 6t^2$.   Find the average velocity and the average acceleration for the interval $t = 1$ to $t = 3$, the units being metres and seconds.

14. A racing car starts from rest and its acceleration after t s is $(k - \frac{1}{5}t)$ m/s^2 until it reaches a velocity of 60 m/s at the end of 1 min.   Find the value of k, and the distance travelled in this first minute.

15. A particle starting from rest at O moves along a straight line OA so that its acceleration after t s is $(24t - 12t^2)$ m/s^2.
   (i) Find when it again returns to O and its velocity then,
   (ii) find its maximum displacement from O during this interval,
   (iii) what is its maximum velocity and its greatest speed during this interval?

16. P and R are two adjacent railway stations, and Q is a signal box on the line between them.   A train which stops at P and R has a velocity of $(\frac{3}{5} + \frac{1}{2}t - \frac{1}{2}t^2)$ km/min at t min past noon, and it passes Q at noon.   Find
   (i) the times of departure from P and arrival at R,
   (ii) an expression for the distance of the train from P in terms of t,
   (iii) the average velocity between P and R, in km/h,
   (iv) the maximum velocity attained, in km/h.

The area under a curve

5.3.   Another important aspect of integration is that it enables us to calculate exactly the areas enclosed by curves.

Let us consider the area enclosed by the axes, the line $x = 3$, and part of the curve $y = 3x^2 + 2$.   This is the area TUVO in Fig. 5.3.

P is the point (x, y) on the curve, PM is its y-coordinate, and the area TPMO we shall call A. Now if we move P along the curve, A increases or decreases as x increases or

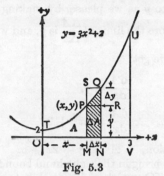

Fig. 5.3

decreases; clearly the size of A depends upon the value of x, and our present aim is to find an expression for A in terms of x.

With the usual notation Q is the point $(x + \Delta x, y + \Delta y)$ adjacent to P, and QN is its y-coordinate. If we move the right-hand boundary of A from PM to QN, we increase x by Δx, and the resulting increase in A, the shaded area PQNM, we call ΔA. In other words ΔA is the increment in A corresponding to the increment Δx in x. It can be seen from Fig. 5.3 that ΔA lies between the areas of the two rectangles PRNM, $y.\Delta x$, and SQNM, $(y + \Delta y).\Delta x$. This may be written *

$$y.\Delta x < \Delta A < (y + \Delta y)\Delta x,$$

and dividing by Δx, which is positive,

$$y < \frac{\Delta A}{\Delta x} < (y + \Delta y).$$

* This statement is called an inequality. $<$ means "is less than"; $>$ means "is greater than". The reader should note in passing that an inequality is reversed by changing the sign of each term. Thus $1 < 2 < 3$, but $-1 > -2 > -3$; this explains the reference to Δx being positive.

Now as $\Delta x \to 0$, $\Delta y \to 0$, and so $(y + \Delta y) \to y$. Thus we find that $\dfrac{\Delta A}{\Delta x}$ lies between y and something which we can make as near to y as we please, by making Δx sufficiently small. Therefore the limit of $\dfrac{\Delta A}{\Delta x}$ is y, and writing the limit of $\dfrac{\Delta A}{\Delta x}$ as $\dfrac{\mathrm{d}A}{\mathrm{d}x}$, we get

$$\frac{\mathrm{d}A}{\mathrm{d}x} = y.$$

$$\therefore \frac{\mathrm{d}A}{\mathrm{d}x} = 3x^2 + 2,$$

and by integration, $A = x^3 + 2x + c.$

If we were to bring in the right-hand boundary of the area A from PM to TO, we should reduce A to zero; that is to say, when $x = 0$, $A = 0$. Substituting these values in the last equation we find that $c = 0$.

$$\therefore \ A = x^3 + 2x,$$

and we have achieved our immediate aim of expressing A in terms of x; now to find the area OTUV. In this case, the right-hand boundary of A has been pushed out from PM to UV, and x is increased to 3.

When $x = 3$, $A = 3^3 + 2.3 = 33,$

$$\therefore \text{ the area TUVO} = 33.$$

Example 2. *Find the area enclosed by the x-axis, the curve $y = 3x^2 + 2$ and the straight lines $x = 3$ and $x = 5$.*

The required area is UWZV in Fig. 5.4, and it may be found as the difference between the areas TWZO and OTUV. Using A as above,

$$\frac{\mathrm{d}A}{\mathrm{d}x} = y = 3x^2 + 2,$$

$$\therefore \ A = x^3 + 2x.$$

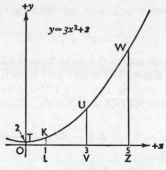

Fig. 5.4

(We have shown above that the constant of integration is zero.)

When $x = 5$,

$$A = 5^3 + 2.5 = 135 \quad \text{(Area TWZO)},$$

and when $x = 3$,

$$A = 3^3 + 2.3 = 33 \quad \text{(Area TUVO)}.$$

$$\therefore \text{ the area UWZV} = 135 - 33 = 102.$$

Qu. 10. Find the area enclosed by the x-axis, the curve $y = 3x^2 + 2$, and the following straight lines:

(i) the y-axis and $x = 4$,   (ii) $x = 1$ and $x = 2$,
(iii) $x = -1$ and $x = 3$,   (iv) $x = -3$ and $x = 3$.

In all the working so far in this chapter we have used the symbol A to denote an area having the y-axis as its left-hand boundary. Suppose that instead we had, in Fig. 5.3, defined a similar area A' having the line $x = 1$ as its left-hand boundary. By the same process of reasoning we should arrive at the result

$$\frac{dA'}{dx} = y = 3x^2 + 2,$$

$$\therefore A' = x^3 + 2x + k.$$

But $A' = 0$ when $x = 1$, and substituting these values we get $k = -3$.

$$\therefore \ A' = x^3 + 2x - 3.$$

Now A' is measured to the right from the line KL ($x = 1$) in Fig. 5.4, and Example 2 might just as well be done using A' instead of A, finding the area UWZV as the difference between the areas KWZL and KUVL. Thus, when $x = 5$,

$$A' = 5^3 + 2.5 - 3 = 135 - 3,$$

and when $x = 3$,

$$A' = 3^3 + 2.3 - 3 = \ 33 - 3.$$

$\therefore$ the area UWZV $= (135 - 3) - (33 - 3) = 102$.

In each solution we have determined the constant of integration; using A, it is zero, and using A', it is -3. But as is clear from the second solution, the constant drops out on subtraction. We could in fact have measured A from any convenient left-hand boundary, and found the area UWZV by subtraction, without evaluating the constant of integration.

We shall from now onwards assume the relationship $\dfrac{\mathrm{d}A}{\mathrm{d}x} = y$ to calculate areas of this nature, and the square bracket notation introduced in §5.2 may now be put to further use, as is illustrated in the next example.

Example 3. *Find the area enclosed by the x-axis, $x = 1$, $x = 3$ and the graph $y = x^3$.* (Fig. 5.5.)

$$\frac{\mathrm{d}A}{\mathrm{d}x} = y = x^3,$$

$$\therefore A = \tfrac{1}{4}x^4 + c.$$

The required area $= \left[\tfrac{1}{4}x^4 + c \right]_1^3,$

$$= (\tfrac{81}{4} + c) - (\tfrac{1}{4} + c),$$
$$= \tfrac{81}{4} + c - \tfrac{1}{4} - c,$$
$$= 20.$$

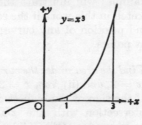

Fig. 5.5

The area evaluated in Example 3 is called *the area under the curve* $y = x^3$ *from* $x = 1$ *to* $x = 3$.

1 and 3 are called, respectively, the *lower* and *upper limits of integration*.

The integral $\tfrac{1}{4}x^4 + c$, involving the arbitrary constant of integration, is called an **indefinite integral**.

When however limits are given, and the integral may be evaluated, e.g. $\left[\tfrac{1}{4}x^4 + c\right]_1^3$, it is called a **definite integral**. Since the constant of integration drops out in a definite integral, it is not necessary to write it in the bracket.

Qu. 11. Evaluate the following definite integrals:

(i) $\left[3x^2 + 2x\right]_{\frac{1}{2}}^1$, (ii) $\left[x^4 - 2x^2\right]_{-1}^2$,

(iii) $\left[x^3 - 3x\right]_{-2}^0$, (iv) $\left[2x^2 + 4x\right]_{-3}^{-1}$,

(v) $\left[x^4 - 2x^3 + x^2 - x\right]_{-2}^0$, (vi) $\left[x^2 + 3x - \dfrac{1}{x^3}\right]_{+\frac{1}{2}}^{+1}$.

Qu. 12. Find the area under $y = \tfrac{1}{2}x$ from $x = 0$ to $x = 10$ by integration. Check by another method.

Qu. 13.　Find the area under

(i) $y = x^2$ from $x = 0$ to $x = 3$,

(ii) $y = 2x^2 + 1$ from $x = 2$ to $x = 5$.

Two further examples will illustrate the advisability of making a rough sketch in this work if the reader is in doubt as to the shape and position of any curve; they also bring out two important points.

Example 4.　*Find the area under the curve* $y = x^2(x - 2)$ (i) *from* $x = 0$ *to* $x = 2$, *and* (ii) *from* $x = 2$ *to* $x = \frac{8}{3}$.

Consideration of the sign of the highest degree term, and the points of intersection with the x-axis, enables an adequate sketch to be made.　(Fig. 5.6.)

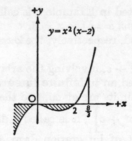

Fig. 5.6

$$\frac{dA}{dx} = y = x^3 - 2x^2$$

$$\therefore A = \tfrac{1}{4}x^4 - \tfrac{2}{3}x^3 + c.$$

(i) The required area $= \left[\tfrac{1}{4}x^4 - \tfrac{2}{3}x^3\right]_0^2,$

$$= (\tfrac{1}{4}.2^4 - \tfrac{2}{3}.2^3) - (0),$$

$$= -1\tfrac{1}{3}.$$

(ii) The required area

$$= \left[\tfrac{1}{4}x^4 - \tfrac{2}{3}x^3\right]_2^{\frac{8}{3}},$$

$$= \left(\frac{1}{4}\cdot\frac{8^4}{3^4} - \frac{2}{3}\cdot\frac{8^3}{3^3}\right) - \left(\frac{1}{4}\cdot2^4 - \frac{2}{3}\cdot2^3\right),$$

$$= (0) - (-1\tfrac{1}{3}),$$

$$= +1\tfrac{1}{3}.$$

Part (i) of this example illustrates that *the area under a curve is negative below the x-axis*. The reader should verify that $\left[\tfrac{1}{4}x^4 - \tfrac{2}{3}x^3\right]_0^{\frac{8}{3}}$ is zero, and now that we have the convention about the sign of an area, we see that this is because it represents the sum of the two areas we have evaluated, numerically equal but of opposite sign.

The reader should now appreciate that a sketch of the relevant curve may help to avoid misleading results arising from perfectly correct calculation.

Qu. 14. Confirm that the total area enclosed by $y = x^2(x - 2)$, the x-axis, $x = 1$ and $x = 3$ is $4\tfrac{1}{2}$.

What is the value of $\left[\tfrac{1}{4}x^4 - \tfrac{2}{3}x^3\right]_1^3$?

Qu. 15. Sketch the curve $y = x(x - 1)(x - 2)$. Find the total area enclosed between this curve and the x-axis.

Example 5. (i) *Find the area under $y = \dfrac{1}{x^2}$ from $x = 1$ to $x = 2$.* (ii) *Can any meaning be attached to the phrase "the area under $y = \dfrac{1}{x^2}$ from $x = -1$ to $x = +2$"?*

(i) $$\frac{dA}{dx} = y = \frac{1}{x^2} = x^{-2},$$

$$\therefore \quad A = -x^{-1} + c.$$

$$\text{The required area} = \left[-\frac{1}{x}\right]_{1}^{2},$$
$$= (-\tfrac{1}{2}) - (-1),$$
$$= -\tfrac{1}{2} + 1,$$
$$= \tfrac{1}{2}.$$

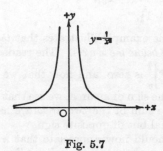

Fig. 5.7

(ii) Fig. 5.7 is a sketch of $y = \dfrac{1}{x^2}$, and we see that if we try to find the area under the graph from $x = -1$ to $x = 2$, between these limits is the value $x = 0$ for which y has no value, and the curve consists of two separate branches. It is possible to go through the motions of evaluating $\left[-\dfrac{1}{x}\right]_{-1}^{+2}$ but the result, $-1\frac{1}{2}$, is meaningless. If we break up the area into two parts and integrate from -1 to 0 and from 0 to 2, in each case we get the meaningless term $\dfrac{1}{0}$.

The second part of Example 5 illustrates that in order that we may calculate the area under a curve, the curve must have no breaks between the limits of x involved; or we may say that *the curve must be continuous throughout the range of integration.*

Exercise 5c

1. Evaluate

(i) $\left[\dfrac{x^4}{4}\right]_{\frac{1}{2}}^{2}$,

(ii) $\left[3x^3 - 4x\right]_{-1}^{+1}$,

(iii) $\left[\frac{1}{3}x^3 - 3x^2 + \frac{1}{2}x\right]_{-2}^{-1}$,

(iv) $\left[x^3 - \dfrac{1}{x^2}\right]_{-4}^{-3}$.

2. Find the area enclosed by $x + 4y - 20 = 0$ and the axes, by integration. Check by another method.

3. Find the areas enclosed by the x-axis, and the following curves and straight lines:

(i) $y = 3x^2$, $x = 1$, $x = 3$,

(ii) $y = x^2 + 2$, $x = -2$, $x = 5$,

(iii) $y = x^2(x - 1)(x - 2)$, $x = -2$, $x = -1$,

(iv) $y = \dfrac{3}{x^2}$, $x = 1$, $x = 6$.

4. Find the area under $y = 4x^3 + 8x^2$ from $x = -2$ to $x = 0$.

5. Sketch the curve $y = x^2 - 5x + 6$ and find the area cut off below the x-axis.

6. Sketch the curve $y = x(x + 1)(2 - x)$, and find the area of each of the two segments cut off by the x-axis.

7. Sketch the following curves and find the areas enclosed by them, and by the x-axis, and the given straight lines:

(i) $y = x(4 - x)$, $x = 5$,

(ii) $y = -x^3$, $x = -2$,

(iii) $y = x^3(x - 1)$, $x = 2$,

(iv) $y = \dfrac{1}{x^2} - 1$, $x = 2$.

8. Find the area of the segment cut off from $y = x^2 - 4x + 6$ by the line $y = 3$.

9. Repeat Qu. 8 for the curve $y = 7 - x - x^2$, and $y = 5$.

10. Find the points of intersection of the following curves and straight lines, and find the area of the segment cut off from each curve by the corresponding straight line:

(i) $y = \frac{1}{2}x^2$, $y = 2x$,

(ii) $y = 3x^2$, $3x + y - 6 = 0$,

(iii) $y = (x + 1)(x - 2)$, $x - y + 1 = 0$.

11. Find the areas enclosed by the following curves and straight lines:

   (i) $y = \frac{1}{2}x^3$, the y-axis, and $y = 32$,

   (ii) $y = x^3 - 1$, the axes and $y = 26$,

   (iii) $y = \dfrac{1}{x^2} - 1$, $y = -1$, $x = \frac{1}{2}$ and $x = 2$.

12. Find the area enclosed by the curves $y = 2x^2$ and $y = 12x^2 - x^3$.

Exercise 5d (Miscellaneous)

1. If $\dfrac{dy}{dx} = \dfrac{3x - 2}{x^3}$ find y in terms of x, if $y = 1$ when $x = 1$.

2. If $\dfrac{dy}{dx} = 2x - \dfrac{1}{x^2}$ and if $y = 1$ when $x = 1$, find y in terms of x.

3. The curve $y = 6 - x - x^2$ cuts the x-axis in two points A and B. By integration find the area enclosed by the x-axis and that portion of the curve which lies between A and B.

4. Sketch the curve $y = x^2 - x - 2$ from $x = -2$ to $x = 3$. Find the area bounded by the curve and the x-axis.

5. Sketch roughly the curve $y = x^2(3 - x)$ between $x = -1$ and $x = 4$. Calculate the area between the portion of the curve in the first quadrant and the axis of x.

6. For the curve $y = 12x - x^3$, find the area between the part of the curve in the first quadrant and the x-axis.

7. The velocity v of a point moving along a straight line is given in terms of the time t by the formula $v = 2t^2 - 9t + 10$, the point being at the origin when $t = 0$. Find expressions in terms of t for the distance from the origin, and the acceleration. Show that the point is at rest twice, and find its distances from the origin at those instants.

8. The velocity v of a point moving along a straight line is connected with the time t by the formula $v = t^2 - 3t + 2$, the units being metres and seconds. If the distance of the point from the origin is 5 m when $t = 1$, find its position and acceleration when $t = 2$.

9. A particle moves in a straight line with a velocity of v m/s after t s, where $v = 3t^2 + 2t$. Find the acceleration at the end of 2 s, and the distance it travels in the 4th second.

10. Find the equation of the curve which passes through the point $(-1, 0)$ and whose gradient at any point (x, y) is $3x^2 - 6x + 4$. Find the area enclosed by the curve, the axis of x and the ordinates $x = 1$ and $x = 2$.

11. Draw in the same figure for values of x from 0 to 6 a rough sketch of the curve $y = 6x - x^2$ and the line $y = 2x$. Calculate the area enclosed by them.

12. The parabola $y = 6x - x^2$ meets the x-axis at O and A. The tangents at O and A meet at T. Show that the curve divides the area of the triangle OAT into two parts in the ratio 2:1.

13. The curve $y = x(x - 1)^2$ touches the x-axis at the point A. B is the point $(2, 2)$ on the curve and N is the foot of the perpendicular from B to the x-axis. Prove that the tangent at B divides the area between the arc AB, BN, and AN in the ratio 11:24.

14. The point P moves in a straight line with an acceleration of $(2t - 4)$ m/s^2 after t s. When $t = 0$, P is at O and its velocity is 3 m/s. Find
    (i) the velocity of P after t s.
    (ii) the value of t when P starts to return to O,
    (iii) the distance of P from O at this moment,

15. A train starts from rest and its acceleration t s after the start is $0 \cdot 1(20 - t)$ m/s^2. What is its speed after 20 s? Acceleration ceases at this instant and the train proceeds at this uniform speed. What is the total distance covered 30 s after the start from rest, to the nearest metre?

16. A particle moves in a straight line with velocity $(7t - t^2 - 6)$ m/s at the end of t s. What is its acceleration when $t = 2$ and when $t = 4$? When $t = 3$ the particle is at A; when $t = 5$ the particle is at B. Find the length of AB. For what values of t is the particle momentarily at rest?

17. A particle, starting from rest, moves along a straight line with a velocity of $(8t - t^2)$ m/s at the end of t s. Find

its velocity when its acceleration vanishes and the distance travelled up to that time. What distance will have been travelled when the velocity vanishes instantaneously?

18. The velocity of a train starting from rest is proportional to t^2, where t is the time which has elapsed since it started. If the distance it has covered at the end of 6 s is 18 m, find the velocity and the acceleration at that instant.

19. A car starts from rest with an acceleration proportional to the time. It travels 9 m in the first 3 s. Calculate its velocity and acceleration at the end of this time. Also find the distance travelled up to the instant when the velocity and acceleration are numerically equal.

20. A particle starts from rest and moves in a straight line. Its speed for the first 3 s is proportional to $(6t - t^2)$, where t is the time in seconds from the commencement of motion, and thereafter it travels with uniform speed at the rate it had acquired at the end of the 3rd second. Prove that the distance travelled in the first 3 s is two-thirds of the distance travelled in the next 3 s.

CHAPTER 6

FURTHER DIFFERENTIATION

Fractional indices

6.1. The rule for differentiating x^n is known to be "multiply the term by the index and decrease the index by 1"—n being a whole number. Fractional indices are dealt with algebraically elsewhere in this book and it is sufficient to assume at this juncture that the rule continues to apply when n is any integer or fraction, positive or negative. The reader is advised to revise §§8.2 and 8.3 on indices at this stage.

Example 1. *Differentiate* (i) $\dfrac{2}{x^3}$, (ii) $\dfrac{1}{\sqrt{x}}$.

(i) Let $\quad y = \dfrac{2}{x^3} = 2x^{-3}$. (ii) Let $\quad y = \dfrac{1}{\sqrt{x}} = x^{-\frac{1}{2}}$.

$$\therefore \frac{dy}{dx} = 2(-3)x^{-4}. \qquad \therefore \frac{dy}{dx} = -\tfrac{1}{2}x^{-\frac{3}{2}}.$$

$$\therefore \frac{d}{dx}\left(\frac{2}{x^3}\right) = \frac{-6}{x^4}. \qquad \therefore \frac{d}{dx}\left(\frac{1}{\sqrt{x}}\right) = -\frac{1}{2\sqrt{x^3}}.$$

Example 2. *Integrate* $\dfrac{3}{\sqrt[3]{x}}$.

If $\qquad\qquad \dfrac{dy}{dx} = \dfrac{3}{\sqrt[3]{x}} = 3x^{-\frac{1}{3}}$,

$$y = 3\,\frac{x^{-\frac{1}{3}+1}}{-\frac{1}{3}+1} + c$$

$$= \frac{9}{2}\sqrt[3]{x^2} + c.$$

109

Qu. 1. Differentiate (i) x^{-4}, (ii) $2x^{-3}$, (iii) $\dfrac{1}{x^2}$, (iv) $\dfrac{4}{x}$, (v) $-\dfrac{2}{x^2}$, (vi) $\dfrac{1}{3x^3}$, (vii) $-\dfrac{1}{x^4}$, (viii) $\dfrac{3}{5x^5}$.

Integrate (i) x^{-3}, (ii) $2x^{-2}$, (iii) $\dfrac{1}{x^2}$, (iv) $\dfrac{2}{x^3}$, (v) $\dfrac{1}{3x^3}$, (vi) $\dfrac{2}{5x^4}$.

Qu. 2. Differentiate (i) $x^{\frac{1}{2}}$, (ii) $2x^{-\frac{1}{2}}$, (iii) $\sqrt{x}$, (iv) $\sqrt[3]{x}$, (v) $\dfrac{1}{\sqrt[3]{x}}$, (vi) $\dfrac{-2}{\sqrt[3]{x}}$, (vii) $2\sqrt{x^3}$, (viii) $\dfrac{2}{3\sqrt{x}}$.

Integrate (i) $x^{-\frac{1}{4}}$, (ii) $2x^{\frac{3}{2}}$, (iii) $\sqrt{x}$, (iv) $\sqrt[3]{x}$, (v) $\dfrac{1}{\sqrt{x}}$, (vi) $\dfrac{1}{\sqrt{x^3}}$.

Function of a function

6.2. The process of differentiating a function has already been dealt with in this book and the reader faced with a simple expression will differentiate it term by term after expansion and know he is quite in order.

If
$$y = (x + 3)^2 = x^2 + 6x + 9,$$

then
$$\frac{\mathrm{d}y}{\mathrm{d}x} = 2x + 6 = 2(x + 3).$$

Quite obviously this expansion process leads to laborious multiplication when something like $(x + 3)^7$ is met. The more venturesome reader might hazard a guess that $7(x + 3)^6$ would be its derivative—and he would be right merely because $x + 3$ has the same gradient as x. Guessing is rather apt to grow indiscriminate, however, and is entirely untrustworthy.

The derivative of $(3x + 2)^4$ is *not* $4(3x + 2)^3$.

The derivative of $(x^2 + 3x)^7$ is *not* $7(x^2 + 3x)^6$.

We want to find a legitimate and foolproof, yet simple, way to differentiate these awkward functions.

Suppose y is a function of t, and t is itself a function of x. If Δy, Δt, and Δx are corresponding small increments in the variables y, t, and x, then

$$\frac{\Delta y}{\Delta x} \equiv \frac{\Delta y}{\Delta t} \cdot \frac{\Delta t}{\Delta x}. \tag{1}$$

When Δy, Δt, and Δx tend to zero,

$$\frac{\Delta y}{\Delta x} \to \frac{dy}{dx}, \quad \frac{\Delta y}{\Delta t} \to \frac{dy}{dt}, \quad \frac{\Delta t}{\Delta x} \to \frac{dt}{dx},$$

and equation (1) becomes

$$\frac{dy}{dx} = \frac{dy}{dt} \cdot \frac{dt}{dx}.$$

This formula enables us to differentiate awkward expressions by allowing us to deal with the difficulties singly, using a substitution.

Example 3. *Differentiate $(3x + 2)^4$.*

Let $y = (3x + 2)^4$ and $t = 3x + 2$, then $y = t^4$.

$$\frac{dt}{dx} = 3, \qquad \frac{dy}{dt} = 4t^3.$$

But

$$\frac{dy}{dx} = \frac{dy}{dt} \cdot \frac{dt}{dx}.$$

$$\therefore \frac{dy}{dx} = 4t^3 \cdot 3.$$

$$\therefore \frac{d}{dx}\{(3x + 2)^4\} = 12(3x + 2)^3.$$

Example 4. *Differentiate $(x^2 + 3x)^7$.*

Let $y = (x^2 + 3x)^7$ and $t = x^2 + 3x$, then $y = t^7$.

$$\frac{dt}{dx} = 2x + 3, \qquad \frac{dy}{dt} = 7t^6.$$

$$\therefore \frac{dy}{dx} = 7t^6(2x + 3).$$

$$\therefore \frac{d}{dx}\{(x^2 + 3x)^7\} = 7(2x + 3)(x^2 + 3x)^6.$$

In the very simple instance of Example 3 a similar method will apply for integration, i.e. $\int(3x + 2)^4 \, dx$ does equal $\frac{1}{5}(3x + 2)^5 . \frac{1}{3}$, but this is an isolated instance. A corresponding division rule in integration does *not* apply. The integration of these awkward "functions of a function" is beyond the compass of this book.

It is not necessary to show the actual substitution, as has been done in the examples above, but it is advisable, until practice has made perfect this art of substitution. The bracket is really treated as a single term—the t of our formula—and then the reader must remember to "multiply by the derivative of the bracket".

Negative and fractional indices are explained algebraically in §8.2 and a knowledge of this is assumed. Differentiation of reciprocals and roots of functions is pure "function of a function" technique.

Example 5. *Differentiate* $\dfrac{1}{1 + \sqrt{x}}$.

Let
$$y = (1 + \sqrt{x})^{-1},$$

$$\therefore \frac{dy}{dx} = -1 . (1 + \sqrt{x})^{-2} . \left\{ \frac{d}{dx}(1 + \sqrt{x}) \right\},$$

$$= -1 . (1 + \sqrt{x})^{-2} . (\tfrac{1}{2} x^{-\frac{1}{2}}).$$

$$\therefore \frac{d}{dx}\left(\frac{1}{1 + \sqrt{x}} \right) = \frac{-1}{2\sqrt{x} . (1 + \sqrt{x})^2}.$$

Example 6. *Differentiate* $\sqrt{(1 + x^2)}$.

Let
$$y = (1 + x^2)^{\frac{1}{2}},$$

$$\therefore \frac{dy}{dx} = \tfrac{1}{2}(1 + x^2)^{-\frac{1}{2}} . 2x.$$

$$\therefore \frac{d}{dx}\{\sqrt{(1 + x^2)}\} = \frac{x}{\sqrt{(1 + x^2)}}.$$

Exercise 6a

1. Differentiate:

(i) $(2x + 3)^2$,   (ii) $2(3x + 4)^4$,   (iii) $(2x + 5)^{-1}$,

(iv) $(3x - 1)^{\frac{3}{2}}$,   (v) $(3 - 2x)^{-\frac{1}{2}}$,   (vi) $(3 - 4x)^{-3}$.

2. Integrate:

(i) $(3x + 2)^3$,   (ii) $(2x + 3)^2$,   (iii) $(3x - 4)^{-2}$,

(iv) $(2x + 3)^{\frac{1}{2}}$.

3. Differentiate:

(i) $\dfrac{1}{(3x + 2)}$,   (ii) $\dfrac{1}{(2x + 3)^2}$,   (iii) $\dfrac{1}{\sqrt{(3x + 1)}}$,

(iv) $\dfrac{1}{(2x - 1)^{\frac{3}{4}}}$.

4. Integrate:

(i) $\dfrac{1}{(2x - 3)^2}$,   (ii) $\dfrac{1}{\sqrt{(3x + 2)}}$,   (iii) $\dfrac{1}{(2x - 1)^{\frac{3}{4}}}$.

5. Differentiate:

(i) $(3x^2 + 5)^3$,   (ii) $(3x^2 + 5x)^2$,   (iii) $(7x^2 - 4)^{\frac{1}{3}}$,

(iv) $(6x^3 - 4x)^{-2}$,   (v) $(3x^2 - 5x)^{-\frac{3}{4}}$.

6. Differentiate:

(i) $\dfrac{1}{(3x^2 + 2)}$,   (ii) $\dfrac{3}{\sqrt{(2 + x^2)}}$,   (iii) $\dfrac{-1}{(1 + \sqrt{x})^2}$,

(iv) $\left(1 - \dfrac{1}{x}\right)^3$,   (v) $\dfrac{1}{(x^2 - 1)^{\frac{1}{3}}}$.

7. Differentiate:

(i) $(3\sqrt{x} - 2x)^3$,   (ii) $\left(\dfrac{2}{\sqrt{x}} - 1\right)^{-1}$,

(iii) $\left(2x^2 - \dfrac{3}{x^2}\right)^{\frac{1}{3}}$,   (iv) $\left(x - \dfrac{1}{x}\right)^{\frac{1}{2}}$.

8. Differentiate:

(i) $\dfrac{1}{x^{\frac{3}{2}} - 1}$,   (ii) $\sqrt{\left(1 - \dfrac{1}{x}\right)}$,

(iii) $\sqrt[3]{(1 - \sqrt{x})}$,   (iv) $\left(\sqrt{x} - \dfrac{1}{\sqrt{x}}\right)^2$.

9. Differentiate:

(i) $\dfrac{1}{(x^2 - 7x)^3}$,

(ii) $\dfrac{1}{(x^2 - \sqrt{x})^2}$,

(iii) $\sqrt{\left(\dfrac{1}{1 - x^2}\right)}$,

(iv) $\left(\dfrac{1}{1 - \sqrt{x}}\right)^2$.

10. Differentiate:

(i) $\sqrt{\left(x^2 - \dfrac{1}{x^2}\right)}$,

(ii) $\dfrac{2}{x + 2\sqrt{x}}$,

(iii) $\left(1 - \dfrac{2}{\sqrt{x}}\right)^{\frac{1}{3}}$,

(iv) $\sqrt{\left(1 - \dfrac{1}{\sqrt{x}}\right)}$.

Rates of change

6.3. *If the volume of a spherical balloon increases by 2 cm³ every second, what is the rate of growth of the radius?*

The solution of this type of problem has obvious calculus possibilities because $\dfrac{\mathrm{d}y}{\mathrm{d}x}$ is the rate of change of y with respect to x, and with the formula of the preceding section we have a ready means of connecting rates of change of dependent variables.

If the radius of the balloon is r, then the volume, $V = \frac{4}{3}\pi r^3$.

The fact we are given is that $\dfrac{\mathrm{d}V}{\mathrm{d}t}$, the rate of change of the volume with respect to time, t, is 2 cm³/s, but

$$\frac{\mathrm{d}V}{\mathrm{d}t} = \frac{\mathrm{d}V}{\mathrm{d}r} \cdot \frac{\mathrm{d}r}{\mathrm{d}t} \quad \text{and} \quad \frac{\mathrm{d}V}{\mathrm{d}r} = 4\pi r^2.$$

which leads to $$\frac{\mathrm{d}r}{\mathrm{d}t} = \frac{2}{4\pi r^2},$$

i.e. the rate of change of the radius is $\dfrac{1}{2\pi r^2}$ cm/s. Any reader will surely at some time have blown up a balloon and noticed that the radius grows much more quickly at the beginning than near the end—sudden though the latter may sometimes be! The rate of change of the radius at any particular time could be calculated when the value of r is known. In the problem chosen, the radius after t s could be calculated from $\frac{4}{3}\pi r^3 = 2t$. The arithmetic is harder than the calculus.

Example 7. *A container in the shape of a right circular cone of height 10 cm and base radius 1 cm is catching the drips from a tap leaking at the rate of 0·1 cm³/s. Find the rate at which the surface area of water is increasing when the water is half-way up the cone.*

Suppose the height of the water at any time is h cm, and that the radius of the surface of water at that time is r cm.

Fig. 0.1

By similar triangles (often used in this type of question),

$$\frac{r}{1} = \frac{h}{10}, \quad \therefore r = \tfrac{1}{10}h.$$

The surface area of water, $A = \pi r^2 = \dfrac{\pi h^2}{100}$ and we wish to find $\dfrac{\mathrm{d}A}{\mathrm{d}t}$ when $h = 5$.

$$\frac{\mathrm{d}A}{\mathrm{d}t} = \frac{\mathrm{d}A}{\mathrm{d}h} \cdot \frac{\mathrm{d}h}{\mathrm{d}t} = \frac{2\pi h}{100} \cdot \frac{\mathrm{d}h}{\mathrm{d}t}. \tag{1}$$

The volume of water, $V = \frac{1}{3}\pi r^2 h = \dfrac{\pi h^3}{300}$,

$$\therefore \frac{\mathrm{d}V}{\mathrm{d}t} = \frac{\mathrm{d}V}{\mathrm{d}h} \cdot \frac{\mathrm{d}h}{\mathrm{d}t} = \frac{3\pi h^2}{300} \cdot \frac{\mathrm{d}h}{\mathrm{d}t}.$$

But we are given that $\dfrac{\mathrm{d}V}{\mathrm{d}t} = 0 \cdot 1$,

$$\therefore \frac{\mathrm{d}h}{\mathrm{d}t} = \frac{\mathrm{d}V}{\mathrm{d}t} \cdot \frac{300}{3\pi h^2} = 0 \cdot 1 \cdot \frac{100}{\pi h^2} = \frac{10}{\pi h^2}. \qquad (2)$$

From (1) and (2) $\dfrac{\mathrm{d}A}{\mathrm{d}t} = \dfrac{2\pi h}{100} \cdot \dfrac{10}{\pi h^2} = \dfrac{1}{5h}$,

and, when $h = 5$, $\dfrac{\mathrm{d}A}{\mathrm{d}t} = \frac{1}{25} = 0 \cdot 04$.

$\therefore$ when the water is halfway up, the rate of change of the surface area $= 0 \cdot 04 \ \text{cm}^2/\text{s}$.

Exercise 6b

1. The side of a cube is increasing at the rate of 6 cm/s. Find the rate of increase of the volume when the length of a side is 9 cm.

2. The area of surface of a sphere is $4\pi r^2$, r being the radius. Find the rate of change of the area in square cm per second when $r = 2$ cm, given that the radius increases at the rate of 1 cm/s.

3. The volume of a cube is increasing at the rate of 2 cm³/s. Find the rate of change of the side of the base when its length is 3 cm.

4. The area of a circle is increasing at the rate of 3 cm²/s. Find the rate of change of the circumference when the radius is 2 cm.

5. At a given instant the radii of two concentric circles are 8 cm and 12 cm. The radius of the outer circle increases at the rate of 1 cm/s and that of the inner at 2 cm/s. Find the rate of change of the area enclosed between the two circles.

6. If $y = (x^2 - 3x)^3$, find $\dfrac{dy}{dt}$ when $x = 2$, given $\dfrac{dx}{dt} = 2$.

7. A hollow right circular cone is held vertex downwards beneath a tap leaking at the rate of 2 cm³/s. Find the rate of rise of water level when the depth is 6 cm given that the height of the cone is 18 cm and its radius 12 cm.

8. An ink blot on a piece of paper spreads at the rate of $\frac{1}{2}$ cm²/s. Find the rate of increase of the radius of the circular blot when the radius is $\frac{1}{2}$ cm.

9. A hemispherical bowl is being filled with water at a uniform rate. When the height of the water is h cm the volume is $\pi(rh^2 - \frac{1}{3}h^3)$ cm³, r cm being the radius of the hemisphere. Find the rate at which the water level is rising when it is half way to the top, given that $r = 6$ and that the bowl fills in 1 min.

10. An inverted right circular cone of vertical angle 120° is collecting water from a tap at a steady rate of 18π cm³/min. Find (i) the depth of the water after 12 min,

(ii) the rate of increase of the depth at this instant.

11. From the formula $v = \sqrt{(60s + 25)}$ the velocity, v, of a body can be calculated when its distance, s, from the origin is known. Find the acceleration when $v = 10$.

12. If $y = \left(x - \dfrac{1}{x}\right)^2$, find $\dfrac{dx}{dt}$ when $x = 2$ given $\dfrac{dy}{dt} = 1$.

13. A rectangle is twice as long as it is broad. Find the rate of change of the perimeter when the breadth of the rectangle is 1 m and its area is changing at the rate of 18 cm²/s, assuming the expansion uniform.

14. A horse-trough has triangular cross-section of height 25 cm and base 30 cm, and is 2 m long. A horse is drinking steadily, and when the water level is 5 cm below the top it is being lowered at the rate of 1 cm/min. Find the rate of consumption in litres per minute.

Products and quotients

6.4. The reader is now able to differentiate quite elaborate functions but no method has been suggested for dealing with the product of two functions.

$(x + 1)^7$ and $(x - 3)^4$, for example, are simple functions and $(x + 1)^7 . (x - 3)^4$ cannot be considered complicated, but if it had to be multiplied out in order to be differentiated term by term the work would be very laborious.

A further brief return to fundamental ideas will produce a formula to help us with functions of this kind.

Let y be the product of two functions u and v of a variable x.

Then $$y = u.v,$$

and $$y + \Delta y = (u + \Delta u)(v + \Delta v),$$

where a small increment Δx in x produces increments Δu in u, Δv in v and Δy in y.

Since $y + \Delta y = uv + v\Delta u + u\Delta v + \Delta u\Delta v$, $y = uv$,

$$\Delta y = v\Delta u + u\Delta v + \Delta u\Delta v.$$

Dividing by Δx,

$$\frac{\Delta y}{\Delta x} = v\frac{\Delta u}{\Delta x} + u\frac{\Delta v}{\Delta x} + \frac{\Delta u}{\Delta x}.\Delta v.$$

As $\Delta x \to 0$, Δu, Δv and Δy also approach 0,

$$\frac{\Delta y}{\Delta x} \to \frac{dy}{dx}, \quad \frac{\Delta u}{\Delta x} \to \frac{du}{dx}, \quad \frac{\Delta v}{\Delta x} \to \frac{dv}{dx}.$$

$$\therefore \quad \frac{dy}{dx} = v\frac{du}{dx} + u\frac{dv}{dx} + \frac{du}{dx}.0.$$

$$\therefore \quad \frac{dy}{dx} = v\frac{du}{dx} + u\frac{dv}{dx}.$$

This formula must be remembered, and this is perhaps most easily done in words,

> "To differentiate the product of two functions, differentiate the first function, leaving the second one alone and then differentiate the second, leaving the first one alone,"

and it is necessary to remember also that, should one of the functions in the product be itself a function of a function, its derivative must be found as carefully as those in §6·2 before insertion in this product formula.

Qu. 3. Use this method to differentiate with respect to x the following functions:

(i) $(x + 1)(x + 2)$, (ii) $(x^2 + 1)x^2$,
(iii) $(x - 2)^2(x^2 - 2)$, (iv) $(x + 1)^2(x + 2)^2$.

check your results by multiplying out and then differentiating.

The most common mistakes made in this type of question are due to careless algebra and so particular attention should be paid to details of simplification.

Example 8. *Differentiate the expression*
$$y = (x^2 - 3)(x + 1)^2$$
and simplify the result.

Let $u = (x^2 - 3)$ and let $v = (x + 1)^2$,

then $\qquad \dfrac{du}{dx} = 2x \quad$ and $\quad \dfrac{dv}{dx} = 2(x + 1)$.

$$\therefore \frac{dy}{dx} = (x + 1)^2 . 2x + (x^2 - 3) . 2(x + 1),$$
$$= 2(x + 1)\{x(x + 1) + (x^2 - 3)\},$$
$$= 2(x + 1)\{2x^2 + x - 3\},$$
$$= 2(x + 1)(2x + 3)(x - 1).$$

Example 9. *Differentiate* $(x^2 + 1)^3(x^3 + 1)^2$.

If $u = (x^2 + 1)^3$ and $v = (x^3 + 1)^2$, then $y = u.v.$

$$\frac{du}{dx} = 3(x^2 + 1)^2 . 2x \quad \text{and} \quad \frac{dv}{dx} = 2(x^3 + 1) . 3x^2,$$

$$\therefore \frac{dy}{dx} = (x^3 + 1)^2 . 6x(x^2 + 1)^2 + (x^2 + 1)^3 . 6x^2(x^3 + 1),$$
$$= 6x(x^3 + 1)(x^2 + 1)^2\{(x^3 + 1) + x(x^2 + 1)\},$$

$$\therefore \frac{d}{dx}\{(x^2 + 1)^3(x^3 + 1)^2\}$$
$$= 6x(x^3 + 1)(x^2 + 1)^2(2x^3 + x + 1).$$

Example 10. *Find the x-coordinates of the stationary points of the curve* $y = (x^2 - 1)\sqrt{(1 + x)}$.

$$y = (x^2 - 1)(x + 1)^{\frac{1}{2}}.$$

$$\therefore \frac{dy}{dx} = (x + 1)^{\frac{1}{2}}.2x + (x^2 - 1).\tfrac{1}{2}(x + 1)^{-\frac{1}{2}},$$

$$= \frac{2(x + 1).2x + (x^2 - 1)}{2(x + 1)^{\frac{1}{2}}},$$

$$= \frac{(x + 1)(4x + x - 1)}{2(x + 1)^{\frac{1}{2}}},$$

$$= \frac{(5x - 1)(x + 1)}{2(x + 1)^{\frac{1}{2}}},$$

$$= \tfrac{1}{2}(5x - 1)(x + 1)^{\frac{1}{2}}.$$

$\therefore$ for stationary points $x = \frac{1}{5}$ or -1.

There is a formula for quotients corresponding to that for products and it is proved in a similar way.

If $y = \dfrac{u}{v}$ then $\dfrac{dy}{dx} = \dfrac{v\dfrac{du}{dx} - u\dfrac{dv}{dx}}{v^2}.$

If the reader wishes to ignore this formula and to deal with the quotient $\dfrac{u}{v}$ as the product uv^{-1} he is at liberty to do so—it is merely a matter of preference.

Example 11. *Differentiate* $\dfrac{(x - 3)^2}{(x + 2)^2}.$

Let $u = (x - 3)^2$ and $v = (x + 2)^2$ then $y = \dfrac{u}{v}.$

$$\frac{du}{dx} = 2(x-3) \quad \text{and} \quad \frac{dv}{dx} = 2(x+2),$$

$$\therefore \frac{dy}{dx} = \frac{(x+2)^2.2(x-3) - (x-3)^2.2(x+2)}{(x+2)^4},$$

$$= \frac{2(x+2)(x-3)\{(x+2)-(x-3)\}}{(x+2)^4}.$$

$$= \frac{2(x-3).5}{(x+2)^3},$$

$$\therefore \frac{d}{dx}\left\{\frac{(x-3)^2}{(x+2)^2}\right\} = \frac{10(x-3)}{(x+2)^3}.$$

Example 12. *Differentiate* $\dfrac{x}{\sqrt{(1+x^2)}}.$

Let $u = x$ and $v = \sqrt{(1+x^2)}.$

Then $\quad \dfrac{du}{dx} = 1 \quad \text{and} \quad \dfrac{dv}{dx} = \dfrac{2x}{2\sqrt{(1+x^2)}}.$

$$\therefore \frac{dy}{dx} = \frac{\sqrt{(1+x^2)}.1 - x.\dfrac{x}{\sqrt{(1+x^2)}}}{(1+x^2)},$$

$$= \frac{1+x^2-x^2}{(1+x^2)^{\frac{3}{2}}}.$$

$$\therefore \frac{d}{dx}\left\{\frac{x}{\sqrt{(1+x^2)}}\right\} = \frac{1}{(1+x^2)^{\frac{3}{2}}}.$$

Qu. 4. Prove the formula for quotients by the Δu, Δv method.

Qu. 5. Differentiate:

(i) $(x^2 + 1)(x + 3)^{-2}$ as a product,

(ii) $\dfrac{x^2 + 1}{(x + 3)^2}$ as a quotient.

Simplify the results and compare them.

* Simplification was an essential part of answering the question in Example 11 and, since the gradient of a function is often needed for a specific purpose, the reader should get into the habit of factorizing and simplifying as far as possible. He will find it necessary, in any case, in order to check his answers with those at the back of the book!

Exercise 6c

Differentiate with respect to x the following functions.

1. $x^2(x + 1)^3.$
2. $x(x^2 + 1)^4.$
3. $(x + 1)^2(x^2 - 1).$

4. $\dfrac{x}{x + 1}.$
5. $\dfrac{1 - x^2}{1 + x^2}.$
6. $\dfrac{x^2 + 1}{(x + 1)^2}.$

7. $(1 + x^2)^2(1 - x^2).$
8. $x^2\left(1 - \dfrac{1}{\sqrt{x}}\right).$

9. $(1 - x^2)^2(1 - x^3).$
10. $(x - 1)\sqrt{(x^2 + 1)}.$

11. $x^2\sqrt{(1 + x^2)}.$
12. $\dfrac{x^2}{\sqrt{(1 + x^2)}}.$
13. $\dfrac{(x - 1)^2}{\sqrt{x}}.$

14. $\dfrac{2x^2 - x^3}{\sqrt{(x^2 - 1)}}.$
15. $\sqrt{(x + 2)}\sqrt{(x + 3)}.$

16. $\dfrac{\sqrt{x}}{\sqrt{(x + 1)}}.$
17. $\dfrac{1 - \sqrt{x}}{1 + \sqrt{x}}.$
18. $\sqrt{\left(\dfrac{1 + x}{2 + x}\right)}.$

19. $\sqrt{(x + 1)}\sqrt{(x + 2)^3}.$
20. $\sqrt{\left\{\dfrac{(x + 1)^3}{x + 2}\right\}}.$

21. $\sqrt{\left\{\dfrac{(1 + x^2)^3}{2 + x^2}\right\}}.$

Implicit functions

6.5. Up to the present we have dealt only with *explicit* functions of x, e.g. $y = x^2 - 5x + \dfrac{4}{x}.$ Here y is given as an expression in x. If, however, y is given *implicitly* by an

---

* Practice in the algebra involved in differentiating a quotient is given in Exercise 8*d*, No. 5.

equation such as $x = y^4 - y - 1$, we cannot express y in terms of x.

Consider an easier case.

If $x = y^2$, $y = x^{\frac{1}{2}}$.

$$\therefore \frac{dy}{dx} = \tfrac{1}{2}x^{-\frac{1}{2}} = \frac{1}{2x^{\frac{1}{2}}} = \frac{1}{2y}.$$

But $\dfrac{dx}{dy} = 2y$, so *in this case*,

$$\frac{dy}{dx} = \frac{1}{\dfrac{dx}{dy}}.$$

Now consider the general case. If x is increased by Δx, y is increased by Δy, but identically,

$$\frac{\Delta y}{\Delta x} = \frac{1}{\dfrac{\Delta x}{\Delta y}}.$$

Now as $\Delta x, \Delta y \to 0$, $\dfrac{\Delta y}{\Delta x} \to \dfrac{dy}{dx}$, $\dfrac{\Delta x}{\Delta y} \to \dfrac{dx}{dy}$.

$$\therefore \frac{dy}{dx} = \frac{1}{\dfrac{dx}{dy}}.$$

When neither x nor y is an explicit function we have to adopt another method. We regard the whole expression as a function of x and differentiate both sides of the equation. Any power of y is a "function of a function", as y is a function of x.

Example 13. *Find the gradient of the curve*

$$x^2 + 2xy - 2y^2 + x = 2$$

at the point $(-4, 1)$.

To find the gradient, differentiate with respect to x.

$$\frac{\mathrm{d}}{\mathrm{d}x}(x^2) + \frac{\mathrm{d}}{\mathrm{d}x}(2xy) - \frac{\mathrm{d}}{\mathrm{d}x}(2y^2) + \frac{\mathrm{d}}{\mathrm{d}x}(x) = \frac{\mathrm{d}}{\mathrm{d}x}(2).$$

$$\therefore 2x + \left(2y + 2x \cdot \frac{\mathrm{d}y}{\mathrm{d}x}\right) - 4y \cdot \frac{\mathrm{d}y}{\mathrm{d}x} + 1 = 0.$$

$$\therefore \frac{\mathrm{d}y}{\mathrm{d}x}(2x - 4y) = -1 - 2x - 2y.$$

When $x = -4$, $y = 1$,

$$\frac{\mathrm{d}y}{\mathrm{d}x}(-8 - 4) = -1 + 8 - 2.$$

$$\therefore \frac{\mathrm{d}y}{\mathrm{d}x} = \frac{+5}{-12}.$$

$\therefore$ the gradient at $(-4, 1)$ is $-\frac{5}{12}$.

Qu. 6.   (a) Differentiate with respect to x:

       (i) x,   (ii) y,   (iii) x^2,   (iv) y^2,

       (v) xy,   (vi) x^2y,   (vii) xy^2.

     (b) Find $\dfrac{\mathrm{d}y}{\mathrm{d}x}$ if $x^2 + y^2 - 6xy + 3x - 2y + 5 = 0$.

Sometimes no attempt is made even to set down an implicit function—both x and y may be given as functions of another variable, a **parameter**. In such cases the gradient is given in terms of the variable parameter.

Example 14. *If $x = t^3 + t^2$, $y = t^2 + t$ find $\dfrac{\mathrm{d}y}{\mathrm{d}x}$ in terms of t.*

$$\frac{\mathrm{d}x}{\mathrm{d}t} = 3t^2 + 2t, \qquad \frac{\mathrm{d}y}{\mathrm{d}t} = 2t + 1.$$

Now $\qquad \dfrac{\mathrm{d}y}{\mathrm{d}x} = \dfrac{\mathrm{d}y}{\mathrm{d}t} \cdot \dfrac{\mathrm{d}t}{\mathrm{d}x}$,   but $\dfrac{\mathrm{d}t}{\mathrm{d}x} = 1 \Big/ \dfrac{\mathrm{d}x}{\mathrm{d}t}$

$$\therefore \frac{\mathrm{d}y}{\mathrm{d}x} = \frac{\mathrm{d}y}{\mathrm{d}t} \div \frac{\mathrm{d}x}{\mathrm{d}t}.$$

$$\therefore \frac{\mathrm{d}y}{\mathrm{d}x} = \frac{2t + 1}{t(3t + 2)}.$$

Qu. 7. Show that the above parametric representation is of the curve $y^3 = x^2 + xy$. Find $\dfrac{dy}{dx}$ for this curve and show that it agrees with the above result.

Example 15. *Find the gradient of the curve* $x = \dfrac{t}{1+t}$, $y = \dfrac{t^3}{1+t}$ *at the point* $(\frac{1}{2}, \frac{1}{2})$.

$$\frac{dx}{dt} = \frac{(1+t).1 - t.1}{(1+t)^2} = \frac{1}{(1+t)^2}.$$

$$\frac{dy}{dt} = \frac{(1+t).3t^2 - t^3.1}{(1+t)^2} = \frac{3t^2 + 2t^3}{(1+t)^2}.$$

$$\therefore \frac{dy}{dx} = 3t^2 + 2t^3.$$

At $(\frac{1}{2}, \frac{1}{2})$ $t = 1$, $\qquad \therefore \dfrac{dy}{dx} = 3 + 2.$

$$\therefore \text{ the gradient at } (\tfrac{1}{2}, \tfrac{1}{2}) \text{ is } 5.$$

Exercise 6d

1. Find the gradient of the ellipse $2x^2 + 3y^2 = 14$ at the points where $x = 1$.

2. Find the x-coordinates of the stationary points of the curve represented by the equation
$$x^3 - y^3 - 4x^2 + 3y = 11x + 4.$$

3. Find the gradient of the ellipse
$$x^2 - 3yx + 2y^2 - 2x = 4$$
at the point $(1, -1)$.

4. Find the gradient of the tangent at the point $(2, 3)$ to the hyperbola $xy = 6$.

5. (i) If $x = t^2$, $y = t^3$ find $\dfrac{dy}{dx}$ in terms of t.

   (ii) If $y = x^{\frac{3}{2}}$ find $\dfrac{dy}{dx}$.

Is there any connection between these two results?

6. At what points are the tangents to the circle
$$x^2 + y^2 - 6y - 8x = 0$$
parallel to the y-axis?

7. Find $\dfrac{dy}{dx}$ when (i) $x^2y^3 = 8$, (ii) $xy(x - y) = 4$.

8. Find $\dfrac{dy}{dx}$, in terms of t, when

(i) $x = at^2$, $y = 2at$;   (ii) $x = (t + 1)^2$, $y = (t^2 - 1)$.

9. If $x = \dfrac{t}{1 - t}$ and $y = \dfrac{t^2}{1 - t}$ find $\dfrac{dy}{dx}$ in terms of t.

10. Find $\dfrac{dy}{dx}$ in terms of x, y when
$$x^2 + y^2 - 2xy + 3y - 2x = 7.$$

11. If $x = \dfrac{2t}{t + 2}$, $y = \dfrac{3t}{t + 3}$, find $\dfrac{dy}{dx}$ in terms of t.

12. Find $\dfrac{dy}{dx}$ in terms of x, y when $3(x - y)^2 = 2xy + 1$.

Small changes

6.6. We have seen that, as $\Delta x \to 0$, $\dfrac{\Delta y}{\Delta x} \to \dfrac{dy}{dx}$. Therefore, if Δx is small,

$$\frac{\Delta y}{\Delta x} \simeq \frac{dy}{dx}.$$

$$\therefore \Delta y \simeq \frac{dy}{dx}\Delta x.$$

Three applications of this formula follow in Examples 16–18.

Example 16. *The side of a square is 5 cm. Find the increase in the area of the square when the side expands 0·01 cm.*

Let the area of the square be A cm² when the side is x cm. Then $A = x^2$.

Now $\Delta A \simeq \dfrac{dA}{dx}\Delta x$ and $\dfrac{dA}{dx} = 2x$,

$$\therefore \Delta A \simeq 2x\Delta x.$$

If $x = 5$ and $\Delta x = 0.01$,

$$\Delta A \simeq 2 \times 5 \times 0.01 = 0.1.$$

∴ the increase in area $\simeq 0.1$ cm².

In this case the increase in area can be found accurately very easily:

$$\Delta A = 5.01^2 - 5^2 = 0.1001.$$

The reader is strongly advised to use the calculus method, for the moment, even if he sees a quicker way, since it is an important introduction to certain topics which he may meet later.

Note that the error in the calculus method is, *in this case* $(0.01)^2 = (\Delta x)^2$.

Example 17. *A 2% error is made in measuring the radius of a sphere. Find the percentage error in surface area.*

Let the surface area be S and the radius be r, then

$$S = 4\pi r^2, \quad \therefore \frac{dS}{dr} = 8\pi r.$$

$$\therefore \Delta S \simeq 8\pi r \Delta r.$$

But the error in r is 2%, therefore $\Delta r = \frac{2}{100} \cdot r$.

$$\therefore \Delta S \simeq 8\pi r \cdot \frac{2r}{100} = \frac{16\pi r^2}{100}.$$

$$\therefore \frac{\Delta S}{S} \simeq \frac{16\pi r^2}{100} \div 4\pi r^2 = \frac{4}{100}.$$

∴ the error in the surface area $\simeq 4\%$.

Example 18. *Find an approximation for $\sqrt{9.01}$.*

Let $y = \sqrt{x}$, so $\dfrac{dy}{dx} = \dfrac{1}{2\sqrt{x}}$.

$$\therefore \Delta y \simeq \frac{1}{2\sqrt{x}} \cdot \Delta x.$$

When $x = 9$, and $\Delta x = 0.01$,

$$\Delta y \simeq \tfrac{1}{6} \times 0.01 \simeq 0.001\ 67.$$
$$\therefore \ \sqrt{9.01} \simeq 3.001\ 67.$$

Exercise 6e

1. The surface area of a sphere is $4\pi r^2$. If the radius of the sphere is increased from 10 cm to 10·1 cm, what is the approximate increase in surface area?

2. An error of 3% is made in measuring the radius of the sphere. Find the percentage error in volume.

3. Find (i) $\sqrt[3]{8.01}$, (ii) $\sqrt{25.1}$ by the method of Example 18.

4. If l cm is the length of a pendulum and t s the time of one complete swing, it is known that $l = kt^2$. If the length of the pendulum is increased by $x\%$, x being small, find the corresponding percentage increase in time of swing.

5. If the pressure and volume of a gas are p and v then Boyle's law states $pv = $ constant (k). If Δp and Δv denote corresponding small changes in p and v express $\dfrac{\Delta p}{p}$ in terms of $\dfrac{\Delta v}{v}$.

6. An error of $2\tfrac{1}{2}\%$ is made in the measurement of the area of a circle. What percentage error results in (i) the radius, (ii) the circumference?

7. The height of a cylinder is 10 cm and its radius is 4 cm. Find the approximate increase in volume when the radius increases to 4·02 cm.

8. One side of a rectangle is three times the other. If the perimeter increases by 2% what is the percentage increase in area?

9. The radius of a closed cylinder is equal to its height. Find the percentage increase in total surface area corresponding to unit percentage increase in height.

10. Find (i) $\sqrt{627}$, (ii) $\sqrt[3]{1005}$, by the method of Example 18.

11. The volume of a sphere increases by 2%. Find the corresponding percentage increase in surface area.

12. As x increases, prove that the area of a circle of radius x and the area of a square of side x increase by the same percentage, providing that the increase in x is small.

13. A hemispherical cap of radius r is placed on the end of a cylinder of radius r and height $3r$. If the radius increases from 8 to 8·1 cm, find the corresponding increase in the volume.

Second derivative

6.7. We know that velocity, v, is the rate of change of displacement, s, with respect to time, t, and may be denoted by $\dfrac{\mathrm{d}s}{\mathrm{d}t}$. Acceleration is the rate of change of velocity with respect to time, and we have up to now denoted this by $\dfrac{\mathrm{d}v}{\mathrm{d}t}$; but $\dfrac{\mathrm{d}}{\mathrm{d}t}(v)$ may also be written as $\dfrac{\mathrm{d}}{\mathrm{d}t}\left(\dfrac{\mathrm{d}s}{\mathrm{d}t}\right)$, and thus acceleration is seen to be the second derivative of s with respect to t.

The second derivative arises in a wide variety of contexts as well as in kinematics, of course, and we need a less cumbersome notation.

$$\dfrac{\mathrm{d}}{\mathrm{d}x}\left(\dfrac{\mathrm{d}y}{\mathrm{d}x}\right) \text{ is written as } \dfrac{\mathrm{d}^2 y}{\mathrm{d}x^2}$$

which is spoken "d two y by d x squared".

Acceleration, $\dfrac{\mathrm{d}^2 s}{\mathrm{d}t^2}$, may be written in yet another way by using the fact that $\dfrac{\mathrm{d}v}{\mathrm{d}t} = \dfrac{\mathrm{d}v}{\mathrm{d}s} \cdot \dfrac{\mathrm{d}s}{\mathrm{d}t} = \dfrac{\mathrm{d}v}{\mathrm{d}s} \cdot v$; thus we have arrived at the following alternative notations,

$$a = \dfrac{\mathrm{d}v}{\mathrm{d}t} = \dfrac{\mathrm{d}^2 s}{\mathrm{d}t^2} = v\dfrac{\mathrm{d}v}{\mathrm{d}s},$$

the last form, $v\dfrac{\mathrm{d}v}{\mathrm{d}s}$, being applicable when velocity or acceleration is a function of s rather than of t.

If $y = \mathrm{f}(x)$, $\dfrac{\mathrm{d}y}{\mathrm{d}x}$ is written as $\mathbf{f'(x)}$, and $\dfrac{\mathrm{d}^2 y}{\mathrm{d}x^2}$ as $\mathbf{f''(x)}$.

Qu. 8. (i) If $y = x^2 - \dfrac{1}{x^3}$, find $\dfrac{dy}{dx}$ and $\dfrac{d^2y}{dx^2}$.

(ii) Given that $v = 3(4 - s^2)^{\frac{1}{2}}$, show that $a = -9s$.

(iii) If $f(x) = \dfrac{x}{x - 1}$, find $f'(x)$ and $f''(x)$.

If $\dfrac{dy}{dx}$ is found in terms of a parameter t, $\dfrac{d^2y}{dx^2}$ requires a differentiation with respect to x, so:

$$\frac{d^2y}{dx^2} = \frac{d}{dt}\left(\frac{dy}{dx}\right) \cdot \frac{dt}{dx} = \frac{d}{dt}\left(\frac{dy}{dx}\right) \div \frac{dx}{dt}.$$

Qu. 9. If $x = a(t^2 - 1)$, $y = 2a(t + 1)$, find $\dfrac{dy}{dx}$ and $\dfrac{d^2y}{dx^2}$.

The miscellaneous questions which follow are in two stages. The first 25 are set in the order of the sections in this chapter in order to give the reader further practice in the fundamentals. The later questions are "mixed" and also include ones involving a knowledge of other sections of this book.

Exercise 6f (Miscellaneous)

1. Differentiate (i) $\dfrac{1}{x^{2n}}$, (ii) x^{n+1}, (iii) $\frac{1}{2}x^{2a-1}$, (iv) $(x^2)^m$,

(v) $\sqrt{x^n}$.

2. Integrate (i) $(x^2)^{k-1}$, (ii) $(x^n)^{-2}$, (iii) $nx^{-\frac{1}{n}-1}$, (iv) x^{-1+k}.

3. Differentiate (i) $\sqrt[n]{x}$, (i) $(\sqrt{x})^n$, (iii) $\dfrac{2x}{x^n}$, (iv) $\dfrac{1}{\sqrt{x^n}}$,

(v) $\dfrac{1}{\sqrt[3]{x^{2n}}}$.

4. Differentiate (i) $\dfrac{2}{\sqrt[n]{x^3}}$, (ii) $\sqrt[3]{\dfrac{1}{x}}$, (iii) $\left(\dfrac{1}{\sqrt{x^n}}\right)^3$, (iv) $\sqrt{(2x^3)}$,

(v) $x^{\frac{n}{n+1}}$.

In nos. 5–12 differentiate with respect to x and simplify.

5. (i) $(x^2 + 3)^4$,　　(ii) $\sqrt{(2x^3 - 3)}$,

(iii) $(\sqrt{x} + 1)^3$,　　(iv) $\left(\dfrac{x}{2} - \dfrac{2}{x}\right)^n$.

6. (i) $\dfrac{1}{x + \sqrt{x}}$,　　(ii) $\dfrac{1}{x^2 - 1}$,

(iii) $\dfrac{1}{(\sqrt{x} - 1)^2}$,　　(iv) $\dfrac{1}{\sqrt[3]{(2x - x^2)}}$.

7. (i) $x^2(x - 1)^3$,　　(ii) $(x + 1)^{\frac{3}{2}}(x - 1)^{\frac{5}{2}}$,

(iii) $(x - 2)^{\frac{1}{2}}(x^2 + 3)$.

8. (i) $\sqrt{\{(x + 1)(x - 2)^3\}}$,　　(ii) $(1 - x^2)\sqrt[3]{(1 - 2x)}$,

(iii) $x\sqrt{(x^2 - 1)}$.

9. (i) $\dfrac{x}{x^2 - 1}$,　　(ii) $\dfrac{x}{\sqrt{(x - 1)}}$,

(iii) $\dfrac{\sqrt{x}}{x - 1}$,　　(iv) $\dfrac{\sqrt{(x - 1)}}{\sqrt{x - 1}}$.

10. (i) $\dfrac{x^2 + 2}{(x + 2)^2}$,　　(ii) $\dfrac{(x - 1)^3}{(x^3 - 1)}$.

11. (i) $\sqrt{\left(\dfrac{x + 1}{x + 2}\right)}$,　　(ii) $\sqrt{\left\{\dfrac{(x + 2)^3}{x - 1}\right\}}$.

12. (i) $\sqrt{\dfrac{x^2 + 1}{x^2 - 1}}$,　　(ii) $\dfrac{(1 - \sqrt{x})^2}{\sqrt{(x^2 - 1)}}$.

13. Find $\dfrac{dy}{dx}$ when $x^2 + 2xy + y^2 = 3$.

14. Find $\dfrac{dy}{dx}$ when $x^2 - 3xy + y^2 - 2y + 4x = 0$.

15. Find $\dfrac{dy}{dx}$ when $3x^2 - 4xy = 7$.

16. If $x = \dfrac{2t}{1 + t^2}$, $y = \dfrac{1 - t^2}{1 + t^2}$ find $\dfrac{dy}{dx}$ in terms of t.

17. If $x = \dfrac{1}{\sqrt{(1 + t^2)}}$, $y = \dfrac{t}{\sqrt{(1 + t^2)}}$ find $\dfrac{dy}{dx}$ in terms of t.

18. If $x = \dfrac{t}{1 - t}$, $y = \dfrac{1 - 2t}{1 - t}$ find $\dfrac{dy}{dx}$.

19. When measuring the area of a circle, 2% error is made. Find the percentage error in the radius.

20. When measuring the dimensions of cubical box 1% error was made—all measurements being too large. Find the percentage error in volume.

21. The circumference of a circle is measured with a piece of string which stretches 1%. What is the percentage error in the area of the circle.

22. Calculate (i) $\sqrt[3]{65}$, (ii) $\sqrt{37}$, without using tables.

23. If $y = 4x^3 - 6x^2 - 9x + 1$, find $\dfrac{dy}{dx}$ and hence find the values of $\dfrac{d^2y}{dx^2}$ when the gradient is zero.

24. If $x = at^2$, $y = 2at$, find $\dfrac{dy}{dx}$ and $\dfrac{d^2y}{dx^2}$ in terms of t.

25. If $f(x) = 8x^3 - 11x^2 - 30x + 9$ for what values of x is $f'(x) = 0$?

---

26. Find the equation of the tangent to the curve $x^2 - y^2 = 9$ at the point $(5, 4)$.

27. Prove from first principles that the derivative of $\dfrac{1}{x^2}$ is $-\dfrac{2}{x^3}$.

28. If $y = \dfrac{x^2}{\sqrt{(x+1)}}$, find $\dfrac{dy}{dx}$ and $\dfrac{d^2y}{dx^2}$.

29. Find what values of x give stationary values of the function $(2x - 3)^2(x - 2)^3$.

30. Differentiate with respect to x (i) $\sec 2x$, (ii) $\sin^2 x$, (iii) $x \cos x$, (iv) $\tan^3 x$, (v) $\sin \sqrt{x}$.

31. A curve is represented parametrically by $x = (t^2 - 1)^2$, $y = t^3$.

Find $\dfrac{dy}{dx}$ and $\dfrac{d^2y}{dx^2}$ in terms of t.

32. The volume of a sphere is increased by 3%. Find the percentage increase in the radius.

33. A curve called the Witch of Agnesi has for its equation $y = \sqrt{\left(\dfrac{3}{x} - 1\right)}$. Find its gradient when $x = \frac{1}{2}$.

34. Show that of all rectangles with given perimeter the square has maximum area.

35. Find the equation of the tangent to the parabola $y^2 = 4x$ which is parallel to the line $y = 3x - 4$. What are the coordinates of the point of contact?

Find also the equation of the normal at this point.

36. The distance, s, of a particle from a point after time t is given by the formula $s^2 = a + bt^2$.

Find the velocity and acceleration in terms of s, t, a, b.

37. Find the equation of the tangent and normal to the curve $x = a \cos^3 t$, $y = a \sin^3 t$.

38. If $R = ar^n$ and an error of $x\%$ is made in measuring r, prove that an error of $nx\%$ will result in R.

39. Find the maximum and minimum values of $x^2\sqrt{(2 - x)}$.

40. Find the equation of the tangent and normal to the cycloid $x = a(2\theta + \sin 2\theta)$, $y = a(1 - \cos 2\theta)$ at any point.

41. If the radius of a spherical soap bubble increases from 1 cm to 1·02 cm, find the approximate increase in volume.

42. If the velocity v is given by the formula $v = \dfrac{u}{1 + ks}$ where u is the initial velocity, s is the distance and k is a constant prove that the acceleration varies as v^3.

43. Differentiate:

(i) $\sin^2 (3x^2 + 4)$, 　　　　(ii) $\tan \sqrt{(x + 1)}$,

(iii) $\dfrac{\sin x}{1 + \cos x}$, 　　　　(iv) $\sqrt{\dfrac{\tan x}{1 - \tan^2 x}}$.

44. If $x^2 + 3xy - y^2 = 3$, find $\dfrac{dy}{dx}$ and $\dfrac{d^2y}{dx^2}$ at the point (1, 1).

45. Differentiate $\sqrt{(x^2 + 1)}$ with respect to x.

Why does the substitution $x = \tan \theta$ in this function and its derivative not give you the derivative of $\sec \theta$?

46. A wire has circular cross-section. It is stretched so that its length is increased by 1%. Assuming that the volume remains constant find the percentage increase in radius.

47. The formula $T = 2\pi \sqrt{\left(\dfrac{l}{g}\right)}$ gives the time of swing of a pendulum. Prove that if a clock pendulum 1 m long is shortened by 1 mm the clock will go about $\frac{3}{4}$ min per day faster.

48. The velocity of a moving point is given by the formula

$$\frac{\mathrm{d}}{\mathrm{d}x}(v^2) = -32x + 64 \quad \text{and} \quad v = 0 \quad \text{when} \quad x = 0.$$

At what other point is $v = 0$? What is the greatest value of v?

49. A rectangular block is of height h cm and has a square base of side x cm. At a certain instant the volume of the block is increasing at the rate of 2% of its value per second and h is increasing at the rate of 3% of its value per second. Is x increasing or decreasing at this instant and at what percentage rate of its value per second?

*50. A triangle ABC, inscribed in a fixed circle, stands on a fixed chord BC, while the vertex A moves. Find an equation connecting Δb and ΔB, and prove that

$$\frac{\Delta b}{\cos B} + \frac{\Delta c}{\cos C} = 0.$$

*51. A rod AB of length a is hinged to a horizontal table at A and turns about A in a vertical plane with angular velocity ω. A luminous point is situated at a height $h(>a)$ vertically above A. Find the length of the shadow when the rod makes an angle θ with the vertical and prove that the length of the shadow is increasing at the rate

$$\frac{\omega a h(h \cos \theta - a)}{(h - a \cos \theta)^2}.$$

52. If $y = \dfrac{x}{\sqrt{(1 + x^2)}}$, prove that

$$(1 + x^2)\frac{\mathrm{d}^2 y}{\mathrm{d}x^2} + 3x\frac{\mathrm{d}y}{\mathrm{d}x} = 0.$$

53. The height of a closed cylinder is 3 cm and remains constant. The radius of its base is 2·5 cm and it is increasing at the rate of 0·01 cm/s. At what rates are (i) the volume, (ii) total surface area of the cylinder increasing?

*54. If $x = \sqrt{(1 - t^2)}$ change the equation

$$x(1 - x^2)\frac{\mathrm{d}^2 y}{\mathrm{d}x^2} + (1 - 3x^2)\frac{\mathrm{d}y}{\mathrm{d}x} = xy$$

into an equation involving $\dfrac{\mathrm{d}^2 y}{\mathrm{d}t^2}$, $\dfrac{\mathrm{d}y}{\mathrm{d}t}$, y and t.

* 50, 51 and 54 are rather more difficult than the others.

CHAPTER 7

FURTHER INTEGRATION

Some standard curves

7.1. In §4.5 we dealt with some simple aids to curve sketching. By this stage, the reader should be thoroughly familiar with some standard curves which will be frequently occurring in the work which follows.

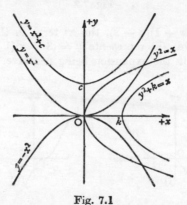

Fig. 7.1

Fig. 7.1 shows some variations on the curve $y = x^2$, which is a *parabola*. The line about which the curve is symmetrical is called the *axis*, and it cuts the curve at the *vertex*. Thus for the curve $y = x^2 + c$, the axis is the y-axis, and the vertex is $(0, c)$. Any equation of the form $y = ax^2 + bx + c$, where a, b, and c are constants (a not being zero), represents a parabola with the axis parallel to the y-axis.

135

Typical shapes of curves for which y is given as a cubic function of x are shown in Fig. 7.2. (i) represents

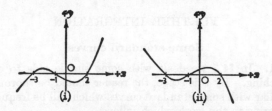

Fig. 7.2

$y = (x + 3)(x + 1)(x - 2)$, the x^3 term in the expansion being positive; (ii) represents $y = (3 + x)(1 + x)(2 - x)$, the x^3 term in the expansion being negative. (See p. 78, Example 7.)

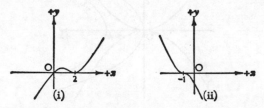

Fig. 7.3

Fig. 7.3 shows (i) $y = x(x - 2)^2$, and (ii) $y = -(x + 1)^3$, illustrating that when the function of x has a squared factor, the curve touches the x-axis; and with a cubed factor, the curve touches *and crosses* the x-axis.

In Fig. 7.4 the curve $y = \dfrac{1}{x}$ is called a *rectangular hyperbola*. It is made up of two separate branches which approach the axes but never meet them.

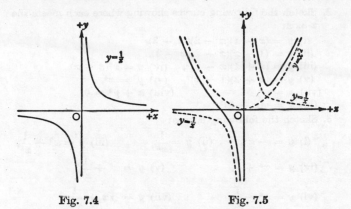

Fig. 7.4   Fig. 7.5

Fig. 7.5 illustrates how a sketch of the curve $y = x^2 + \dfrac{1}{x}$ may be built up by adding the y-coordinates of the two known curves $y = x^2$ and $y = \dfrac{1}{x}$.

The integration of x^n (n fractional)

7.2. In Chapter 6 the differentiation of x^n was assumed to include cases where n is a fraction, and so we can now integrate powers of x with fractional indices. Thus, if

$$\frac{dy}{dx} = \sqrt{x} = x^{\frac{1}{2}},$$

$$y = \frac{x^{\frac{3}{2}}}{\frac{3}{2}} + c = \tfrac{2}{3}x^{\frac{3}{2}} + c.$$

Exercise 7a

1. Sketch the following curves:

(i) $y = 4x^2$,   (ii) $y = -x^2 + 9$,

(iii) $y - 1 = x^2$,   (iv) $x = -y^2$,

(v) $x - y^2 + 4 = 0$,   (vi) $2x + y^2 + 16 = 0$.

2. Sketch the following curves showing where each meets the x-axis:

    (i) $y = (x - 1)(x - 2)(x - 3)$,

    (ii) $y = (1 - x)(x - 2)(x - 3)$,

    (iii) $y = (x + 1)(x - 2)^2$,   (iv) $y = x^2(3 - x)$,

    (v) $y = (x + 2)(1 - x)^2$,   (vi) $y^2 = x^6$,

  (vii) $x = y^3$,   (viii) $x + y^3 = 0$,

    (ix) $x = y(y - 3)^2$.

3. Sketch the following curves:

    (i) $y = - x^4$,   (ii) $y = \dfrac{1}{x^2}$,   (iii) $y = x^2 + \dfrac{1}{x^2}$,

    (iv) $y = x^3 + \dfrac{1}{x}$,   (v) $y = x^3 + \dfrac{1}{x^2}$,

    (vi) $y = x^2 - \dfrac{1}{x}$,   (vii) $y = \sqrt{x} + \dfrac{1}{\sqrt{x}}$.

4. Integrate with respect to x:

    (i) $x^{\frac{1}{3}}$,   (ii) $\sqrt[4]{x}$,   (iii) $2x^{\frac{1}{5}}$,

    (iv) $k . \sqrt[3]{x}$,   (v) $x^{-\frac{1}{2}}$,   (vi) $\dfrac{1}{\sqrt[3]{x}}$,

    (vii) $x^{-\frac{1}{6}}$,   (viii) $\dfrac{2}{\sqrt[5]{x}}$,   (ix) $\sqrt[3]{x^2}$,

    (x) $x^{\frac{7}{3}}$,   (xi) $(\sqrt{x})^3$,   (xii) $x^{-\frac{4}{3}}$,

  (xiii) $x^{\frac{1}{a}}$,   (xiv) $\dfrac{1}{\sqrt[n]{x}}$,

    (xv) $\dfrac{x^3 + 2x^2 - 3x}{\sqrt{x}}$,   (xvi) $\sqrt{x} + \dfrac{2}{\sqrt{x}}$,

  (xvii) $(\sqrt{x} + 2)(\sqrt{x} - 3)$, (xviii) $\sqrt{(x + 2)}$,

  (xix) $x\sqrt{(x^2 - 3)}$.

5. Evaluate the following:

    (i) $\left[x^{-\frac{1}{2}}\right]_1^4$,   (ii) $\left[x^{\frac{3}{2}} + 2x^{\frac{1}{2}}\right]_4^9$,   (iii) $\left[\tfrac{2}{3}(x + 4)^{\frac{3}{2}}\right]_0^5$.

Area as the limit of a sum

7.3. We have already discussed the use of integration in finding the area under a curve (§5.3). The word *integration* implies the putting together of parts to make up a whole, and this fundamental aspect of the process is brought out in the following alternative approach to the area under a curve.

Suppose that we wish to find the area under the curve in Fig. 7.6 from $x = 0$ to $x = 3$. We divide this area into three equal strips by the lines $x = 1$ and $x = 2$.

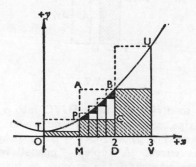

Fig. 7.6

The required area TUVO lies between the sum of the areas of the three shaded "inside" rectangles, and the sum of the three "outside" rectangles bounded at the top by the broken lines; for example, the middle strip PBDM lies between the areas PCDM and ABDM.

We shall for the time being confine our attention to the "inside" rectangles; the sum of these falls short of the required area by the sum of PBC and the two corresponding areas. We now divide TUVO into 12 strips (for clarity only 4 of these are shown in Fig. 7.6). The sum of the 12 "inside" rectangles is clearly a better approximation to the area under the curve, since an error such as PBC has been reduced to a much smaller error represented by the 4 black roughly triangular areas. Thus by taking a sufficient number of strips (in other words, by making the width of each strip sufficiently small) we can make the sum of the areas of the "inside" rectangles as near as we please to the area under the curve.

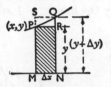

Fig. 7.7

If we were to divide the area TUVO into a very large number of strips, then a typical one would be PQNM (Fig. 7.7), where P(x, y) and Q($x + \Delta x$, $y + \Delta y$) are two points on the curve. A typical "inside" rectangle is PRNM, of area $y\Delta x$, and the process of increasing the number of strips is the same as letting $\Delta x \to 0$. The required area TUVO is found by adding all the "inside" rectangular areas $y\Delta x$ between $x = 0$ and $x = 3$, and then finding the *limit* of this sum as $\Delta x \to 0$. Using the symbol Σ to denote "the sum of",

$$\text{as } \Delta x \to 0, \sum_{x=0}^{x=3} y\Delta x \to \text{ the area TUVO.}$$

*Hence area TUVO = the *limit*, as $\Delta x \to 0$, of $\displaystyle\sum_{x_{1}=0}^{x=3} y\Delta x.$

---

* For simplicity we have confined our attention to the "inside" rectangles. Fig. 7.7 also shows a typical "outside" rectangle SQNM of area $(y + \Delta y)\Delta x$; as $\Delta x \to 0, \displaystyle\sum_{x=0}^{x=3} (y + \Delta y)\Delta x$ tends to the same limit.

Example 1. *Calculate the area under $y = x + 1$ from $x = 0$ to $x = 10$.*

Divide the area into n strips of equal width parallel to Oy (Fig. 7.8); the width of each strip will be $\dfrac{10}{n}$. To find the sum of the areas of the inner shaded rectangles we must first calculate their heights.

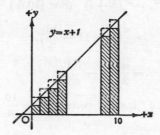

Fig. 7.8

For the three smallest,

$$\text{when } x = 0, \qquad y = x + 1 = 1,$$
$$\text{when } x = \frac{10}{n}, \qquad y = \frac{10}{n} + 1,$$
$$\text{when } x = 2 \cdot \frac{10}{n}, \qquad y = \frac{20}{n} + 1;$$

and for the largest,

$$\text{when } x = 10 - \frac{10}{n}, \quad y = 11 - \frac{10}{n}.$$

The sum of the areas of the inner rectangles is

$$\left\{ \frac{10}{n} \cdot 1 + \frac{10}{n}\left(\frac{10}{n} + 1\right) + \frac{10}{n}\left(\frac{20}{n} + 1\right) + \ldots + \frac{10}{n}\left(11 - \frac{10}{n}\right) \right\}$$
$$= \frac{10}{n}\left\{ 1 + \left(\frac{10}{n} + 1\right) + \left(\frac{20}{n} + 1\right) + \ldots + \left(11 - \frac{10}{n}\right) \right\}$$

The dots have been used to signify the terms corresponding to all the intermediate rectangles; we know that there are as many terms in the curly brackets as there are strips, namely n, and they form an A.P. with common difference $\dfrac{10}{n}$. We can now sum the terms in the brackets using the formula

$$S_n = \frac{n}{2}(a + l) \quad \text{(See §10.6.)}$$

$$= \frac{n}{2}\left(1 + 11 - \frac{10}{n}\right),$$

$$= \frac{n}{2}\left(12 - \frac{10}{n}\right).$$

$\therefore$ the sum of the "inside" rectangles $= \dfrac{10}{n} \cdot \dfrac{n}{2}\left(12 - \dfrac{10}{n}\right)$

$$= 60 - \frac{50}{n}.$$

As $n \to \infty$, the limit of the sum is 60,

$\therefore$ the area under $y = x + 1$ from $x = 0$ to $x = 10$ is 60.

Qu. 1. Calculate the sum of the areas of the n "outside" rectangles in Example 1, and find the limit of this sum as $n \to \infty$.

The integral notation

7.4. Example 1 could be done by integration. Before doing this, we introduce the symbol $\int(\ldots)dx$ to denote integration with respect to x. The symbol $\int$, which is an elongated S, for "sum", is a reminder that integration is essentially summation.

The area under $y = x + 1$ from $x = 0$ to $x = 10$ is

$$\int_0^{10} y\,dx = \int_0^{10} (x + 1)\,dx,$$

$$= \left[\tfrac{1}{2}x^2 + x \right]_0^{10},$$
$$= (\tfrac{1}{2} \cdot 10^2 + 10) - (0),$$
$$= 60.$$

For indefinite integrals, where there are no limits, a similar notation is used. Thus

$$\int (3x^2 + 4)\, dx = x^3 + 4x + c.$$

Qu. 2. Find the following indefinite integrals:

(i) $\int (3x - 4)\, dx$,

(ii) $\int \dfrac{8x^5 - 3x}{x^3}\, dx$,

(iii) $\int \sqrt[7]{x}\, dx$,

(iv) $\int (2\sqrt{t} - 3)(1 - \sqrt{t})\, dt$.

Qu. 3. Evaluate the following definite integrals:

(i) $\int_{\frac{1}{2}}^{1} (60t - 16t^2)\, dt$,

(ii) $\int_1^2 \dfrac{1}{2x^4}\, dx$,

(iii) $\int_1^4 \dfrac{(y + 3)(y - 3)}{\sqrt{y}}\, dy$.

We have shown above that when $y = x + 1$, the limit of $\displaystyle\sum_{x=0}^{x=10} y\Delta x$, as $\Delta x \to 0$, is identical with, and is more readily evaluated as $\displaystyle\int_0^{10} y\, dx$.

We shall now assume that for any curve which is continuous between $x = a$ and $x = b$, the area under the curve from $x = a$ to $x = b$ is

the limit, as $\Delta x \to 0$, of $\displaystyle\sum_{x=a}^{x=b} y\Delta x = \int_a^b y\, dx.$*

---

* The reader may be interested to note the parallel between this statement and that concerning gradient, namely the limit, as $\Delta x \to 0$, of $\dfrac{\Delta y}{\Delta x} = \dfrac{dy}{dx}$.

The reader should in future think of every area bounded by a curve as a summation, first writing down the area of one of the typical strips, or *elements of area*, into which it is most conveniently divided, and then evaluating the limit of the sum of those strips by integration. A convenient way of laying out the working is shown in the following examples; these extend the work of Chapter 5 in the following ways:

(i) by using elements of area parallel to the x-axis, we may integrate with respect to y;

(ii) by finding the element of area cut off between two curves we may evaluate in only one step the area enclosed between them.

Example 2. *Find the area enclosed by* $y = 4x - x^2$, $x = 1$, $x = 2$ *and the x-axis.* (Fig. 7.9.)

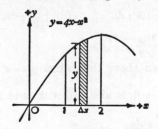

Fig. 7.9

The element of area is $y\Delta x = (4x - x^2)\Delta x$.

$$\therefore \text{ the required area} = \int_1^2 (4x - x^2)\,dx,$$
$$= \left[2x^2 - \tfrac{1}{3}x^3\right]_1^2,$$
$$= (8 - \tfrac{8}{3}) - (2 - \tfrac{1}{3}),$$
$$= 3\tfrac{2}{3}.$$

Example 3. *Find the area enclosed by that part of $y = x^2$ for which x is positive, the y-axis, $y = 1$ and $y = 4$.*

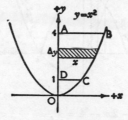

Fig. 7.10

The required area is ABCD in Fig. 7.10. The equation may be written $x = \pm \sqrt{y}$, and for the part of the curve with which we are concerned $x = + \sqrt{y} = + y^{\frac{1}{2}}$.

The element of area is $x\Delta y$.

$$\therefore \text{ the required area} = \int_1^4 x \, dy,$$
$$= \int_1^4 y^{\frac{1}{2}} \, dy,$$
$$= \left[\tfrac{2}{3} y^{\frac{3}{2}} \right]_1^4,$$
$$= (\tfrac{2}{3}.8) - (\tfrac{2}{3}),$$
$$= 4\tfrac{2}{3}.$$

Example 4. *Find the area enclosed between the curves $y = 4 - x^2$ and $y = x^2 - 2x$.*

We must first sketch the curves, and to find the limits of integration we must find the x-coordinates of the points of intersection.

When $\qquad x^2 - 2x = 4 - x^2,$

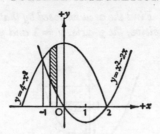

Fig. 7.11

$$2x^2 - 2x - 4 = 0,$$

$$x^2 - x - 2 = 0,$$

$$(x - 2)(x + 1) = 0,$$

$$x = -1 \quad \text{or} \quad +2.$$

The element of area is shown shaded in Fig. 7.11.

If we write $\qquad Y = 4 - x^2$
and $\qquad\qquad\qquad y = x^2 - 2x,$
the element of area is $(Y - y)\Delta x.$

∴ the required area

$$= \int_{-1}^{+2} (Y - y)\, \mathrm{d}x,$$

$$= \int_{-1}^{+2} \{(4 - x^2) - (x^2 - 2x)\}\, \mathrm{d}x,$$

$$= \int_{-1}^{+2} (4 + 2x - 2x^2)\, \mathrm{d}x,$$

$$= \left[4x + x^2 - \tfrac{2}{3}x^3 \right]_{-1}^{+2},$$

$$= (4.2 + 2^2 - \tfrac{2}{3}.2^3) - (-4 + 1 + \tfrac{2}{3}),$$

$$= 8 + 4 - 5\tfrac{1}{3} + 4 - 1\tfrac{2}{3},$$

$$= 9.$$

Exercise 7b

1. Find the following integrals:

$$\text{(i)} \int x(x-3)\,\mathrm{d}x, \qquad\qquad \text{(ii)} \int \frac{2(x-1)}{x^3}\,\mathrm{d}x,$$

$$\text{(iii)} \int \left(at^2 + b + \frac{c}{t^2}\right)\,\mathrm{d}t,$$

$$\text{(iv)} \int \left(x^4 - \sqrt[3]{x^2} + 2 - \frac{1}{x^2}\right)\,\mathrm{d}x,$$

$$\text{(v)} \int \left(y + \frac{1}{\sqrt{y}}\right)\left(y + \frac{1}{y}\right)\,\mathrm{d}y, \quad \text{(vi)} \int \frac{(s+1)^2}{\sqrt[3]{s}}\,\mathrm{d}s.$$

2. Evaluate:

$$\text{(i)} \int_{-2}^{+3} (v^2 + 3)\,\mathrm{d}v, \qquad\qquad \text{(ii)} \int_{1}^{4} (y^2 + \sqrt{y})\,\mathrm{d}y,$$

$$\text{(iii)} \int_{0}^{1} \sqrt{x}(x+2)\,\mathrm{d}x, \qquad \text{(iv)} \int_{1}^{2} \left(3 + \frac{1}{t^2} + \frac{1}{t^4}\right)\,\mathrm{d}t,$$

$$\text{(v)} \int_{1}^{9} \left(\sqrt{x} + \frac{1}{\sqrt{x}}\right)\,\mathrm{d}x, \qquad \text{(vi)} \int_{4}^{11} \sqrt{(x+5)}\,\mathrm{d}x.$$

3. Find the area under each of the following curves between the given limits:

(i) $y = x^2 + 3$, $x = -1$ to $x = 2$,

(ii) $y = x^2(3 - x)$, $x = 4$ to $x = 5$,

(iii) $y = x^2 + \dfrac{1}{x^2}$, $x = \frac{1}{2}$ to $x = 1$.

4. Find the area enclosed by the y-axis and the following curves and straight lines:

(i) $x = y^2$, $y = 3$; \qquad\qquad (ii) $y = x^3$, $y = 1$, $y = 8$;

(iii) $x - y^2 - 3 = 0$, $y = -1$, $y = 2$;

(iv) $x = +\dfrac{1}{\sqrt{y}}$, $y = 2$, $y = 3$.

5. Find the area enclosed by each of the following curves and the y-axis:

(i) $x = (y-1)(y-4)$ (Why is this negative?),

(ii) $x = 3y - y^2$, \qquad\qquad (iii) $x = y(y-2)^2$.

6. Find the area enclosed by $y^2 = 4x$ and the straight lines $x = 1$ and $x = 4$.

7. Find the area enclosed by $y^2 = x + 9$ and the y-axis, by taking an element of area (i) parallel to the y-axis, and (ii) parallel to the x-axis.

8. Find the area enclosed by $9x^2 + y - 16 = 0$ and the x-axis, by integrating (i) with respect to x, and (ii) with respect to y.

9. Calculate the areas enclosed by

   (i) $y = \dfrac{1}{x^2}$, $y = 1$ and $y = 4$;

   (ii) $x = \dfrac{1}{y^2}$, $y = 1$, $y = 4$, and $x = 0$.

10. Find the area of the segment cut off from each of the following curves by the given straight line.

   (i) $y = x^2 - 2x + 2$, $y = 5$;
   (ii) $y = x^2 - 6x + 9$, $y = 1$;
   (iii) $y = -x^2 + 3x - 4$, $y = -4$;
   (iv) $y = x(x - 2)$, $y = x$;
   (v) $y = 4 - 3x - x^2$, $2x + y + 2 = 0$;
   (vi) $y = x^2 - 6x + 2$, $x + y - 2 = 0$.

11. Find the area enclosed by each of the following pairs of curves:

   (i) $y = x(x - 1)$ and $y = x(2 - x)$,
   (ii) $y = x(x + 3)$ and $y = x(5 - x)$,
   (iii) $y = x^2 - 5x$ and $y = 3x^2 - 6x$,
   (iv) $y^2 = 4x$ and $x^2 = 4y$,
   (v) $y = x^2 - 3x - 7$ and $y = 5 - x - x^2$,
   (vi) $y = 2x^2 + 7x + 3$ and $y = 9 + 4x - x^2$.

12. Find the area of the segment cut off from $y = \dfrac{1}{x^2}$ by $10x + 4y - 21 = 0$, given that one of the points of intersection of the straight line and the curve is $(-\frac{2}{5}, \frac{25}{4})$.

Solids of revolution

7.5. If we take a triangular piece of cardboard ABC with a right angle at B, and rotate it through 360 degrees about

AB, we sweep out the volume of a right circular cone
(Fig. 7.12).   The cone can thus be thought of as the **solid
of revolution** generated by rotating the area ABC about
the line AB.

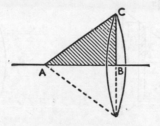

Fig. 7.12

Qu. 4.   State the solid generated by rotating through 360
degrees:

    (i) the above $\triangle$ABC (a) about BC, (b) about AC,
    (ii) the area of a semi-circle about the bounding diameter,
    (iii) a quadrant of a circle about a boundary radius,
    (iv) the area of a circle centre (3, 3) radius 1, about the y-axis,
    (v) a rectangle about one of its sides.

The method of calculating the volume of a solid of revolution is best illustrated by discussing an example; the ideas
involved are the same as those of §7.3.

Example 5.   *Find the volume of the solid generated by
rotating about the x-axis the area under* $y = \frac{3}{4}x$ *from* $x = 0$ *to*
$x = 4$.

A typical element of area under $y = \frac{3}{4}x$ is $y\Delta x$, shown
shaded in Fig. 7.13; rotating this area about the x-axis we
generate the typical *element of volume*.   (Fig. 7.14.)   This
is approximately a cylinder of thickness Δx, but one
circular face has radius y, the other, radius $y + \Delta y$, and

the volume lies between that of an " inside " cylinder $\pi y^2 \Delta x$, and an " outside " cylinder $\pi (y + \Delta y)^2 \Delta x$.

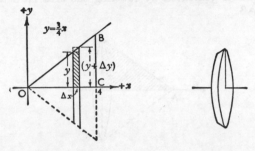

Fig. 7.13   Fig. 7.14

The sum of the volumes of all the "inside" (or "outside") cylinders is an approximation to the volume required, and, by making Δx sufficiently small, we can make this sum approach as close as we please to the volume of the solid of revolution, which may therefore be written as the limit, as $\Delta x \to 0$, of $\sum\limits_{x=0}^{x=4} \pi y^2 \Delta x$.

This may be evaluated as $\displaystyle\int_0^4 \pi y^2 \, \mathrm{d}x$; thus the solution of this example may be presented as follows.

The element of volume $= \pi y^2 \Delta x = \pi \dfrac{9x^2}{16} \Delta x.$

$$\therefore \text{ the required volume } = \int_0^4 \pi \frac{9x^2}{16} \, \mathrm{d}x,$$

$$= \left[\pi \frac{3x^3}{16} \right]_0^4,$$

$$= \pi \frac{3 . 4^3}{16},$$

$$= 12\pi.$$

Qu. 5. Find the volume of the solid generated by rotating about the x-axis

(i) the area under $y = x^2$ from $x = 1$ to $x = 2$,

(ii) the area under $y = x^2 + 1$ from $x = -1$ to $x = +1$.

· The volumes of solids generated by rotating areas about the y-axis may be evaluated by integration with respect to y. This, and other aspects of this work, are illustrated by the following examples.

Example 6. *Find the volume of the solid generated by rotating about the y-axis the area in the first quadrant enclosed by* $y = x^2$, $y = 1$, $y = 4$ *and the y-axis.*   (Fig. 7.15.)

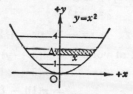

Fig. 7.15

The element of volume $= \pi x^2 \Delta y = \pi y \Delta y$.

$\therefore$ the required volume $= \displaystyle\int_1^4 \pi y \ \mathrm{d}y$,

$$= \left[\tfrac{1}{2}\pi y^2 \right]_1^4,$$

$$= \tfrac{1}{2} . \pi . 16 - \tfrac{1}{2}\pi,$$

$$= \frac{15\pi}{2}.$$

Example 7. *The area of the segment cut off by $y = 5$ from the curve $y = x^2 + 1$ is rotated about $y = 5$; find the volume generated.*   (Fig. 7.16.)

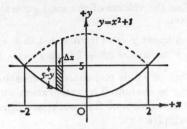

Fig. 7.16

The points of intersection occur when

$$x^2 + 1 = 5$$
$$x^2 = 4$$
$$x = -2 \quad \text{or} \quad +2.$$

The element of volume $= \pi(5 - y)^2\Delta x,$
$$= \pi(5 - x^2 - 1)^2\Delta x,$$
$$= \pi(16 - 8x^2 + x^4)\Delta x.$$

∴ the required volume

$$= \int_{-2}^{+2} \pi(16 - 8x^2 + x^4) \, dx,$$
$$= \left[\pi(16x - \tfrac{8}{3}x^3 + \tfrac{1}{5}x^5)\right]_{-2}^{+2},$$
$$= \pi(32 - 21\tfrac{1}{3} + 6\tfrac{2}{5}) - \pi(-32 + 21\tfrac{1}{3} - 6\tfrac{2}{5}),$$
$$= 34\tfrac{2}{15}\pi.$$

Example 8. *The area of the segment cut off by $y = 5$ from the curve $y = x^2 + 1$ is rotated about the x-axis; find the volume generated.* (Fig. 7.17.)

The solid generated is a cylinder fully open at each end, but with the internal diameter decreasing towards the middle; its volume is found by subtracting the volume of the cavity from the volume of the solid cylinder of the same external dimensions.

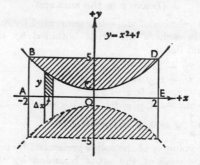

Fig. 7.17

The required volume = the volume generated by rotation, about the x-axis. of the rectangle ABDE . . . (1),

minus   the volume generated by rotation, about the x-axis, of the area under $y = x^2 + 1$ from $x = -2$ to $x = +2$, i.e. ABCDE . . . (2).

Volume (1) = $\pi r^2 h = \pi.5^2.4 = 100\pi$.

Element of Volume (2) = $\pi y^2 \Delta x$,
$$= \pi(x^4 + 2x^2 + 1)\Delta x.$$

$\therefore$ Volume (2) = $\displaystyle\int_{-2}^{+2} \pi(x^4 + 2x^2 + 1)\ dx,$

$$= \left[\pi\left(\frac{x^5}{5} + \frac{2}{3}x^3 + x\right)\right]_{-2}^{+2},$$

$$= \pi(6\tfrac{2}{5} + 5\tfrac{1}{3} + 2) - \pi(-6\tfrac{2}{5} - 5\tfrac{1}{3} - 2),$$

$$= 27\tfrac{7}{15}\pi.$$

$\therefore$ the required volume = $100\pi - 27\tfrac{7}{15}\pi,$

$$= 72\tfrac{8}{15}\pi.$$

Exercise 7c

(Leave π in the answers)

1. Find the volumes of the solids generated by rotating about the x-axis each of the areas bounded by the following curves and lines:

   (i) $x + 2y - 12 = 0$, $x = 0$, $y = 0$;

   (ii) $y = x^2 + 1$, $y = 0$, $x = 0$, $x = 1$;

   (iii) $y = \sqrt{x}$, $y = 0$, $x = 2$;

   (iv) $y = x(x - 2)$, $y = 0$;   (v) $y = x^2(1 - x)$, $y = 0$;

   (vi) $y = \dfrac{1}{x}$, $y = 0$, $x = 1$, $x = 4$.

2. Find the volumes of the solids generated by rotating about the y-axis each of the areas bounded by the following curves and lines:

   (i) $y = 2x - 4$, $y = 2$, $x = 0$;

   (ii) $x = \sqrt{(y - 1)}$, $x = 0$, $y = 4$;

   (iii) $x - y^2 - 2 = 0$, $x = 0$, $y = 0$, $y = 3$:

   (iv) $y^2 = x + 4$, $x = 0$;

   (v) $y = 1 - x^3$, $x = 0$, $y = 0$;

   (vi) $xy = 1$, $x = 0$, $y = 2$, $y = 5$.

3. Find the volumes of the solids generated when each of the areas enclosed by the following curves and lines is rotated about the given line:

   (i) $y = x$, $x = 0$, $y = 2$, about $y = 2$;

   (ii) $y = \sqrt{x}$, $y = 0$, $x = 4$, about $x = 4$;

   (iii) $y^2 = x$, $x = 0$, $y = 2$, about $y = 2$;

   (iv) $y = 2 - x^2$, $y = 1$, about $y = 1$;

   (v) $y = x^3 - 2x^2 + 3$, $y = 3$, about $y = 3$;

   (vi) $y = \dfrac{1}{x^2}$, $y = 4$, $x = 1$, about $y = 4$.

4. Repeat No. 3 for the following areas:

   (i) $x - 3y + 3 = 0$, $x = 0$, $y = 2$, about the x-axis;

   (ii) $x - y^2 - 1 = 0$, $x = 2$, about the y-axis;

   (iii) $y^2 = 4x$, $y = x$, about $y = 0$;

   (iv) $y = \dfrac{1}{x}$, $y = 1$, $x = 2$, about $y = 0$.

5. Obtain, by integration, the formula for the volume of a right circular cone of base radius r, height h. (Consider the area enclosed by $y = \dfrac{h}{r}x$, $x = 0$ and $y = h$.)

6. The equation of a circle centre the origin and radius r is $x^2 + y^2 = r^2$. By considering the area of this circle cut off in the first quadrant being rotated about either the x- or y-axis, deduce the formula for the volume of a sphere radius r.

7. A hemispherical bowl of internal radius 13 cm contains water to a maximum depth of 8 cm. Find the volume of the water.

8. A gold-fish bowl is a glass sphere of inside diameter 20 cm. Calculate the volume of water it contains when the maximum depth is 18 cm.

9. A wall vase has one plane face, and its volume is equivalent to that generated when the area enclosed by $x = \tfrac{1}{64}y^3 + 1$, the y-axis and $y = 8$ is rotated through 2 right angles about the y-axis, the units being cm. Calculate its volume.

10. The area under $y = \tfrac{1}{6}x^2 + 1$ from $x = 0$ to $x = 3$, and the area enclosed by $y = 0$, $y = 2$, $x = 3$, and $x = 4$, are rotated about the y-axis, and the solid generated represents a metal ash tray, the units being cm. Calculate the volume of metal.

11. The area enclosed by $y = x^2 - 6x + 18$ and $y = 10$ is rotated about $y = 10$. Find the volume generated.

12. The area enclosed by $y = x^2 + \dfrac{1}{x}$, the x-axis and $x = -2$, is rotated about the x-axis; find the volume generated.

13. The area enclosed by $y = \dfrac{4}{x^2}$, $y = 1$ and $y = 4$ is rotated about the x-axis; find the volume generated.

14. The area enclosed by $y = x^2 - 6x + 18$ and $y = 10$ is rotated about the y-axis; find the volume generated. [Take an element of area parallel to the x-axis of length $(x_2 - x_1)$; express the typical element of volume in terms of y by using the fact that x_1 and x_2 are the roots of $x^2 - 6x + (18 - y) = 0$; see §8.6.]

15. Repeat No. 14 for the area enclosed by $4y = 4x^2 - 20x + 25$ and $4y = 9$.

Centre of gravity

7.6. The reader who has dealt with this topic in mechanics will be familiar with the fact that, for a system of bodies whose centres of gravity lie in a plane, taking moments about any line in the plane,

the moment of their total weight acting at the centre of gravity of the system = the sum of the moments of the weight of each body.

If n bodies of weight w_1, w_2, w_3, ... w_n have their centres of gravity at (x_1, y_1), (x_2, y_2), (x_3, y_3) ... (x_n, y_n) respectively, writing the coordinates of the centre of gravity of the system as $(\bar{x}, \bar{y})$, and taking moments about the y-axis,

$$\bar{x}(w_1 + w_2 + w_3 + \ldots + w_n)$$
$$= x_1 w_1 + x_2 w_2 + x_3 w_3 + \ldots + x_n w_n.$$

Using the Σ notation,

$$\bar{x} . \Sigma w = \Sigma xw.$$

Similarly, taking moments about the x-axis,

$$\bar{y} . \Sigma w = \Sigma yw.$$

If, instead of separate bodies, we consider the elements of area of a uniform lamina, then Σxw and Σyw become the sums of the moments of the weights of the elements about the axes, and these can be evaluated by integration.

Example 9. *Find the centre of gravity of a uniform lamina whose shape is the area bounded by $y^2 = 4x$ and $x = 9$.*

By symmetry the centre of gravity lies on the x-axis, hence $\bar{y} = 0$.

Consider the lamina as made up of strips parallel to the y-axis, then if the weight per unit area is ρ, a typical element (Fig. 7.18) at a distance x from the y-axis has weight $\rho . 2y . \Delta x$ and its moment about the y-axis is $x . 2\rho y \Delta x$.

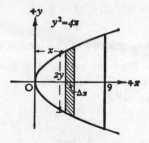

Fig. 7.18

The sum of the moments of the weights of the elements

$$= \sum_{x=0}^{x=9} 2\rho xy \Delta x,$$

and the limit of this, as $\Delta x \to 0$, is evaluated as

$$2\rho \int_0^9 xy \; dx.$$

The weight of the whole lamina

$= \rho \times$ twice the area under $y = 2x^{\frac{1}{2}}$ from $x = 0$ to $x = 9$

$= 2\rho \int_0^9 y \; dx.$

Since $\bar{x} . \Sigma w = \Sigma xw,$

$$\therefore \; \bar{x} . 2\rho \int_0^9 y \; dx = 2\rho \int_0^9 xy \; dx,$$

$$\therefore \; \bar{x} . \int_0^9 y \; dx = \int_0^9 xy \; dx.$$

But $y = 2x^{\frac{1}{2}}$,

$$\therefore \; \bar{x} . \int_0^9 x^{\frac{1}{2}} \; dx = \int_0^9 x^{\frac{3}{2}} \; dx,$$

$$\therefore \; \bar{x} . \left[\tfrac{2}{3} x^{\frac{3}{2}} \right]_0^9 = \left[\tfrac{2}{5} x^{\frac{5}{2}} \right]_0^9,$$

$$\therefore \bar{x}.\tfrac{2}{3}.3^3 = \tfrac{2}{5}.3^5,$$

$$\therefore \bar{x} = \tfrac{27}{5}.$$

∴ the centre of gravity of the lamina is at $(\tfrac{27}{5}, 0)$.

Example 10. *Find the centre of gravity of a uniform lamina whose shape is the area bounded by* $y = x^2$, *the x-axis and* $x = 4$. (Fig. 7.19.)

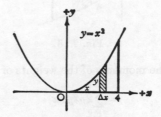

Fig. 7.19

Let the weight per unit area be ρ.

Taking moments about the y-axis,

$$\bar{x}.\rho\int_0^4 y\,\mathrm{d}x = \rho\int_0^4 xy\,\mathrm{d}x,$$

$$\therefore \bar{x}.\int_0^4 y\,\mathrm{d}x = \int_0^4 xy\,\mathrm{d}x.$$

But $y = x^2$,

$$\therefore \bar{x}.\int_0^4 x^2\,\mathrm{d}x = \int_0^4 x^3\,\mathrm{d}x,$$

$$\therefore \bar{x}.\left[\tfrac{1}{3}x^3\right]_0^4 = \left[\tfrac{1}{4}x^4\right]_0^4,$$

$$\therefore \bar{x}.\tfrac{1}{3}.4^3 = \tfrac{1}{4}.4^4.$$

$$\therefore \bar{x} = 3.$$

The centre of gravity of the element is at its mid-point, thus the moment of its weight about the x-axis is $\tfrac{1}{2}y.\rho y\Delta x$.

Taking moments about the x-axis,

$$\bar{y} . \rho \int_0^4 y \, dx = \rho \int_0^4 \tfrac{1}{2} y^2 \, dx,$$

$$\therefore \ \bar{y} . \int_0^4 y \, dx = \int_0^4 \tfrac{1}{2} y^2 \, dx,$$

$$\therefore \ \bar{y} . \int_0^4 x^2 \, dx = \int_0^4 \tfrac{1}{2} x^4 \, dx,$$

$$\therefore \ \bar{y} \left[\tfrac{1}{3} x^3 \right]_0^4 = \left[\tfrac{1}{10} x^5 \right]_0^4,$$

$$\therefore \ \bar{y} . \tfrac{1}{3} . 4^3 = \tfrac{1}{10} . 4^5,$$

$$\therefore \ \bar{y} = \tfrac{24}{5}.$$

$\therefore$ the centre of gravity of the lamina is at $(3, \tfrac{24}{5})$.

Qu. 6. Find the centre of gravity of the lamina whose area is bounded by

(1) $y^2 = x$ and $x = 2$,   (ii) $y = \sqrt{x}$, $y = 0$ and $x = 2$.

The centre of gravity of a solid of revolution may be found in the same way, since the centre of gravity of each element of volume lies in the plane of the axes.

Example 11. *Find the centre of gravity of the solid generated by rotating about the x-axis the area under $y = x$ from $x = 0$ to $x = 3$.* (Fig. 7.20.)

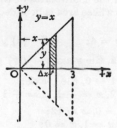

Fig. 7.20

The solid is a cone, vertex O, and axis Ox. By symmetry, the centre of gravity lies on the x-axis.

Let the weight per unit volume be ρ.

The centre of gravity of the element of volume is on the x-axis, thus the moment of its weight about the y-axis is $x \cdot \rho \pi y^2 \Delta x$.

Taking moments about the y-axis,

$$\bar{x} \cdot \rho \int_0^3 \pi y^2 \, dx = \int_0^3 x \cdot \rho \pi y^2 \, dx.$$

$$\therefore \; \bar{x} \int_0^3 y^2 \, dx = \int_0^3 x y^2 \, dx.$$

But $y = x$,

$$\therefore \; \bar{x} \int_0^3 x^2 \, dx = \int_0^3 x^3 \, dx,$$

$$\therefore \; \bar{x} \cdot \left[\tfrac{1}{3} x^3 \right]_0^3 = \left[\tfrac{1}{4} x^4 \right]_0^3,$$

$$\therefore \; \bar{x} \cdot \tfrac{1}{3} \cdot 3^3 = \tfrac{1}{4} \cdot 3^4,$$

$$\therefore \; \bar{x} = \tfrac{9}{4}.$$

$\therefore$ the centre of gravity of the cone is at $(\tfrac{9}{4}, 0)$.

Exercise 7d

In Nos. 1 to 3, find the coordinates of the centre of gravity of the uniform lamina whose area is bounded by the given straight lines and curves.

1.   (i) $y^2 = 9x$, $x = 4$;   (ii) $y = \tfrac{1}{4}x^2$, $y = 1$;

    (iii) $y^2 = 4 - x$, the y-axis;   (iv) $y = \dfrac{1}{x^2}$, $y = 1$, $y = 4$.

2.   (i) $x + y^2 - 1 = 0$, the y-axis;
    (ii) $y = x^2 + 2$, $x = -1$, $x = +1$ and $y = 0$.

3.   (i) $y = \tfrac{1}{8}x$, $y = 0$, $x = 12$;
    (ii) $x = 2\sqrt{y}$, $y = 1$, $x = 0$;
    (iii) $y = x^2$, $y = 0$, $x = 3$;   (iv) $y = x^3$, $y = 0$, $x = 2$.

4. Find the centres of gravity of the solids of revolution generated when the areas bounded by the following straight lines and curves are rotated about the given axes:

   (i) $x + 3y - 6 = 0$, $x = 0$, $y = 0$, about the x-axis;

   (ii) $y = 2\sqrt{x}$, $y = 0$, $x = 4$, about the x-axis;

   (iii) $y^2 = 4x$, $y = 4$, $x = 0$, about the y-axis;

   (iv) $y = x^2(2 - x)$, $y = 0$, about the x-axis;

   (v) $y = \dfrac{1}{x^2}$, $y = 0$, $x = 1$, $x = 2$, about the x-axis;

   (vi) $y = x^3$, $y = 1$, $y = 8$, $x = 0$, about the y-axis.

5. By considering the solid generated by rotating, about the x-axis, the area enclosed by $y = \dfrac{r}{h}x$, the x-axis and $x = h$, deduce the position of the centre of gravity of a right circular cone.

6. $x^2 + y^2 = r^2$ is the equation of the circle centre the origin, radius r. By considering the solid generated by rotating about either axis the area of one quadrant, deduce the distance of the centre of gravity of a solid hemisphere from its plane surface.

7. A gold-fish bowl consists of a sphere of inside radius 10 cm. If it contains water to a maximum depth of 16 cm, find the height of the centre of gravity of the water above the lowest point.

8. A uniform lamina is of the shape of the quadrant of the circle $x^2 + y^2 = r^2$ cut off by the positive axes. Find the coordinates of its centre of gravity.

Exercise 7e (Miscellaneous)

1. Calculate $\displaystyle\int_{-1}^{1} x(x^2 - 1)\, dx$.

   Find the area bounded by the curve $y = x(x^2 - 1)$ and the x-axis (i) between $x = -1$ and $x = 0$, and (ii) between $x = 0$ and $x = 1$.

2. Find the area between the curve $y = x(x - 1)^2(2 - x)$ and the portion of the x-axis between $x = 1$ and $x = 2$.

3. The line $y = \frac{1}{2}x + 1$ meets the curve $y = \frac{1}{4}(7x - x^2)$ at the points A and B. Calculate the coordinates of A and B

and the length of the line **AB**.　Prove that the segment of
the curve cut off by the line has an area $1\frac{1}{3}$.

4. The area enclosed between the line $x = 1$, the x-axis, the
   line $x = 3$ and the line $3x - y + 2 = 0$, is rotated through
   four right angles about the x-axis.　Find the volume
   generated.

5. Solids of revolution are generated by rotating
   (i) about the x-axis the area bounded by the arc of the
       curve $y = 2x^2$ between $(0, 0)$ and $(2, 8)$, the line
       $x = 2$ and the x-axis;
   (ii) about the y-axis the area bounded by the same arc,
        the line $y = 8$ and the y-axis.

   Calculate the volumes of the two solids so formed.

6. The corners of a trapezium are at the points $(0, 2)$, $(2, 2)$,
   $(0, 4)$, $(3, 4)$.　Find the volume of the solid formed by
   revolving the area about the y-axis.

7. Sketch the curve $y = x^2(1 - x)$.　The area between the
   curve and the part of the x-axis from $x = 0$ to $x = 1$ is
   rotated about the x-axis.　Find the volume swept out.

8. The portion of the parabola $y = \frac{2}{3}\sqrt{x}$ between $x = \frac{1}{4}$ and
   $x = 2$ is revolved about the x-axis so as to obtain a para-
   bolic cup with a circular base and top.　Show that the
   volume of the cup is approximately $2\cdot75$.

9. Find the equation of the tangent to the curve $y = x - \dfrac{1}{x}$
   at the point $(1, 0)$.

   The area between the curve, the x-axis ánd the ordinate
   $x = 2$ is rotated about the x-axis.　Prove that the volume
   thus obtained is $\frac{5}{6}\pi$.

10. The curve $y = x^2 + 4$ meets the axis of y at the point **A**,
    and **B** is the point on the curve where $x = 2$.　Find the
    area between the arc **AB**, the axes, and the line $x = 2$.

    If this area is revolved about the x-axis, prove that the
    volume swept out is approximately 188.　(Take π as $3\cdot14$.)

11. The area bounded by the x-axis, the line $x = 1$, the line
    $x = 4$, and the curve $y^2 = 4x^3$ is rotated about the x-axis.
    Find the volume of the resulting solid.　　　　(C)

12. A cylindrical hole of radius 4 cm is cut from a sphere of
    radius 5 cm, the axis of the cylinder coinciding with a

diameter of the sphere. Prove that the volume of the remaining portion of the sphere is 36π cm^3.

13. Find the area bounded by the curve $y = 3x^2 - x^3$ and the x-axis. Find the x-coordinate of the centre of gravity of this area.

14. Find the area bounded by the x-axis and the arc of the curve $y = x^2(x - 1)(3 - x)$ from $x = 1$ to $x = 3$. Find also the x-coordinate of the centre of gravity of this area.

15. Find the area and the x-coordinate of the centre of gravity of the lamina whose edges are formed by the lines $x = 0$, $y = 0$, and the part of the curve $y = (1 - x)(5 + 4x + x^2)$ which is cut off by these lines in the first quadrant.

16. (i) Find the area bounded by the curve $y = x^2$, the axis of x and the ordinates $x = 1$ and $x = 2$.
    (ii) Find the x and y coordinates of the centre of gravity of this area.

17. Find the coordinates of the centre of gravity of the area enclosed by the x-axis and the curve $y = x^2(3 - x)$.

18. Find the area bounded by the curve $y = (x + 1)(x - 2)^2$ and the x-axis from $x = -1$ to $x = 2$. Also find the x-coordinate of the centre of gravity of this area.

19. Find the coordinates of the centre of gravity of the area bounded by the curve $y = 5x^2$ and the straight lines $x = 1$, $x = 2$, $y = 0$.

20. The area bounded by the arc of the curve $y = x(3 - x)$ between the points where $x = 0$ and $x = 2$, the x-axis, and the line $x = 2$, is rotated about the x-axis. Find the volume of the solid of revolution so generated, and the x-coordinate of its centre of gravity.

21. An area in the first quadrant is bounded by the ellipse $4x^2 + 9y^2 = 36$ and the axes. This area is rotated through four right angles about the x-axis. Find
    (i) the volume generated,
    (ii) the x-coordinate of the centre of gravity of this volume.

22. The triangle formed by the three straight lines $y = 0$, $y = 2x$, and $y = 3 - x$ is rotated about the side $y = 0$.
    Find the volume of the solid so generated, and the x-coordinate of its centre of gravity.

23. Plot sufficient points to enable you to make a sketch of the curve $y^2 = x(x - 2)^2$ from $x = 0$ to $x = 2$. Calculate the volume of the solid formed by revolving the enclosed area about the x-axis, and find the distance of the centre of gravity of the solid from the y-axis.

24. The curve $y^2 = x^2(2 - x)$ cuts the x-axis at the points given by $x = 0$ and $x = 2$. The area enclosed by the x-axis and the curve between these two points is rotated about the axis of x so as to form a solid of revolution. Find the volume of this solid and the x-coordinate of its centre of gravity.

CHAPTER 8

SOME USEFUL TOPICS IN ALGEBRA

Surds

8.1. It is not immediately obvious that

$$\frac{3\sqrt{5}}{2\sqrt{7}}, \quad \frac{\sqrt{45}}{\sqrt{28}}, \quad \frac{3}{14}\sqrt{35}, \quad \frac{15}{2\sqrt{35}}, \quad \frac{3}{2}\sqrt{\frac{5}{7}}, \quad \frac{\sqrt{45}}{2\sqrt{7}}, \quad \frac{3\sqrt{5}}{\sqrt{28}}, \quad \sqrt{\frac{45}{28}}$$

all represent the same number. Again, it may not be clear on first sight that $\dfrac{1}{\sqrt{2}-1}$ and $\sqrt{2}+1$ are equal.

Since square roots frequently occur in trigonometry and coordinate geometry, it is useful to be able to recognize a number when it is written in different ways, and the purpose of this section is to give the reader practice in this.

The reader may have calculated an approximate value of $\sqrt{2} = 1\cdot414\ 213\ 562\ldots$ as an exercise in arithmetic, and he may know that this decimal does not terminate or recur. The ancient Greeks did not use decimals, but they discovered that $\sqrt{2}$ could not be expressed as a fraction of two integers (whole numbers). Such a root ($\sqrt{3}$, $\sqrt{5}$, $\sqrt[3]{6}$ are other examples) is called a **surd**. In general, a number which cannot be expressed as a fraction of two integers is called an **irrational** number.

It should be noted that by $\sqrt{2}$ is meant the *positive* number whose square is 2, thus $\sqrt{2} = 1\cdot414\ldots$, and the negative square root of 2 is written as $-\sqrt{2}$. The same applies to other numbers.

Qu. 1. Square: (i) $\sqrt{2}$, (ii) $\sqrt{6}$, (iii) $\sqrt{a}$, (iv) $\sqrt{(ab)}$, (v) $3\sqrt{2}$, (vi) $4\sqrt{5}$, (vii) $2\sqrt{a}$, (viii) $\sqrt{2}.\sqrt{3}$, (ix) $\sqrt{5}.\sqrt{7}$, (x) $\sqrt{2}.\sqrt{8}$, (xi) $\sqrt{12}.\sqrt{3}$, (xii) $\sqrt{a}.\sqrt{b}$.

Note that the answers to parts (iv) and (xii) are the same, therefore $\sqrt{(ab)} = \sqrt{a}.\sqrt{b}$. This result will be used in the next example.

Example 1. *Write $\sqrt{63}$ as the simplest possible surd.*
The factors of 63 are $3^2.7$.

$$\therefore \quad \sqrt{63} = \sqrt{(3^2.7)} = \sqrt{3^2}.\sqrt{7} = 3\sqrt{7}.$$

Example 2. *Express $6\sqrt{5}$ as a simple square root.*
$$6\sqrt{5} = \sqrt{36}.\sqrt{5} = \sqrt{(36.5)} = \sqrt{180}.$$

Example 3. *Simplify $\sqrt{50} + \sqrt{2} - 2\sqrt{18} + \sqrt{8}$.*

$$\begin{aligned}
\sqrt{50} + \sqrt{2} - 2\sqrt{18} + \sqrt{8} \\
= 5\sqrt{2} + \sqrt{2} - 2.3\sqrt{2} + 2\sqrt{2}, \\
= 8\sqrt{2} - 6\sqrt{2}, \\
= 2\sqrt{2}.
\end{aligned}$$

It is usual not to write surds in the denominator of a fraction when this can be avoided. The process of clearing *irrational* numbers is called **rationalization.**

Example 4. *Rationalize the denominators of* (i) $\dfrac{1}{\sqrt{2}}$, (ii) $\dfrac{1}{3 - \sqrt{2}}$.

(i) Multiply numerator and denominator by $\sqrt{2}$. Thus

$$\frac{1}{\sqrt{2}} = \frac{1}{\sqrt{2}}.\frac{\sqrt{2}}{\sqrt{2}} = \frac{\sqrt{2}}{2}.$$

(ii) Multiply numerator and denominator by the denominator with the sign of $\sqrt{2}$ changed:

$$\frac{1}{3 - \sqrt{2}} = \frac{1}{3 - \sqrt{2}}.\frac{3 + \sqrt{2}}{3 + \sqrt{2}} = \frac{3 + \sqrt{2}}{9 - 2} = \frac{1}{7}(3 + \sqrt{2}).$$

Exercise 8a (Oral)

1. Square:

    (i) $\sqrt{5}$,   (ii) $\sqrt{\tfrac{1}{2}}$,   (iii) $4\sqrt{3}$,

    (iv) $\tfrac{1}{2}\sqrt{2}$,   (v) $\sqrt{\dfrac{a}{b}}$,   (vi) $\sqrt{3}\cdot\sqrt{5}$,

    (vii) $\sqrt{3}\cdot\sqrt{7}$,   (viii) $\dfrac{\sqrt{p}}{\sqrt{q}}$,   (ix) $\dfrac{1}{2\sqrt{p}}$,

    (x) $\dfrac{3\sqrt{a}}{\sqrt{(2b)}}$.

2. Express in terms of the simplest possible surds:

    (i) $\sqrt{8}$,   (ii) $\sqrt{12}$,   (iii) $\sqrt{27}$,   (iv) $\sqrt{50}$,
    (v) $\sqrt{45}$,   (vi) $\sqrt{1210}$,   (vii) $\sqrt{75}$,   (viii) $\sqrt{32}$,
    (ix) $\sqrt{72}$,   (x) $\sqrt{98}$,   (xi) $\sqrt{60}$,   (xii) $\sqrt{512}$.

3. Express as square roots:

    (i) $3\sqrt{2}$,   (ii) $2\sqrt{3}$,   (iii) $4\sqrt{5}$,   (iv) $2\sqrt{6}$,

    (v) $3\sqrt{8}$,   (vi) $6\sqrt{6}$,   (vii) $8\sqrt{2}$,   (viii) $10\sqrt{10}$,

    (ix) $\dfrac{\sqrt{2}}{2}$,   (x) $\dfrac{\sqrt{3}}{3}$,   (xi) $\dfrac{\sqrt{2}}{2\sqrt{3}}$,   (xii) $\dfrac{2}{\sqrt{6}}$.

4. Rationalize the denominators of the following fractions:

    (i) $\dfrac{1}{\sqrt{5}}$,   (ii) $\dfrac{1}{\sqrt{7}}$,   (iii) $-\dfrac{1}{\sqrt{2}}$,

    (iv) $\dfrac{2}{\sqrt{3}}$,   (v) $\dfrac{3}{\sqrt{6}}$,   (vi) $\dfrac{1}{2\sqrt{2}}$,

    (vii) $-\dfrac{3}{2\sqrt{3}}$,   (viii) $\dfrac{9}{4\sqrt{6}}$,   (ix) $\dfrac{1}{\sqrt{2}+1}$,

    (x) $\dfrac{1}{2-\sqrt{3}}$,   (xi) $\dfrac{1}{4-\sqrt{10}}$,   (xii) $\dfrac{2}{\sqrt{6}+2}$,

    (xiii) $\dfrac{1}{\sqrt{5}-\sqrt{3}}$,   (xiv) $\dfrac{3}{\sqrt{6}-\sqrt{5}}$,   (xv) $\dfrac{1}{3-2\sqrt{2}}$,

    (xvi) $\dfrac{1}{3\sqrt{2}-2\sqrt{3}}$.

Exercise 8b

1. Simplify:

(i) $\sqrt{8} + \sqrt{18} - 2\sqrt{2}$,   (ii) $\sqrt{75} + 2\sqrt{12} - \sqrt{27}$,

(iii) $\sqrt{28} + \sqrt{175} - \sqrt{63}$,   (iv) $\sqrt{1000} - \sqrt{40} - \sqrt{90}$,

(v) $\sqrt{512} + \sqrt{128} + \sqrt{32}$,

(vi) $\sqrt{24} - 3\sqrt{6} - \sqrt{216} + \sqrt{294}$.

2. Given that $\sqrt{2} = 1\cdot414\ldots$ and $\sqrt{3} = 1\cdot732\ldots$, evaluate correct to 3 significant figures:

(i) $\sqrt{648}$,   (ii) $\sqrt{5\cdot12}$,   (iii) $\dfrac{1}{\sqrt{3} - \sqrt{2}}$,

(iv) $(3 + \sqrt{2})^2$,   (v) $\sqrt{\tfrac{1}{2}} - \sqrt{\tfrac{1}{8}}$,   (vi) $\sqrt{0\cdot0675}$.

3. Express in the form $A + B\sqrt{C}$:

(i) $\dfrac{2}{3 - \sqrt{2}}$,   (ii) $(\sqrt{5} + 2)^2$,

(iii) $(1 + \sqrt{2})(3 - 2\sqrt{2})$,   (iv) $(\sqrt{3} - 1)^2$,

(v) $(1 - \sqrt{2})(3 + 2\sqrt{2})$,   (vi) $\sqrt{\tfrac{1}{2}} + \sqrt{\tfrac{1}{4}} + \sqrt{\tfrac{1}{8}}$,

(vii) $\sqrt{\tfrac{1}{3}} - \sqrt{\tfrac{1}{27}}$,   (viii) $\dfrac{1}{\sqrt{5}} + \sqrt{\dfrac{1}{125}}$,

(ix) $\dfrac{\sqrt{3} + 2}{2\sqrt{3} - 1}$,   (x) $\dfrac{\sqrt{5} + 1}{\sqrt{5} - 1}$,

(xi) $\dfrac{\sqrt{8} + 3}{\sqrt{18} + 2}$,   (xii) $\sqrt{3} + 2 + \dfrac{1}{\sqrt{3} - 2}$.

4. Rationalize the denominators of:

(i) $\dfrac{\sqrt{3} + \sqrt{2}}{\sqrt{3} - \sqrt{2}}$,   (ii) $\dfrac{\sqrt{5} + 1}{\sqrt{5} - \sqrt{3}}$,

(iii) $\dfrac{2\sqrt{2} - \sqrt{3}}{\sqrt{2} + \sqrt{3}}$,   (iv) $\dfrac{\sqrt{2} + 2\sqrt{5}}{\sqrt{5} - \sqrt{2}}$,

(v) $\dfrac{\sqrt{6} + \sqrt{3}}{\sqrt{6} - \sqrt{3}}$,   (vi) $\dfrac{\sqrt{10} + 2\sqrt{5}}{\sqrt{10} + \sqrt{5}}$.

5. Express in surd form and rationalize the denominators:

(i) $\dfrac{1}{1 + \cos 45°}$,   (ii) $\dfrac{2}{1 - \cos 30°}$,   (iii) $\dfrac{1 + \tan 60°}{1 - \tan 60°}$,

(iv) $\dfrac{1 + \tan 30°}{1 - \tan 30°}$,   (v) $\dfrac{1 + \sin 45°}{1 - \sin 45°}$,   (vi) $\dfrac{1}{(1 - \sin 45°)^2}$.

Indices

8.2. It is assumed that the reader knows the three laws of indices for positive integers:

(i) $a^m \times a^n = a^{m+n}$;

(ii) $a^m \div a^n = a^{m-n}$, $(m > n)$;

(iii) $(a^m)^n = a^{mn}$.

We shall now assume that these laws hold for *any* indices, and see what meaning must be assigned to fractional and negative indices as a result of this assumption.

Fractional indices

We know that $4^3 = 4 \times 4 \times 4$, but so far $4^{\frac{1}{2}}$ has not been given any meaning. If fractional indices are to be used, clearly it is an advantage if they are governed by the laws of indices. This being so, what meaning should be given to $4^{\frac{1}{2}}$? By the first law of indices

$$4^{\frac{1}{2}} \times 4^{\frac{1}{2}} = 4^1 = 4.$$

Therefore $4^{\frac{1}{2}}$ is defined as the square root of 4 (to avoid ambiguity we take it to be the positive square root) and so $4^{\frac{1}{2}} = 2$. Similarly, $a^{\frac{1}{2}} = \sqrt{a}$.

To see what value should be given to $8^{\frac{1}{3}}$, consider

$$8^{\frac{1}{3}} \times 8^{\frac{1}{3}} \times 8^{\frac{1}{3}} = 8^1 = 8.$$

Therefore $8^{\frac{1}{3}}$ is defined as $\sqrt[3]{8}$, which is 2. Similarly, $a^{\frac{1}{3}} = \sqrt[3]{a}$.

In general, taking n factors of $a^{\frac{1}{n}}$,

$$a^{\frac{1}{n}} . a^{\frac{1}{n}} \dots a^{\frac{1}{n}} = a,$$

so that

$$a^{\frac{1}{n}} = \sqrt[n]{a}.$$

Next consider $8^{\frac{2}{3}}$. We know that $8^{\frac{1}{3}} = 2$, so

$$8^{\frac{2}{3}} = 8^{\frac{1}{3}} \times 8^{\frac{1}{3}} = 2 \times 2 = 4.$$

Therefore we must take $8^{\frac{2}{3}}$ to be the square of the cube root of 8, and in general $a^{\frac{m}{n}}$ must be taken to be the mth power of $\sqrt[n]{a}$ (or the nth root of a^m), and we may write

$$a^{\frac{m}{n}} = \sqrt[n]{a^m}.$$

Qu. 2. Find the values of (i) $9^{\frac{1}{2}}$, (ii) $27^{\frac{1}{3}}$, (iii), $27^{\frac{2}{3}}$, (iv) $4^{\frac{1}{2}}$, (v) $4^{\frac{3}{2}}$, (vi) $9^{\frac{5}{2}}$, (vii) $8^{\frac{4}{3}}$, (viii) $16^{\frac{3}{4}}$.

Zero and negative indices

8.3. So far 2^0 has been given no meaning. Again it is desirable for it to be given a meaning consistent with the laws of indices, so we divide 2^1 by 2^1 using the second law:

$$2^1 \div 2^1 = 2^0.$$

But $2^1 \div 2^1 = 1$, so 2^0 must be taken to be 1. In the same way, $a^n \div a^n = a^0$, so

$$a^0 = 1 \quad (a \neq 0).$$

Qu. 3. Why does the above not hold for $a = 0$?

To find what meaning must be given to 2^{-1}, divide 2^0 by 2^1, using the second law of indices:

$$2^0 \div 2^1 = 2^{-1}.$$

But $2^0 \div 2^1 = 1 \div 2 = \frac{1}{2}$, therefore we must take 2^{-1} to be $\frac{1}{2}$.

Similarly, $\qquad 2^{-3} = 2^0 \div 2^3 = \dfrac{1}{2^3}.$

Thus 2^{-3} is the reciprocal of 2^3.

In the same way

$$a^{-n} = \frac{1}{a^n},$$

that is, a^{-n} is the reciprocal of a^n.

Example 5. *Find the value of* $\left(\dfrac{27}{8}\right)^{-\frac{2}{3}}$.

Using the last result, $\left(\dfrac{27}{8}\right)^{-\frac{2}{3}} = \left(\dfrac{8}{27}\right)^{\frac{2}{3}}$.

Taking the cube root, $\left(\dfrac{8}{27}\right)^{\frac{2}{3}} = \left(\dfrac{2}{3}\right)^{2}$.

$$\therefore \left(\dfrac{27}{8}\right)^{-\frac{2}{3}} = \dfrac{4}{9}.$$

Example 6. *Simplify* $\dfrac{(1 + x)^{\frac{1}{2}} - \frac{1}{2}x(1 + x)^{-\frac{1}{2}}}{1 + x}$.

Multiply numerator and denominator by $2(1 + x)^{\frac{1}{2}}$.

$$\dfrac{(1 + x)^{\frac{1}{2}} - \frac{1}{2}x(1 + x)^{-\frac{1}{2}}}{1 + x} = \dfrac{2(1 + x) - x}{2(1 + x)^{\frac{3}{2}}},$$

$$= \dfrac{2 + x}{2(1 + x)^{\frac{3}{2}}}.$$

Exercise 8c (Oral)

1. Find the values of:

(i) $25^{\frac{1}{2}}$,   (ii) $27^{\frac{1}{3}}$,   (iii) $64^{\frac{1}{6}}$,   (iv) $49^{\frac{1}{2}}$,

(v) $\left(\dfrac{1}{4}\right)^{\frac{1}{2}}$,   (vi) $1^{\frac{1}{4}}$,   (vii) $(-8)^{\frac{1}{3}}$,   (viii) $(-1)^{\frac{1}{5}}$,

(ix) $8^{\frac{4}{3}}$,   (x) $27^{\frac{2}{3}}$,   (xi) $25^{\frac{3}{2}}$,   (xii) $49^{\frac{2}{3}}$,

(xiii) $\left(\dfrac{1}{4}\right)^{\frac{3}{2}}$,   (xiv) $\left(\dfrac{4}{9}\right)^{\frac{1}{2}}$,   (xv) $\left(\dfrac{27}{8}\right)^{\frac{1}{3}}$,   (xvi) $\left(\dfrac{16}{81}\right)^{\frac{1}{4}}$.

2. Find the values of:

(i) 7^{0},   (ii) 3^{-1},   (iii) 5^{0},   (iv) 4^{-1},

(v) 2^{-3},   (vi) $\left(\dfrac{1}{2}\right)^{-1}$,   (vii) $\left(\dfrac{1}{3}\right)^{-2}$,   (viii) $\left(\dfrac{4}{9}\right)^{0}$,

(ix) 3^{-3},   (x) $(-6)^{-1}$,   (xi) $\left(-\dfrac{1}{6}\right)^{0}$,   (xii) $\left(\dfrac{2}{3}\right)^{-2}$,

(xiii) $\left(-\dfrac{1}{2}\right)^{-2}$,　　　　　(xiv) $\dfrac{1}{3^{-1}}$,

(xv) $\dfrac{2^{-1}}{3^{-2}}$,　　　　　　　　(xvi) $\dfrac{2^0 \cdot 3^{-2}}{5^{-1}}$.

3. Find the values of:

(i) $8^{-\frac{1}{3}}$,　　　(ii) $8^{-\frac{2}{3}}$,　　　(iii) $4^{-\frac{1}{2}}$,　　　(iv) $4^{-\frac{3}{2}}$,

(v) $27^{-\frac{2}{3}}$,　　(vi) $\left(\dfrac{1}{4}\right)^{-\frac{1}{2}}$,　(vii) $\left(\dfrac{1}{8}\right)^{-\frac{1}{3}}$,　(viii) $\left(\dfrac{1}{27}\right)^{-\frac{2}{3}}$,

(ix) $\left(\dfrac{4}{9}\right)^{-\frac{1}{2}}$,　(x) $\left(\dfrac{8}{27}\right)^{-\frac{1}{3}}$,　(xi) $\left(\dfrac{16}{81}\right)^{-\frac{1}{4}}$,　(xii) $\left(\dfrac{27}{8}\right)^{-\frac{4}{3}}$.

Exercise 8d

1. Find the values of:

(i) $256^{\frac{1}{2}}$,　　(ii) $1296^{\frac{1}{4}}$,　　(iii) $64^{\frac{1}{3}}$,　　　(iv) $216^{\frac{1}{3}}$,

(v) $(2\frac{1}{4})^{\frac{1}{2}}$,　(vi) $(1\frac{7}{9})^{\frac{1}{2}}$,　(vii) $8^{-\frac{1}{3}}$,　　(viii) $4^{-\frac{3}{2}}$,

(ix) $64^{-\frac{2}{3}}$,　(x) $81^{-\frac{1}{4}}$,　(xi) $\left(\dfrac{121}{16}\right)^{\frac{1}{2}}$,　(xii) $\left(\dfrac{1}{16}\right)^{-\frac{3}{4}}$,

(xiii) $\left(\dfrac{8}{27}\right)^{\frac{2}{3}}$,　(xiv) $1\cdot331^{\frac{1}{3}}$,　(xv) $0\cdot04^{-\frac{1}{2}}$,　(xvi) $\dfrac{4^{-\frac{3}{2}}}{8^{-\frac{2}{3}}}$.

2. Find the values of:

(i) $\dfrac{16^{\frac{1}{4}} \cdot 4^{\frac{1}{3}}}{8}$,　　　(ii) $\dfrac{27^{\frac{1}{2}} \cdot 243^{\frac{1}{2}}}{243^{\frac{4}{5}}}$,　　(iii) $\dfrac{32^{\frac{2}{4}} \cdot 16^0 \cdot 8^{\frac{2}{4}}}{128^{\frac{3}{2}}}$,

(iv) $\dfrac{6^{\frac{1}{2}} \cdot 96^{\frac{1}{4}}}{216^{\frac{1}{4}}}$,　　(v) $\dfrac{12^{\frac{1}{3}} \cdot 6^{\frac{1}{3}}}{81^{\frac{1}{6}}}$,　　(vi) $\dfrac{8^{\frac{1}{6}} \cdot 4^{\frac{1}{3}}}{32^{\frac{1}{6}} \cdot 16^{\frac{1}{12}}}$.

3. Simplify:

(i) $8^n \times 2^{2n} \div 4^{3n}$,　　　(ii) $3^{n+1} \times 9^n \div 27^{\frac{2}{3}n}$,

(iii) $16^{\frac{3}{4}n} \div 8^{\frac{5}{6}n} \times 4^{n+1}$,　　(iv) $9^{-\frac{1}{2}n} \times 3^{n+2} \times 81^{-\frac{1}{4}}$,

(v) $6^{\frac{1}{2}n} \times 12^{n+1} \times 27^{-\frac{1}{2}n} \div 32^{\frac{1}{2}n}$,

(vi) $10^{\frac{1}{3}n} \times 15^{\frac{1}{2}n} \times 6^{\frac{1}{6}n} \div 45^{\frac{1}{3}n}$.

4. Simplify:

(i) $\dfrac{x^{-\frac{2}{3}} \cdot x^{\frac{1}{2}}}{x^{\frac{1}{6}}}$,

(ii) $\dfrac{\sqrt{(xy)} \cdot x^{\frac{1}{3}} \cdot 2y^{\frac{1}{4}}}{(x^{10}y^9)^{\frac{1}{12}}}$,

(iii) $\dfrac{x^{2n+1} \cdot x^{\frac{1}{2}}}{\sqrt{x^{3n}}}$,

(iv) $\dfrac{x^{3n+1}}{x^{2n+2\frac{1}{2}} \cdot \sqrt{x^{2n-3}}}$,

(v) $\dfrac{x^{p+\frac{1}{2}q} \cdot y^{2p-q}}{(xy^2)^p \cdot \sqrt{x^q}}$,

(vi) $\dfrac{x^{-\frac{2}{3}} \cdot y^{-\frac{1}{3}}}{(x^4y^2)^{-\frac{1}{6}}}$.

5. Simplify:

(i) $\dfrac{x^2(x^2+1)^{-\frac{1}{2}} - (x^2+1)^{\frac{1}{2}}}{x^2}$,

(ii) $-\dfrac{\frac{1}{2}x(1-x)^{-\frac{1}{2}} + (1-x)^{\frac{1}{2}}}{x^2}$,

(iii) $\dfrac{\frac{1}{2}x^{\frac{1}{2}}(1+x)^{-\frac{1}{2}} - \frac{1}{2}x^{-\frac{1}{2}}(1+x)^{\frac{1}{2}}}{x}$,

(iv) $\dfrac{(1+x)^{\frac{1}{3}} - \frac{1}{3}x(1+x)^{-\frac{2}{3}}}{(1+x)^{\frac{2}{3}}}$,

(v) $\dfrac{\sqrt{(1-x)}\frac{1}{2}(1+x)^{-\frac{1}{2}} + \frac{1}{2}(1-x)^{-\frac{1}{2}}\sqrt{(1+x)}}{1-x}$.

Logarithms

8.4. Readers will be familiar with the use of logarithms for multiplication and division, but there are certain properties of logarithms that are useful in more advanced work. Having just considered indices, this is the appropriate place for logarithms because *a logarithm is an index*.

In four-figure tables it is found that the logarithm of 3, to the base 10, is 0·4771. This means that $10^{0\cdot4771} = 3$, working to four significant figures. The statement "the logarithm of 3, to the base 10, is 0·4771" is abbreviated to $\log_{10} 3 = 0\cdot4771$.

Similarly, $10^{0 \cdot 9031} = 8$,

which may be expressed as

$$\log_{10} 8 = 0 \cdot 9031.$$

Now $2^3 = 8$, and this statement may also be written in logarithmic notation. Here the base is 2 and the index (i.e. logarithm) is 3, thus $\log_2 8 = 3$.

Qu. 4. What are the bases and logarithms in the following statements?

(i) $10^2 = 100$,   (ii) $10^{1 \cdot 6021} = 40$,   (iii) $9 = 3^2$,

(iv) $4^3 = 64$,   (v) $1 = 2^0$,   (vi) $8 = \left(\dfrac{1}{2}\right)^{-3}$,

(vii) $a^b = c$.

Exercise 8e (Oral)

1. Express the following statements in logarithmic notation:

(i) $2^4 = 16$,   (ii) $27 = 3^3$,   (iii) $125 = 5^3$,

(iv) $10^6 = 1\,000\,000$,   (v) $1728 = 12^3$,

(vi) $64 = 16^{\frac{3}{2}}$,   (vii) $10^4 = 10\,000$,

(viii) $4^0 = 1$,   (ix) $0 \cdot 01 = 10^{-2}$,   (x) $\frac{1}{2} = 2^{-1}$,

(xi) $9^{\frac{3}{2}} = 27$,   (xii) $8^{-\frac{2}{3}} = \frac{1}{4}$,

(xiii) $81 = \left(\dfrac{1}{3}\right)^{-4}$,   (xiv) $e^0 = 1$,

(xv) $16^{-\frac{1}{4}} = \frac{1}{2}$,   (xvi) $\left(\dfrac{1}{8}\right)^0 = 1$,

(xvii) $27 = 81^{\frac{3}{4}}$,   (xviii) $4 = \left(\dfrac{1}{16}\right)^{-\frac{1}{2}}$,

(xix) $\left(-\dfrac{2}{3}\right)^2 = \dfrac{4}{9}$,   (xx) $(-3)^{-1} = -\frac{1}{3}$,

(xxi) $c = a^5$,   (xxii) $a^3 = b$,

(xxiii) $p^q = r$,   (xxiv) $a = b^c$.

2. Express in index notation:

(i) $\log_2 32 = 5$,   (ii) $\log_3 9 = 2$,

(iii) $2 = \log_5 25$,   (iv) $\log_{10} 100\,000 = 5$,

(v) $7 = \log_2 128$,   (vi) $\log_9 1 = 0$,

(vii) $-2 = \log_3 \frac{1}{9}$,   (viii) $\log_4 2 = \frac{1}{2}$,

(ix) $\log_e 1 = 0$,   (x) $\log_{27} 3 = \frac{1}{3}$,

(xi) $2 = \log_a x$,   (xii) $\log_3 a = b$,

(xiii) $\log_a 8 = c$,   (xiv) $y = \log_x z$,

(xv) $p = \log_q r$.

3. Evaluate:

(i) $\log_2 64$,   (ii) $\log_{10} 100$,   (iii) $\log_{10} 10^7$,

(iv) $\log_a a^2$,   (v) $\log_8 2$,   (vi) $\log_4 1$,

(vii) $\log_{27} 3$,   (viii) $\log_{\frac{3}{2}} \frac{4}{9}$,   (ix) $\log_5 125$,

(x) $\log_{0.1} 10$,   (xi) $\log_e e^9$,   (xii) $\log_e \dfrac{1}{e}$.

8.5. Two numbers can be multiplied by adding their logarithms and divided by subtracting them. The rules are familiar, but it is worth while proving them as an example of logarithmic notation.

Qu. 5. Write in logarithmic notation:
$$a = c^x, \quad b = c^y, \quad ab = c^{x+y}, \quad a/b = c^{x-y}.$$

Deduce that

(i) $\log_c a + \log_c b = \log_c ab$,

(ii) $\log_c a - \log_c b = \log_c (a/b)$.

The logarithm of the nth power of a number is obtained by multiplying its logarithm by n. A method of proving this rule is suggested in the next question.

Qu. 6. Write in logarithmic notation: $a = c^x$, $a^n = c^{nx}$. Deduce that (iii) $\log_c a^n = n \log_c a$.

These three results are used in the next example.

Example 7. *Express* $\log_{10} \dfrac{a^2 b^3}{100\sqrt{c}}$ *in terms of* $\log_{10} a$, $\log_{10} b$, $\log_{10} c$.

First note that $\sqrt{c} = c^{\frac{1}{2}}$.

Using the two rules of Qu. 5,

$$\log_{10} \frac{a^2 b^3}{100 c^{\frac{1}{2}}} = \log_{10} a^2 + \log_{10} b^3 - \log_{10} 100 - \log_{10} c^{\frac{1}{2}}.$$

Then by the rule of Qu. 6, and writing $\log_{10} 100 = 2$,

$$\log_{10} \frac{a^2 b^3}{100 c^{\frac{1}{2}}} = 2 \log_{10} a + 3 \log_{10} b - 2 - \tfrac{1}{2} \log_{10} c.$$

Example 8. *Simplify* $\dfrac{\log_{10} 125}{\log_{10} 25}$.

[Note that 125 and 25 are both powers of 5, so their logarithms can be expressed in terms of $\log_{10} 5$.]

$$\frac{\log_{10} 125}{\log_{10} 25} = \frac{\log_{10} 5^3}{\log_{10} 5^2} = \frac{3 \log_{10} 5}{2 \log_{10} 5} = \frac{3}{2}.$$

Example 9. *Use tables to find the value of* $\log_2 7$.

Write $x = \log_2 7$, then $2^x = 7$. Since $2^x = 7$, their logarithms to the base 10 are equal, therefore

$$\log_{10} 2^x = \log_{10} 7.$$
$$\therefore x \log_{10} 2 = \log_{10} 7.$$
$$\therefore x = \frac{\log_{10} 7}{\log_{10} 2} = \frac{0 \cdot 8451}{0 \cdot 3010}.$$

[Note that $\log_{10} 7$ has to be *divided* by $\log_{10} 2$. This means that, if logarithms are used to perform the division, the logarithm of $\log_{10} 2$ will be subtracted from the logarithm of $\log_{10} 7$.]

0·8451	$\bar{1}$·9270
0·3010	$\bar{1}$·4786
2·808	0·4484

Therefore $\log_2 7 = 2 \cdot 808.$

Exercise 8f

Logarithms are to the base 10, unless stated otherwise.

1. Express in terms of $\log a$, $\log b$, $\log c$:

(i) $\log ab$, (ii) $\log \dfrac{a}{c}$, (iii) $\log \dfrac{1}{b}$,

(iv) $\log a^2 b^{\frac{3}{2}}$, (v) $\log \dfrac{1}{b^4}$, (vi) $\log \dfrac{a^{\frac{1}{3}}b^4}{c^3}$,

(vii) $\log \sqrt{a}$, (viii) $\log \sqrt[3]{b}$, (ix) $\log \sqrt{(ab)}$,

(x) $\log (10a)$, (xi) $\log \dfrac{1}{100b^2}$, (xii) $\log \sqrt{\left(\dfrac{a}{b}\right)}$,

(xiii) $\log \sqrt{\left(\dfrac{ab^3}{c}\right)}$, (xiv) $\log \dfrac{b\sqrt{a}}{\sqrt[3]{c}}$,

(xv) $\log \sqrt{\left(\dfrac{10a}{b^5 c}\right)}$.

2. Express as single logarithms:

(i) $\log 2 + \log 3$, (ii) $\log 18 - \log 9$,

(iii) $\log 4 + 2 \log 3 - \log 6$,

(iv) $3 \log 2 + 2 \log 3 - 2 \log 6$,

(v) $\log c + \log a$, (vi) $\log x + \log y - \log z$,

(vii) $2 \log a - \log b$, (viii) $2 \log a + 3 \log b - \log c$,

(ix) $\frac{1}{2} \log x - \frac{1}{2} \log y$, (x) $\log p - \frac{1}{3} \log q$,

(xi) $2 + 3 \log a$, (xii) $1 + \log a - \frac{1}{2} \log b$,

(xiii) $2 \log a - 3 - \log 2c$, (xiv) $3 \log x - \frac{1}{2} \log y + 1$.

3. Simplify:

(i) $\log 1000$, (ii) $\frac{1}{2} \log_3 81$, (iii) $\frac{1}{6} \log_2 64$,

(iv) $- \log_2 \frac{1}{2}$, (v) $\frac{1}{3} \log 8$, (vi) $\frac{1}{2} \log 49$,

(vii) $- \frac{1}{2} \log 4$, (viii) $3 \log 3 - \log 27$,

(ix) $5 \log 2 - \log 32$, (x) $\dfrac{\log 8}{\log 2}$,

(xi) $\dfrac{\log 81}{\log 9}$, (xii) $\dfrac{\log 49}{\log 343}$.

4. Solve the equations:

(i) $2^x = 5$, (ii) $3^x = 2$, (iii) $3^{4x} = 4$,

(iv) $2^x \cdot 2^{x+1} = 10$, (v) $\left(\dfrac{1}{2}\right)^x = 6$, (vi) $\left(\dfrac{2}{3}\right)^x = \dfrac{1}{16}$.

5. Evaluate, taking $\log \pi = 0.4971$ and $e = 2.718$:

   (i) $\log_2 9$,   (ii) $\log_{12} 6$,   (iii) $\log_3 \pi$,
   (iv) $\log_e 10$,   (v) $\log_e \pi$,   (vi) $\log_3 \frac{1}{2}$.

6. Show that $\log_a b = 1/\log_b a$,
   (i) using the result $\log_a b \cdot \log_b c = \log_a c$,
   (ii) from first principles.

7. Evaluate:

   (i) $2.56^{1.21}$,   (ii) $1.57^{0.576}$,   (iii) $2.718^{3.142}$,
   (iv) $0.561^{\frac{2}{3}}$,   (v) $0.513^{\frac{3}{2}}$,   (vi) $0.0057^{1.39}$.

Roots of quadratic equations

8.6. It is often useful to be able to obtain information about the roots of an equation without actually solving it. For instance, if α and β are the roots of the equation $3x^2 + x - 1 = 0$, the value of $\alpha^2 + \beta^2$ can be found without first finding the values of α and β. This is done by finding the values of $\alpha + \beta$ and $\alpha\beta$, and expressing $\alpha^2 + \beta^2$ in terms of $\alpha + \beta$ and $\alpha\beta$.

If α and β are roots of the equation $ax^2 + bx + c = 0$, then

$$\alpha + \beta = -\frac{b}{a}, \qquad \alpha\beta = \frac{c}{a}.$$

The equation whose roots are α and β may be written

$$(x - \alpha)(x - \beta) = 0.$$
$$\therefore \ x^2 - \alpha x - \beta x + \alpha\beta = 0.$$
$$\therefore \ x^2 - (\alpha + \beta)x + \alpha\beta = 0. \qquad (1)$$

But α and β are also the roots of the equation

$$ax^2 + bx + c = 0,$$

which may be written

$$x^2 + \frac{b}{a}x + \frac{c}{a} = 0, \qquad (2)$$

Now equations (1) and (2), having the same roots, must be precisely the same equation, written in two different ways, since the coefficients of x^2 are both 1. Therefore

(i) the coefficients of x must be equal,

$$\therefore \ \alpha + \beta = -\frac{b}{a}.$$

(ii) the constant terms must be equal,

$$\therefore \ \alpha\beta = \frac{c}{a}.$$

NOTE.—If it is required to write down an equation whose roots are known, equation (1) gives it in a convenient form. It may be written:

$x^2 -$ (sum of the roots)$x +$ (product of the roots) $= 0$. (3)

Qu. 7. Write down the sums and products of the roots of the following equations:

(i) $3x^2 - 2x - 7 = 0$,   (ii) $5x^2 + 11x + 3 = 0$,
(iii) $2x^2 + 5x = 1$,   (iv) $2x(x + 1) = x + 7$.

Qu. 8. Write down equations, the sums and products of whose roots are respectively

(i) $7, 12$;   (ii) $3, -2$;
(iii) $-\frac{1}{2}, -\frac{3}{2}$;   (iv) $\frac{2}{3}, 0$.

Qu. 9. Write down the sum and product of the roots of the equation $3x^2 + 9x + 7 = 0$. Show that the equation has no real root.

Example 10. *The roots of the equation* $3x^2 + 4x - 5 = 0$ *are* α, β. *Find the values of* (i) $\dfrac{1}{\alpha} + \dfrac{1}{\beta}$, (ii) $\alpha^2 + \beta^2$.

Both $\dfrac{1}{\alpha} + \dfrac{1}{\beta}$ and $\alpha^2 + \beta^2$ can be expressed in terms of $\alpha + \beta$ and $\alpha\beta$.

$$\alpha + \beta = -\tfrac{4}{3}, \qquad \alpha\beta = -\tfrac{5}{3}.$$

(i) $\dfrac{1}{\alpha} + \dfrac{1}{\beta} = \dfrac{\beta + \alpha}{\alpha\beta} = \dfrac{-\frac{4}{3}}{-\frac{5}{3}} = \tfrac{4}{5}.$

(ii) $\alpha^2 + \beta^2 = \alpha^2 + 2\alpha\beta + \beta^2 - 2\alpha\beta,$

$$= (\alpha + \beta)^2 - 2\alpha\beta = (-\tfrac{4}{3})^2 - 2(-\tfrac{5}{3}),$$

$$\therefore \ \alpha^2 + \beta^2 = \frac{16}{9} + \frac{10}{3} = \frac{46}{9}.$$

Alternatively, since α and β are roots of the equation $3x^2 + 4x - 5 = 0$,

$$3\alpha^2 + 4\alpha - 5 = 0,$$
$$3\beta^2 + 4\beta - 5 = 0.$$

Adding,

$$3(\alpha^2 + \beta^2) + 4(\alpha + \beta) - 10 = 0.$$
$$\therefore \ 3(\alpha^2 + \beta^2) - \tfrac{16}{3} - 10 = 0.$$
$$\therefore \ \alpha^2 + \beta^2 = \tfrac{16}{9} + \tfrac{10}{3} = \tfrac{46}{9}.$$

Example 11. *The roots of the equation $2x^2 - 7x + 4 = 0$ are α, β. Find an equation with integral coefficients whose roots are $\dfrac{\alpha}{\beta}, \dfrac{\beta}{\alpha}$.*

Since α, β are the roots of the equation $2x^2 - 7x + 4 = 0$, we have

$$\alpha + \beta = \tfrac{7}{2}, \qquad \alpha\beta = 2.$$

Then the required equation may be formed from equation (3) above, if the sum and product of $\dfrac{\alpha}{\beta}, \dfrac{\beta}{\alpha}$ are expressed in terms of $\alpha + \beta$ and $\alpha\beta$.

$$\frac{\alpha}{\beta} + \frac{\beta}{\alpha} = \frac{\alpha^2 + \beta^2}{\alpha\beta} = \frac{(\alpha + \beta)^2 - 2\alpha\beta}{\alpha\beta}$$
$$= \frac{\frac{49}{4} - 4}{2} = \tfrac{33}{8}.$$

Therefore the sum of the roots is $\tfrac{33}{8}$.

$$\frac{\alpha}{\beta} \cdot \frac{\beta}{\alpha} = 1.$$

Therefore the product of the roots is 1.

Therefore the equation with roots $\dfrac{\alpha}{\beta}$, $\dfrac{\beta}{\alpha}$ is

$$x^2 - \tfrac{33}{8}x + 1 = 0.$$

Multiplying through by 8, in order to obtain integral coefficients, the required equation is

$$8x^2 - 33x + 8 = 0.$$

Symmetrical functions

8.7. The functions of α and β that have been used in this chapter all show a certain symmetry. Consider, for example,

$$\alpha + \beta, \quad \alpha\beta, \quad \frac{1}{\alpha} + \frac{1}{\beta}, \quad \alpha^2 + \beta^2, \quad \frac{\alpha}{\beta} + \frac{\beta}{\alpha}.$$

Notice that if α and β are interchanged:

$$\beta + \alpha, \quad \beta\alpha, \quad \frac{1}{\beta} + \frac{1}{\alpha}, \quad \beta^2 + \alpha^2, \quad \frac{\beta}{\alpha} + \frac{\alpha}{\beta},$$

the resulting functions are the same. When a function of α and β is unchanged when α and β are interchanged, it is called a **symmetrical** function of α and β. Such functions occurring in this chapter may be expressed in terms of $\alpha + \beta$ and $\alpha\beta$, as in the next example.

Example 12. *Express in terms of* $\alpha + \beta$ *and* $\alpha\beta$:
(i) $\alpha^3 + \beta^3$, (ii) $(\alpha - \beta)^2$.

(i) α^3 and β^3 occur in the expansion of $(\alpha + \beta)^3$.

$$(\alpha + \beta)^3 = \alpha^3 + 3\alpha^2\beta + 3\alpha\beta^2 + \beta^3.$$

$$\therefore \;\; \alpha^3 + \beta^3 = (\alpha + \beta)^3 - 3\alpha^2\beta - 3\alpha\beta^2.$$

$$\therefore \;\; \alpha^3 + \beta^3 = (\alpha + \beta)^3 - 3\alpha\beta(\alpha + \beta).$$

(ii) $\qquad\qquad (\alpha - \beta)^2 = \alpha^2 - 2\alpha\beta + \beta^2.$

α^2 and β^2 occur in the expansion of $(\alpha + \beta)^2$.

$$(\alpha + \beta)^2 = \alpha^2 + 2\alpha\beta + \beta^2.$$

$$\therefore \;\; (\alpha - \beta)^2 = (\alpha + \beta)^2 - 4\alpha\beta.$$

Exercise 8g

1. Find the sums and products of the roots of the following equations:

(i) $2x^2 - 11x + 3 = 0$,   (ii) $2x^2 + x - 1 = 0$,

(iii) $3x^2 = 7x + 6$,   (iv) $x^2 + x = 1$,

(v) $t(t - 1) = 3$,   (vi) $y(y + 1) = 2y + 5$,

(vii) $x + \dfrac{1}{x} = 4$,   (viii) $\dfrac{1}{t} + \dfrac{1}{t + 1} = \dfrac{1}{2}$.

2. Find equations, with integral coefficients, the sums and products of whose roots are respectively:

(i) 3, 4;   (ii) -5, 6;   (iii) $\frac{2}{3}$, $-\frac{5}{4}$;

(iv) $-\frac{7}{3}$, 0;   (v) 0, -7;   (vi) $1 \cdot 2$, $0 \cdot 8$;

(vii) $-\frac{1}{3}$, $\frac{1}{36}$;   (viii) $-2 \cdot 5$, $-1 \cdot 6$.

3. The roots of the equation $2x^2 + 3x - 4 = 0$ are α, β. Find the values of

(i) $\alpha^2 + \beta^2$,   (ii) $\dfrac{1}{\alpha} + \dfrac{1}{\beta}$,

(iii) $(\alpha + 1)(\beta + 1)$,   (iv) $\dfrac{\beta}{\alpha} + \dfrac{\alpha}{\beta}$.

4. If the roots of the equation $3x^2 - 5x + 1 = 0$ are α, β, find the values of

(i) $\alpha\beta^2 + \alpha^2\beta$,   (ii) $\alpha^2 - \alpha\beta + \beta^2$,

(iii) $\alpha^3 + \beta^3$,   (iv) $\dfrac{\alpha^2}{\beta} + \dfrac{\beta^2}{\alpha}$.

5. The equation $4x^2 + 8x - 1 = 0$ has roots α, β. Find the values of

(i) $\dfrac{1}{\alpha^2} + \dfrac{1}{\beta^2}$,   (ii) $(\alpha - \beta)^2$,

(iii) $\alpha^3\beta + \alpha\beta^3$,   (iv) $\dfrac{1}{\alpha^2\beta} + \dfrac{1}{\alpha\beta^2}$.

6. If the roots of the equation $x^2 - 5x - 7 = 0$ are α, β, find equations whose roots are

(i) α^2, β^2;   (ii) $\alpha + 1, \beta + 1$;   (iii) $\alpha^2\beta, \alpha\beta^2$.

7. The roots of the equation $2x^2 - 4x + 1 = 0$ are α, β. Find equations with integral coefficients whose roots are

(i) $\alpha - 2$, $\beta - 2$;   (ii) $\dfrac{1}{\alpha}$, $\dfrac{1}{\beta}$;   (iii) $\dfrac{\alpha}{\beta}$, $\dfrac{\beta}{\alpha}$.

8. Find an equation, with integral coefficients, whose roots are the squares of the roots of the equation

$$2x^2 + 5x - 6 = 0.$$

9. The roots of the equation $x^2 + 6x + q = 0$ are α and $\alpha - 1$. Find the value of q.

10. The roots of the equation $x^2 - px + 8 = 0$ are α and $\alpha + 2$. Find two possible values of p.

11. The roots of the equation $x^2 + 2px + q = 0$ differ by 2. Show that $p^2 = 1 + q$.

12. If the roots of the equation $ax^2 + bx + c = 0$ are α, β, find expressions for

(i) $\alpha^2\beta + \alpha\beta^2$,   (ii) $\alpha^2 + \beta^2$,   (iii) $\alpha^3 + \beta^3$,

(iv) $\dfrac{1}{\alpha} + \dfrac{1}{\beta}$,   (v) $\dfrac{\alpha}{\beta} + \dfrac{\beta}{\alpha}$,   (vi) $\alpha^4 + \beta^4$,

in terms of a, b, c.

13. The equation $ax^2 + bx + c = 0$ has roots α, β. Find equations whose roots are

(i) $-\alpha$, $-\beta$;   (ii) $\alpha + 1$, $\beta + 1$;

(iii) α^2, β^2;   (iv) $-\dfrac{1}{\alpha}$, $-\dfrac{1}{\beta}$;

(v) $\alpha - \beta$, $\beta - \alpha$;   (vi) $2\alpha + \beta$, $\alpha + 2\beta$.

14. Prove that, if the difference between the roots of the equation $ax^2 + bx + c = 0$ is 1, then $a^2 = b^2 - 4ac$.

15. Prove that, if one root of the equation $ax^2 + bx + c = 0$ is twice the other, then $2b^2 = 9ac$.

16. Prove that, if the sum of the squares of the roots of the equation $ax^2 + bx + c = 0$ is 1, then $b^2 = 2ac + a^2$.

17. Prove that, if the sum of the reciprocals of the roots of the equation $ax^2 + bx + c = 0$ is 1, then $b + c = 0$. If, in addition, one root of the equation is twice the other, use the result of No. 15 to find one set of values of a, b, c. Solve the equation.

18. In the equation $ax^2 + bx + c = 0$, make the substitutions:

(i) $x = y - 1$,   (ii) $x = y^2$,   (iii) $x = \sqrt{y}$,

and simplify the equations.

If the roots of the equation $ax^2 + bx + c = 0$ are α, β, what are the roots of the three equations in y? [Express y in terms of x, and give your answers in terms of α, β.]

19. If the roots of the equation $ax^2 + bx + c = 0$ are α, β, make substitutions, as in No. 18, to find equations whose roots are

(i) $\alpha + 2, \beta + 2$;   (ii) $\dfrac{1}{\alpha}, \dfrac{1}{\beta}$;   (iii) $1 \pm \sqrt{\alpha}, 1 \pm \sqrt{\beta}$.

20. Find a, b which satisfy the identity

$$x^2 + 5x + 9 \equiv (x + a)^2 + b.$$

What is the value of $x^2 + 5x + 9$ when $x = -a$? What is the least value of the expression?

21. Express $2x^2 + 8x + 9$ in the form $2(x + a)^2 + b$, and find the least value of the expression.

22. Express $1 - x - x^2$ in the form $b - (x + a)^2$ and find the maximum value of the expression.

23. Find, without using calculus, the greatest values of the functions

(i) $5 - 2x - x^2$,   (ii) $12x - 3x^2 - 8$.

24. Find, without using calculus, the least values of

(i) $x^2 - 6x + 10$,   (ii) $2x^2 - 3x + 5$.

25. If a is positive, show that the minimum value of the function $ax^2 + bx + c$ is $\dfrac{4ac - b^2}{4a}$.

The remainder theorem

8.8. If we divide $x^3 - 3x^2 + 6x + 5$ by $x - 2$:

$$
\require{enclose}
\begin{array}{r}
x^2 - x + 4 \\
x - 2 \enclose{longdiv}{x^3 - 3x^2 + 6x + 5} \\
\underline{x^3 - 2x^2} \\
-x^2 + 6x \\
\underline{-x^2 + 2x} \\
4x + 5 \\
\underline{4x - 8} \\
13,
\end{array}
$$

the result may be expressed in the identity

$$x^3 - 3x^2 + 6x + 5 \equiv (x - 2)(x^2 - x + 4) + 13.$$

Here $x^2 - x + 4$ is called the *quotient* and 13 the *remainder*.

The remainder theorem gives a method of finding the remainder without going through the process of division.

Suppose it is required to find the remainder when $x^4 - 5x + 6$ is divided by $x - 2$. If the division were performed, we could write

$$x^4 - 5x + 6 \equiv (x - 2) \times \text{quotient} + \text{remainder}.$$

Now if we put $x = 2$ in this identity we obtain

$$16 - 10 + 6 = 0 \times \text{quotient} + \text{remainder}.$$
$$\therefore \text{ the remainder} = 12.$$

Applying this process to any such expression divided by $x - a$, we may write

$$\text{expression} \equiv (x - a) \times \text{quotient} + \text{remainder}.$$

Putting $x = a$ in this identity, it follows that

the remainder = the value of the expression when $x = a$.

This is the remainder theorem, but the formal statement of it is left until the next section.

Qu. 10. For what type of expression is the above method valid?

The function notation

8.9. The phrase "the value of the expression when $x = a$" is too long for convenience, and an abbreviation is used. Just as when drawing graphs some expression in x (or function of x) is called y, so now some expression in x is called $f(x)$. (This does *not* mean f multiplied by x, but is an abbreviation for a function of x.)

Example 13. *If* $f(x) \equiv x^2 - 3x + 4$, *find* (i) $f(1)$, (ii) $f(-2)$, (iii) $f(a)$, (iv) $f(a + h)$.

(i) f(1) means the value of the function when $x = 1$.

$$\therefore \ f(1) = 1^2 - 3 \times 1 + 4 = 2.$$

(ii) To find f(-2), put $x = -2$ in the expression:

$$f(-2) = (-2)^2 - 3 \times (-2) + 4 = 14.$$

(iii) $\qquad\qquad\qquad f(a) = a^2 - 3a + 4.$

(iv) $\qquad f(a + h) = (a + h)^2 - 3(a + h) + 4.$

Qu. 11. If $f(x) \equiv x^3 + 3x - 4$, find (i) f(1), (ii) f(0), (iii) f(-1), (iv) f(-2), (v) f(a).

The notation may now be used to state the remainder theorem. *If a polynomial* f(x) *is divided by* $x - a$, *the remainder is* f(a).

Example 14. *Find the remainder when*

$$x^5 - 4x^3 + 2x + 3$$

is divided by (i) $x - 1$, (ii) $x + 2$.

Let $f(x) \equiv x^5 - 4x^3 + 2x + 3$, then

(i) the remainder when f(x) is divided by $x - 1$ is

$$f(1) = 1 - 4 + 2 + 3 = 2.$$

(ii) the remainder when f(x) is divided by $x + 2$ is

$$f(-2) = -32 + 32 - 4 + 3 = -1.$$

Example 15. *Find the remainder when* $4x^3 - 6x + 5$ *is divided by* $2x - 1$.

As $2x - 1$ is not in the form $x - a$, imagine the division to have been performed, then

$$4x^3 - 6x + 5 \equiv (2x - 1) \times \text{quotient} + \text{remainder}.$$

Putting $x = \frac{1}{2}$ in this identity,

$$\tfrac{1}{2} - 3 + 5 = 0 \times \text{quotient} + \text{remainder}.$$

Therefore the remainder is $2\frac{1}{2}$.

Example 16. *Factorize the expression*

$$2x^3 + 3x^2 - 32x + 15.$$

Let $f(x) \equiv 2x^3 + 3x^2 - 32x + 15.$

[$x - a$ will be a factor of $f(x)$ only if there is no remainder on division, i.e. if $f(a) = 0$.]

$f(1) = 2 + 3 - 32 + 15 \neq 0.$ $\therefore x - 1$ is not a factor.

$f(-1) = -2 + 3 + 32 + 15 \neq 0.$ $\therefore x + 1$ is not a factor.

$x - 2$ and $x + 2$ cannot be factors, as 2 is not a factor of the constant term 15.

$f(3) = 54 + 27 - 96 + 15 = 0.$ $\therefore x - 3$ is a factor.

On division (or by inspection),

$$2x^3 + 3x^2 - 32x + 15 \equiv (x - 3)(2x^2 + 9x - 5).$$

Therefore the factors of $2x^3 + 3x^2 - 32x + 15$ are

$$(x - 3)(x + 5)(2x - 1).$$

Example 17.   *When the expression* $x^5 + 4x^2 + ax + b$ *is divided by* $x^2 - 1$, *the remainder is* $2x + 3$. *Find the values of a and b.*

Suppose the division to have been performed, then

$x^5 + 4x^2 + ax + b \equiv (x^2 - 1) \times$ quotient $+ 2x + 3.$

Putting $x = 1$,

$$1 + 4 + a + b = 2 + 3.$$

Putting $x = -1$,

$$-1 + 4 - a + b = -2 + 3.$$

These equations may be rewritten

$$a + b = 0,   -a + b = -2.$$

Adding,

$$2b = -2.   \therefore b = -1   \text{and}   a = 1.$$

Exercise 8h

1. Find the values of $f(0)$, $f(1)$, $f(-1)$, $f(2)$, $f(-2)$ when
   (i) $f(x) \equiv x^3 + 3x^2 - 4x - 12$,
   (ii) $f(x) \equiv 3x^3 - 2x - 1$,

(iii) $f(x) \equiv x^5 + 2x^4 + 3x^3$,

(iv) $f(x) \equiv x^4 - 4x^2 + 3$.

State one factor of each expression.

2. Find the remainders when

  (i) $x^3 + 3x^2 - 4x + 2$ is divided by $x - 1$,

  (ii) $x^3 - 2x^2 + 5x + 8$ is divided by $x - 2$,

  (iii) $x^5 + x - 9$ is divided by $x + 1$,

  (iv) $x^3 + 3x^2 + 3x + 1$ is divided by $x + 2$,

  (v) $4x^3 - 5x + 4$ is divided by $2x - 1$,

  (vi) $4x^3 + 6x^2 + 3x + 2$ is divided by $2x + 3$.

3. Find the values of a in the expressions below when the following conditions are satisfied:

  (i) $x^3 + ax^2 + 3x - 5$ has remainder $- 3$ when divided by $x - 2$,

  (ii) $x^3 + x^2 + ax + 8$ is divisible by $x - 1$,

  (iii) $x^3 + x^2 - 2ax + a^2$ has remainder 8 when divided by $x - 2$,

  (iv) $x^4 - 3x^2 + 2x + a$ is divisible by $x + 1$,

  (v) $x^3 - 3x^2 + ax + 5$ has remainder 17 when divided by $x - 3$,

  (vi) $x^5 + 4x^4 - 6x^2 + ax + 2$ has remainder 6 when divided by $x + 2$.

4. Show that $2x^3 + x^2 - 13x + 6$ is divisible by $x - 2$, and hence find the other factors of the expression.

5. Show that $12x^3 + 16x^2 - 5x - 3$ is divisible by $2x - 1$ and find the factors of the expression.

6. Factorize

  (i) $x^3 - 2x^2 - 5x + 6$,   (ii) $x^3 - 4x^2 + x + 6$,

  (iii) $2x^3 + x^2 - 8x - 4$,   (iv) $2x^3 + 5x^2 + x - 2$,

  (v) $2x^3 + 11x^2 + 17x + 6$,   (vi) $2x^3 - x^2 + 2x - 1$.

7. Find the values of a and b if $ax^4 + bx^3 - 8x^2 + 6$ has remainder $2x + 1$ when divided by $x^2 - 1$.

8. The expression $px^4 + qx^3 + 3x^2 - 2x + 3$ has remainder $x + 1$ when divided by $x^2 - 3x + 2$. Find the values of p and q.

9. The expression $ax^2 + bx + c$ is divisible by $x - 1$, has remainder 2 when divided by $x + 1$, and has remainder 8 when divided by $x - 2$. Find the values of a, b, c.

10. $x - 1$ and $x + 1$ are factors of the expression

$$x^3 + ax^2 + bx + c,$$

and it leaves a remainder of 12 when divided by $x - 2$. Find the values of a, b, c.

Exercise 8i (Miscellaneous)

1. Write in terms of the simplest possible surds:

    (i) $\sqrt{180} + \sqrt{1125} - \sqrt{1280}$,   (ii) $\dfrac{3\sqrt{2} - 4}{3 - 2\sqrt{2}}$,

    (iii) $(\sqrt{3} + \sqrt{2})^3 + (\sqrt{3} - \sqrt{2})^3$.

2. Given that $\sqrt{2} = 1 \cdot 414$ and $\sqrt{3} = 1 \cdot 732$, evaluate correct to three significant figures:

    (i) $\sqrt{48} + \sqrt{72} + \sqrt{12 \cdot 5}$,   (ii) $\dfrac{2}{\sqrt{3} - \sqrt{2}}$,

    (iii) $\sqrt{\tfrac{1}{8}} + \sqrt{\tfrac{1}{12}}$.

3. Express in the form $a + b\sqrt{2}$:

    (i) $\dfrac{3 + \sqrt{2}}{3 - \sqrt{2}}$,   (ii) $(3 + \sqrt{2})(5 - \sqrt{2})$.

4. (i) Find the values of $8^{-\frac{2}{3}}$, $\left(\dfrac{4}{9}\right)^{\frac{3}{2}}$, $512^{-\frac{2}{3}}$.

    (ii) Solve the equation $x^{\frac{2}{3}} - 5x^{\frac{1}{3}} + 6 = 0$.

5. (i) Multiply $x^{\frac{2}{3}} + 2x^{\frac{1}{3}} + 1$ by $x^{\frac{1}{3}} - 2$.

    (ii) Divide $x^{\frac{3}{2}} - 2x^{\frac{1}{2}} - 2x^{-\frac{1}{2}} + x^{-\frac{3}{2}}$ by $x^{\frac{1}{2}} + x^{-\frac{1}{2}}$.

Check your answers by substituting $x = 8$ in (i) and $x = 4$ in (ii).

6. Without using tables, find the values of

    (i) $\dfrac{12^{\frac{3}{2}} \times 10^{\frac{1}{4}}}{27^{\frac{1}{8}} \times 18^{\frac{1}{4}}}$,   (ii) $\log_{10} 75 + 2 \log_{10} 2 - \log_{10} 3$.

7. Given that $\log_{10} 2 = 0 \cdot 301\ 030$ and $\log_{10} 3 = 0 \cdot 477\ 121$, find, without using tables, the values correct to five places of decimals of:

(i) $\log_{10} 12$,   (ii) $2 \log_{10} 21 - \log_{10} 98$,   (iii) $\log_{10} \sqrt[3]{60}$.

8. (i) Express $\log_{10} \dfrac{100a^2}{b^3\sqrt{c}}$ in terms of $\log_{10} a$, $\log_{10} b$, $\log_{10} c$.

(ii) Given that $\log_{10} 5 = 0{\cdot}698\,970\,0$, find correct to six decimal places the value of $\log_{10} 40$.

9. Taking $\log_{10} 2 = 0{\cdot}301\,030\,0$ and $\log_{10} 3 = 0{\cdot}477\,121\,3$, find the values of (i) $\log_{10} 5$, (ii) $\log_{10} 18$, (iii) $\log_{10} 1{\cdot}5$, correct to six decimal places.

10. (i) Without using tables, solve the equation $9^x = 27^{\frac{3}{4}}$.

(ii) Use tables to find the values, correct to three significant figures, of (a) $0{\cdot}0596^{\frac{2}{3}}$, (b) the value of x given by $2^x = 9$.

11. On a slide rule the distance from mark "1" to mark "n" is proportional to $\log_{10} n$. If the distance from the mark "1" to the mark "10" is 25 cm, calculate the distances

(i) from the mark "1" to the mark "2",

(ii) from the mark "2" to the mark "3".

12. Find the sum and product of the roots of the equation $3x^2 + 5x - 1 = 0$. Also find the equation whose roots are the squares of the roots of this equation.

13. Find the values of m for which the equation

$$x^2 + (m + 3)x + 4m = 0$$

has equal roots. For what value of m are the roots equal and opposite?

14. If α and β are the roots of the equation $2x^2 - 5x - 1 = 0$, find

(i) the value of $\alpha^2 + \beta^2$,

(ii) an equation with integral coefficients whose roots are $\dfrac{1}{\alpha}$ and $\dfrac{1}{\beta}$.

15. Find the range of values of k for which the equation $x^2 + 3x + 1 = kx$ has real roots.

16. Show that the equation $x^2 + 2(k - 3)x = k^2$ cannot have equal roots for any real value of k.

17. Find constants a, b, c such that

$$2x^2 - 6x + 5 \equiv a(x + b)^2 + c.$$

Hence

(i) show that $y = 2x^2 - 6x + 5$ is always positive,

(ii) find the least value of y,

(iii) sketch the graph of y.

18. Express $7 - 5x - 2x^2$ in the form $a - b(x + c)^2$. Hence show that the greatest value of the expression is $10\frac{1}{8}$.

19. What are the values of a and b if $x - 3$ and $x + 7$ are factors of the quadratic $ax^2 + 12x + b$?

20. Show that $3x^3 + x^2 - 8x + 4$ is zero when $x = \frac{2}{3}$, and hence factorize the expression.

21. What is the value of a if $2x^2 - x - 6$, $3x^2 - 8x + 4$ and $ax^3 - 10x - 4$ have a common factor?

22. Factorize the expression $3x^3 - 11x^2 - 19x - 5$.

23. If the expression $ax^4 + bx^3 - x^2 + 2x + 3$ has remainder $3x + 5$ when it is divided by $x^2 - x - 2$, find the values of a and b.

24. Find the values of p and q which make

$$x^4 + 6x^3 + 13x^2 + px + q$$

a perfect square.

CHAPTER 9

PERMUTATIONS AND COMBINATIONS

9.1. This chapter aims at teaching a method of approach to certain problems involving arrangements and selections. In the course of the work, a notation is introduced, and a formula is obtained for use in the proof of the binomial theorem (Chapter 11).

Example 1. *From a pack of playing cards, the Ace, King, Queen, Jack, and Ten of Spades are taken. In how many ways can three of these five cards be placed in a row from left to right?*

The first card can be any one of the five, viz.:

$$A; \quad K; \quad Q; \quad J; \quad 10.$$

When the first card has been placed, there are four cards left to chose from, and so the possible ways of placing the first two cards are:

A K,	A Q,	A J,	A 10;
K A,	K Q,	K J,	K 10;
Q A,	Q K,	Q J,	Q 10;
J A,	J K,	J Q,	J 10;
10 A,	10 K,	10 Q,	10 J.

Thus, for *each* of the 5 ways of choosing the first, there are 4 ways in which the second card may be chosen; therefore there are 5×4 (i.e. 20) ways of choosing the first two cards.

Now for each of the 20 ways of placing the first two cards, there are 3 cards left to choose from (e.g. if the first two cards were A K, the third could be Q, J, or 10); therefore there are 20×3 ways of placing the third card.

Thus, three cards chosen from the Ace, King, Queen, Jack, and Ten of Spades may be placed in a row from left to right in 60 different ways.

Example 2. *Three schools have teams of six or more runners in a cross-country race. In how many ways can the first six places be taken by the three schools, if there are no dead heats?*

First it should be made clear that there is no question of the individuality of the runners, but only which school each of the first six runners belongs to.

The first place can be taken by any of the 3 schools.

When the first runner has come in, the second place can be taken by any of the 3 schools, so the first two places can be taken in 3×3, or 3^2, ways.

Similarly, the third place can be taken by any of the 3 schools, so the first three places can be taken in $3^2 \times 3$, or 3^3, ways.

Continuing the argument for the fourth, fifth and sixth places, it follows that the first six places may be taken in 3^6, or 729, ways by the three schools.

Example 3. *How many even numbers, greater than 2000, can be formed with the digits 1, 2, 4, 8, if each digit may be used only once in each number?*

If the number is greater than 2000, the first digit can be chosen in 3 ways (viz.: 2, 4, or 8).

Then, whichever has been chosen to be the first digit, there are 2 ways in which the last digit may be chosen, in order to make the number even. Thus there are 3×2 ways of choosing the first and last digits.

When the first and last digits have been chosen, there are 2 digits, either of which may be the second digit of the number. Thus there are $3 \times 2 \times 2$ ways of choosing the first, last, and second digit.

Now, when three digits have been chosen, there is only 1

left to fill the remaining place, and so there are
$3 \times 2 \times 2 \times 1$, i.e. 12, even numbers greater than 2000
which may be formed from the digits 1, 2, 4, 8, without
repetitions.

The following table is useful for showing the argument
briefly:

Position of digit	First	Last	Second	Third
Number of possibilities	3	2	2	1

It is to be understood, in this and later tables, that the
choice is made in the order of the first line.

Exercise 9a

1. Ten boys are running a race.   In how many ways can the
   first three places be filled, if there are no dead heats?
2. In how many ways can four letters of the word BRIDGE
   be arranged in a row, if no letter is repeated?
3. Five letters from the word SHILLING are arranged in a
   row.   Find the number of ways in which this can be done,
   when the first letter is I and the last is L,
     (i) if no letter may be repeated,
     (ii) if each letter may occur as many times as it does in
   SHILLING.
4. A man, who works a five-day week, can travel to work on
   foot, by cycle or by bus.   In how many ways can he
   arrange a week's travelling to work?
5. How many five-figure odd numbers can be made from the
   digits 1, 2, 3, 4, 5, if no digit is repeated?
6. A girl has two hats, four skirts and three coats.   How
   many different costumes, consisting of hat, coat, and skirt,
   can she make out of these?
7. In a class of thirty boys, one prize is awarded for Latin,
   another for French, and a third for mathematics.   In how
   many ways can the recipients be chosen?

8. A man has five flags. In how many ways can he fly them one above the other?

9. On the dial of a telephone there are ten numbered holes, one of which is marked 0. If 0 is dialled first, the exchange is rung. In how many ways can a number of three digits be dialled, without ringing the exchange?

10. There are sixteen books on a shelf. In how many ways can these be arranged if twelve of them are volumes of a history, and must be kept together, in order?

11. A typist has six envelopes and six letters. In how many ways can she place one letter in each envelope without getting every letter in the right envelope?

12. A London telephone number consisted of three letters followed by four figures. How many numbers could be made by arranging the symbols in **BYR 2556**?

13. In how many ways can the letters of the word **NOTATION** be arranged?

14. How many odd numbers, greater than 500 000, can be made from the digits 2, 3, 4, 5, 6, 7, without repetitions?

15. Three letters from the word **RELATION** are arranged in a row. In how many ways can this be done? How many of these contain exactly one vowel?

16. Seven men and six women are to be seated in a row on a platform. In how many ways can they be arranged if no two men sit next to each other? In how many ways can the arrangement be made if there are six men and six women, subject to the same restriction?

17. A man stays three days at a hotel and the menu is the same for breakfast each day. He may have any one of three types of egg dish, or two types of fish, or meat. In how many ways can he order his three breakfasts if he does not have egg two days running nor repeat any dish?

18. A boy has five blue marbles, four green marbles and three red marbles. In how many ways can he arrange four of them in a row, if the marbles of any one colour are indistinguishable?

19. I have fifteen books of three different sizes, five of each. In how many ways can I arrange them on my shelf if I keep books of the same size together?

20. Four men and their wives sit on a bench. In how many ways can they be arranged if
  (i) there is no restriction,
  (ii) each man sits next to his wife?

The factorial notation

9.2. There are times when a problem on arrangements leads to an answer involving a product of more factors than it is convenient to write down. The next example shows how this may arise.

Example 4. *In how many ways can the cards of one suit, from a pack of playing cards, be placed in a row?*

Position of card in row	First	Second	...	Twelfth	Thirteenth
Number of possibilities	13	12	...	2	1

The table abbreviates the type of argument used in the last three examples, and it leads to the conclusion that the cards of one suit can be placed in a row in

$$13 \times 12 \times 11 \times 10 \times 9 \times 8 \times 7 \times 6 \times 5 \times 4 \times 3 \times 2 \times 1 \text{ ways.}$$

To shorten the answer, the product could be evaluated, giving 6 227 020 800; but it is easier to write

$$13!$$

(which is read, "factorial thirteen", or by some, "thirteen shriek"!). Thus,

$$7! = 7 \times 6 \times 5 \times 4 \times 3 \times 2 \times 1 = 5040,$$

and similarly for any other positive integer.

The factorial notation will be used freely in this chapter and Chapter 11, and the reader should become thoroughly used to it before going on to the next section.

Example 5. (i) *Evaluate* $\dfrac{9!}{2!7!}$,

(ii) *Write* $40 \times 39 \times 38 \times 37$ *in factorial notation.*

(i) Written in full,

$$\frac{9!}{2!7!} = \frac{9 \times 8 \times 7 \times 6 \times 5 \times 4 \times 3 \times 2 \times 1}{2 \times 1 \times 7 \times 6 \times 5 \times 4 \times 3 \times 2 \times 1},$$

$$= \frac{9 \times 8}{2 \times 1}, \quad \text{on cancelling } 7!$$

$$\therefore \frac{9!}{2!7!} = 36.$$

(ii) $\quad 40 \times 39 \times 38 \times 37$

$$= 40 \times 39 \times 38 \times 37 \times \frac{36 \times 35 \times \ldots \times 2 \times 1}{36 \times 35 \times \ldots \times 2 \times 1},$$

$$\therefore \ 40 \times 39 \times 38 \times 37 = \frac{40!}{36!}.$$

Exercise 9b

1. Evaluate:

(i) $3!$,   (ii) $4!$,   (iii) $5!$,   (iv) $\dfrac{10!}{8!}$,

(v) $\dfrac{7!}{4!}$,   (vi) $\dfrac{12!}{9!}$,   (vii) $\dfrac{11!}{7!4!}$,   (viii) $\dfrac{6!2!}{8!}$,

(ix) $(2!)^2$,   (x) $\dfrac{6!}{(3!)^2}$,   (xi) $\dfrac{10!}{3!7!}$,   (xii) $\dfrac{10!}{2!3!5!}$.

2. Express in factorial notation:

(i) $6 \times 5 \times 4$,   (ii) 10×9,

(iii) $12 \times 11 \times 10 \times 9$,   (iv) $n(n-1)(n-2)$,

(v) $(n+2)(n+1)n$,   (vi) $\dfrac{10 \times 9}{2 \times 1}$,

(vii) $\dfrac{7 \times 6 \times 5}{3 \times 2 \times 1}$,   (viii) $\dfrac{52 \times 51 \times 50}{3 \times 2 \times 1}$,

$\text{(ix)} \ \dfrac{n(n-1)}{2 \times 1},$ $\qquad \text{(x)} \ \dfrac{(n+1)n(n-1)}{3 \times 2 \times 1},$

$\text{(xi)} \ \dfrac{2n(2n-1)}{2 \times 1},$ $\qquad \text{(xii)} \ n(n-1)\ldots(n-r+1).$

3. Express in factors:

    (i) $20! + 21!$,   (ii) $26! - 25!$,

    (iii) $14! - 2(13!)$,   (iv) $15! + 4(14!)$,

    (v) $(n+1)! + n!$,   (vi) $(n-1)! - (n-2)!$,

    (vii) $n! + 2(n-1)!$,

    (viii) $(n+2)! + (n+1)! + n!$.

4. Simplify:

    (i) $\dfrac{15!}{11!4!} + \dfrac{15!}{12!3!}$,   (ii) $\dfrac{21!}{7!14!} + \dfrac{21!}{8!13!}$,

    (iii) $\dfrac{16!}{9!7!} + \dfrac{2.16!}{10!6!} + \dfrac{16!}{11!5!}$,   (iv) $\dfrac{35!}{16!19!} + \dfrac{3.35!}{17!18!}$,

    (v) $\dfrac{n!}{(n-r)!r!} + \dfrac{n!}{(n-r+1)!(r-1)!}$,

    (vi) $\dfrac{n!}{(n-r)!r!} + \dfrac{2.n!}{(n-r+1)!(r-1)!} + $

$$\dfrac{n!}{(n-r+2)!(r-2)!}.$$

Permutations

9.3. In Example 4, it was found that 13 playing cards could be placed in a row in 13! ways. If we consider n unlike objects placed in a row, using the same method:

Position of object in row:	1st	2nd	...	$(n-1)$th	nth
Number of possibilities:	n	$n-1$	...	2	1

we find that they may be arranged in $n!$ ways.

The arrangements of the n objects are called **permutations.** Thus

ABC, ACB, BCA, BAC, CAB, CBA,

are the 3! permutations of the three letters A, B, C.

Again, in Example 1, it was found that 3 cards chosen from 5 unlike cards could be arranged in 60 ways. This might be expressed by saying that there are 60 permutations of 3 cards chosen from 5 unlike cards.

A permutation is an arrangement of a number of objects in a particular order. In practice, the order may be in space, such as from left to right in a row; or it may be in time, such as reaching the winning post in a race, or dialling on a telephone.

How many permutations are there of r objects chosen from n unlike objects?

The method is indicated in the table below.

Order of choice of object:	1st	2nd	3rd	...	$(r-1)$th	rth
Number of possibilities:	n	$(n-1)$	$(n-2)$	...	$(n-r+2)$	$(n-r+1)$

Thus there are

$$n(n-1)(n-2) \ldots (n-r+2)(n-r+1)$$

permutations of the objects. But

$$n(n-1)(n-2) \ldots (n-r+2)(n-r+1)$$

$$= \frac{n(n-1)(n-2) \ldots (n-r+2)(n-r+1).(n-r) \ldots 2.1}{(n-r) \ldots 2.1},$$

$$= \frac{n!}{(n-r)!}.$$

Therefore there are $\dfrac{n!}{(n-r)!}$ permutations of r objects chosen from n unlike objects, if r is less than n.

(We have already found that there are $n!$ permutations of n unlike objects.)

For brevity, the number of permutations of r objects chosen from n unlike objects is written ^nP_r. Therefore

$$^nP_r = \frac{n!}{(n-r)!}, \quad (r < n),$$

and $$^nP_n = n!$$

Example 6. *There are 20 books on a shelf, but the red covers of two of them clash, and they must not be put together. In how many ways can the books be arranged?*

This is best tackled by finding out the number of ways in which the two books *are* together, and subtracting this from the number of ways in which the 20 books can be arranged if there is no restriction.

Suppose the two red books are tied together, then there are 19 objects, which can be arranged in 19! ways. Now if the order of the two red books is reversed, there will again be 19! arrangements; so that there are $2 \times 19!$ ways of arranging the books with the red ones next to each other.

With *no* restriction, 20 books can be arranged in 20! ways; therefore the number of arrangements in which the red books are not together is $20! - 2 \times 19! = 18 \times 19!$

Example 7. *In how many ways can 8 people sit at a round table?*

Since the table is round, the position of people relative to the *table* is of no consequence. Thus, supposing they sit down, and then all move one place to the left, the arrangement is still the same.

Therefore one person may be considered to be fixed, and the other 7 can then be arranged about him in 7! ways.

Thus there are 5040 ways in which 8 people can sit at a round table.

Example 8. *In how many ways can the letters of the word* BESIEGE *be arranged?*

First, give the three E's suffixes: $BE_1SIE_2GE_3$. Then, treating the E's as different, the 7 letters may be arranged in 7! ways.

Now, in every distinct arrangement, the 3 E's may be rearranged amongst themselves in 3! ways, without altering the positions of the B, S, I, or G; for instance, SEIBEEG would have been counted 3! times in the 7! arrangements as

$$SE_1IBE_2E_3G, \quad SE_2IBE_3E_1G, \quad SE_3IBE_1E_2G,$$
$$SE_1IBE_3E_2G, \quad SE_2IBE_1E_3G, \quad SE_3IBE_2E_1G.$$

Therefore the number of distinct arrangements of the letters in BESIEGE is $\dfrac{7!}{3!} = 840$.

In the next exercise there are some examples which are best tackled from first principles, like the next example.

Example 9. *How many even numbers, greater than* 50 000, *can be formed with the digits* 3, 4, 5, 6, 7, 0, *without repetitions?*

Compared with Example 3, page 193, there are two extra difficulties: that the number can have either 5 or 6 digits, and that the number cannot begin with 0. Therefore the problem is split up into four parts:

(i) Numbers with 5 digits, the first digit being even.

Position of digit in number:	1st	5th	2nd	3rd	4th
Number of possibilities:	1	2	4	3	2

This gives
$$1 \times 2 \times 4 \times 3 \times 2 = 48 \text{ possibilities.}$$

(ii) Numbers with 5 digits, the first digit being odd.

Position of digit in number:	1st	5th	2nd	3rd	4th
Number of possibilities:	2	3	4	3	2

This gives

$$2 \times 3 \times 4 \times 3 \times 2 = 144 \text{ possibilities.}$$

(iii) Numbers with 6 digits, the first digit being even.

Position of digit in number:	1st	6th	2nd	3rd	4th	5th
Number of possibilities:	2	2	4	3	2	1

This gives

$$2 \times 2 \times 4 \times 3 \times 2 \times 1 = 96 \text{ possibilities.}$$

(iv) Numbers with 6 digits, the first digit being odd.

Position of digit in number:	1st	6th	2nd	3rd	4th	5th
Number of possibilities:	3	3	4	3	2	1

This gives

$$3 \times 3 \times 4 \times 3 \times 2 \times 1 = 216 \text{ possibilities.}$$

Therefore the total number of possibilities is

$$48 + 144 + 96 + 216 = 504.$$

Exercise 9c

1. Seven boys and two girls are to sit together on a bench. In how many ways can they arrange themselves so that the girls do not sit next to each other?
2. Eight women and two men are to sit at a round table. In how many ways can they be arranged? If, however, the two men sit directly opposite each other, in how many ways can the ten people be arranged?

3. How many arrangements can be made of the letters in the word TROTTING? In how many of these are the N and the G next to each other?

4. On a bookshelf, four books are bound in leather and sixteen in cloth. If the books are to be arranged so that the leather-bound ones are together, in how many ways can this be done? If, in addition, the cloth-bound books are to be kept together, in how many ways can the shelf be arranged?

5. There is room for ten books on a bedside table, but there are fifteen to choose from. Of these, however, a Bible and a book of ghost stories must go at the ends. In how many ways can the books be arranged?

6. Ten beads of different colours are arranged on a ring. If a salesman claims that no two of his rings are the same, what is the greatest number of rings he could have? (A ring can be turned over.)

7. In his cowhouse, a farmer has seven stalls for cows, and four for calves. If he has ten cows and five calves, in how many ways can he arrange the animals in his cowhouse?

8. At a conference of five powers, each delegation consists of three members. If each delegation sits together, with their leader in the middle, in how many ways can the members be arranged at a round table?

9. How many numbers, divisible by 5, can be made with the digits 2, 3, 4, 5, no digit being used more than once in each number?

10. In a cricket team, the captain has settled the first four places in the batting order, and has decided that the four bowlers will occupy the last four places. In how many ways can the batting order be made out?

11. How many arrangements can be made of the letters in the word TERRITORY?

12. A man has ten ornaments for his mantelpiece, and of these the clock must go in the centre. If there is only room for seven ornaments altogether, how many arrangements can be made on the mantelpiece?

13. How many odd numbers, greater than 60 000, can be made

from the digits 5, 6, 7, 8, 9, 0, if no number contains any digit more than once?

14. A code word consists of three letters, followed by two digits. How many code words can be made, if no letter nor digit is repeated in any code word?

15. How many numbers of five digits can be made from the digits 1, 2, 3, 4, 5, 6, 7, 8, 9, when each number contains exactly one even digit and no digit more than once?

16. A bridge player holds five spades, four hearts, two diamonds and two clubs. If he keeps the cards of each suit together, in how many ways can he arrange the cards he holds

(i) if the suits are in the above order,

(ii) if the suits may be arranged in any order?

17. Find the number of ways in which the letters of ISOSCELES can be arranged if the two E's are separated.

18. Find how many numbers greater than 400 000 can be made, using all the digits of 416 566.

19. In how many ways can four red beads, three green beads, and five beads of different colours be strung on a circular wire?

20. Six natives and two foreigners are seated in a compartment of a railway carriage with four seats either side. In how many ways can the passengers seat themselves if

(i) the foreigners do not sit opposite each other,

(ii) the foreigners do not sit next to each other?

Combinations

9.4. In the last section, attention was given to permutations, where the order of a set of objects was of importance; but in other circumstances, the order of selection is irrelevant. If, for instance, eight tourists find there is only room for five of them at a hotel, they will be chiefly interested in which five of them stay there, rather than in any order of arrangement.

When a selection of objects is made, with no regard being paid to order, it is referred to as a **combination**. Thus,

ABC, ACB, CBA, are different permutations, but they are the same combination of letters.

Example 10. *In how many ways can 13 cards be selected from a pack of 52 playing cards?*

First of all, suppose that thirteen cards from the pack are laid on a table in an order from left to right. From the last section, it follows that this can be done in $^{52}P_{13}$ (i.e. $52!/39!$) ways.

Now each combination of cards can be arranged in $13!$ ways, therefore

the number of permutations

$$= 13! \times \text{(the number of combinations)},$$

$$\therefore \frac{52!}{39!} = 13! \times \text{(the number of combinations)}.$$

Therefore the number of combinations of 13 cards chosen from a pack of playing cards is $\dfrac{52!}{39! \, 13!}$.

In how many ways can r objects be chosen from n unlike objects?

In §9.3 it was shown that there are $n!/(n-r)!$ permutations of r objects chosen from n unlike objects.

Now each combination of r objects can be arranged in $r!$ ways, therefore

the number of permutations

$$= r! \times \text{(the number of combinations)},$$

$$\therefore \frac{n!}{(n-r)!} = r! \times \text{(the number of combinations)}.$$

Therefore the number of combinations of r objects chosen from n unlike objects is

$$\frac{n!}{(n-r)! \, r!}.$$

For brevity, the number of combinations of r objects chosen from n unlike objects is written ^nC_r, thus

$$^nC_r = \frac{n!}{(n-r)!\,r!}.$$

^nC_r is also sometimes written as $_nC_r$ and $\binom{n}{r}$.

Qu. 1. What are the values of (i) $^8C_3$, $^8C_5$; (ii) $^{10}C_6$, $^{10}C_4$?

Qu. 2. In how many ways can $n - r$ objects be chosen from n unlike objects?

Qu. 3. Show that $^nC_r = {}^nC_{n-r}$.

Example 11. *A mixed hockey team containing 5 men and 6 women is to be chosen from 7 men and 9 women. In how many ways can this be done?*

Five men can be selected from 7 men in $^7C_5$ ways, and 6 women can be selected from 9 women in $^9C_6$ ways.

Now for each of the $^7C_5$ ways of selecting the men, there are $^9C_6$ ways of selecting the women, therefore there are $^7C_5 \times {}^9C_6$ ways of selecting the team.

$$^7C_5 \times {}^9C_6 = \frac{7!}{2!5!} \cdot \frac{9!}{3!6!},$$

$$= 21 \times 84.$$

Therefore the team can be chosen in 1764 ways.

Exercise 9d

1. Evaluate: (i) $^{10}C_2$, (ii) $^6C_4$, (iii) $^7C_3$, (iv) $^9C_5$, (v) $^8C_4$. Express in factors: (vi) $^nC_2$, (vii) $^nC_3$, (viii) $^nC_{n-2}$, (ix) $^{n+1}C_2$, (x) $^{n+1}C_{n-1}$.

2. In how many ways can a cricket team be selected from thirteen players?

3. There are ten possible players for the VI to represent a tennis club, and of these the captain and the secretary must be in the team. In how many ways can the team be selected?

4. Ten boxes each hold one white ball and one coloured ball, every colour being different. Find the number of ways in which one ball may be taken from each box if half those taken are white.

5. Nine people are going to travel in two taxis. The larger has five seats, and the smaller has four. In how many ways can the party be split up?

6. A girl wants to ask eight friends to tea, but there is only room for four of them. In how many ways can she choose whom to invite if two of them are sisters and must not be separated? (Consider two cases, (i) when both sisters are invited, (ii) when neither sister is invited.)

7. In a game of mixed hockey there are ten married couples and two spinsters playing. In how many ways can the two teams be made up, if no husband may play against his wife?

8. A ferry which holds ten people carries a party of thirteen men and seven women across a river. Find the number of ways in which the party may be taken across if all the women go on the first trip.

9. Twelve people each spin a coin. Find the number of ways in which exactly five heads may be obtained.

10. Two punts each hold six people. In how many ways can a party of six boys and six girls divide themselves so that there are equal numbers of boys and girls in each punt?

11. In how many ways can eight white and four black draughtsmen be arranged in a pile?

12. A committee of six is to be formed from nine women and three men. In how many ways can the members be chosen so as to include at least one man?

13. Ten men are present at a club. In how many ways can four be chosen to play bridge if two men refuse to sit at the same table?

14. A man is allowed to take six volumes to a desert island. He is going to choose these from eleven books, one of which contains two volumes, which he will take or leave together. Find the number of ways in which he can make his choice.

15. Four people are to play bridge and four others are to play whist. Find the number of ways in which they may be chosen if eleven people are available.

16. A party of twelve is to dine at three tables at a hotel. In how many ways may they be split up if each table holds four?

17. Twelve people are to travel by three cars, each of which holds four. Find the number of ways in which the party may be divided if two people refuse to travel in the same car.

18. A committee of ten is to be chosen from nine men and six women. In how many ways can it be formed if at least four women are to be on the committee?

19. In how many ways can eleven men be chosen to represent a cricket club if they are selected from seven Englishmen, six Welshmen and five Scots, and if at least one of each nationality must be in the team?

Exercise 9e (Miscellaneous)

1. In the absence of the chairman, a committee of three vice-chairmen and four ordinary members is to sit on a platform. In how many ways can they be arranged if one of the vice-chairmen sits in the middle?

2. In how many ways can a committee of four men and three women be formed from seven men and eight women?

3. Show that the number of ways of choosing six objects from fourteen unlike objects is equal to the number of ways of choosing five objects from fifteen unlike objects.

4. How many arrangements can be made of the letters in THIRTIETH?

5. In how many ways can a committee of eight be arranged at a round table? In how many of these does the chairman sit between the secretary and the treasurer?

6. How many circular rings can be formed from seven differently coloured beads? In how many of these are the red and the blue beads separated?

7. In how many ways can a boy arrange in a row six balls from seven cricket balls, six tennis balls and five squash balls?

8. Find the number of diagonals of a polygon of n sides.

9. How many five-figure numbers can be made from the digits of 10 242?

10. In how many ways can ten books be arranged on a shelf if four of them are kept together?

11. In how many ways can a man who has ten chairs put five in one room, three in a second and two in a third?

12. How many odd numbers, greater than 600 000, can be made from the digits 5, 6, 7, 8, 9, 0,
    (i) if repetitions are not allowed,
    (ii) if repetitions are allowed?

13. How many arrangements can be made with the letters of **LEATHERETTE**?

14. In how many ways can four mince-pies, three jam tarts, and three cakes be given to ten children if each receives one?

15. In how many ways can a committee of nine be formed from ten men and their wives, if no husband serves on it with his wife?

16. There are six ornaments on my mantelpiece. In how many ways can I put three more on it without changing the order of those already there?

17. How many mixed hockey teams may be made from six married couples, one bachelor and three spinsters, if no wife will play without her husband?

18. A man has ten pieces of clothing to dispose of. In how many ways can he do this if he gives away at least two articles and sells the rest?

19. Eight boys and two girls sit on a bench. If the girls may sit neither at the ends nor together, in how many ways can they be arranged?

20. In how many arrangements of the letters of **REVERSE** are the V and S separated?

21. In how many ways is it possible to select one or more letters from those in **INSIPIDITY**?

22. Four men and their wives, four bachelors and four spinsters are travelling in two eight-seat compartments of a train, one of which is a smoking compartment and the other is not. In how many ways can the party be split up if no wife is separated from her husband?

23. A painter has to paint the doors of twelve new council houses and has sufficient paint to do five green and three

yellow. If he is given paint of only one colour—blue, green, or yellow—for the remaining doors, in how many ways can the twelve doors be painted ?

24. In how many ways can a lift holding eight passengers carry a party of thirteen up a building in two journeys.

25. How many numbers of five or six digits can be formed from the digits 1, 2, 2, 2, 3, 4 ?

26. In how many ways is it possible to select six letters, including at least one vowel, from the letters of (i) INCOMPUTABLE, (ii) FLABELLIFOR M ?

CHAPTER 10

SERIES

10.1. The reader should examine the following sets of numbers. Each set is written down in a definite order, and there is a simple rule by which the terms are obtained. Such a set of terms is called a **sequence**.

Qu. 1. Write down the next two terms in each of the following sequences:

(i) $1, 3, 5, 7, \ldots$ (ii) $2, 5, 8, 11, \ldots$

(iii) $1, 2, 4, 8, \ldots$ (iv) $\frac{1}{3}, \frac{1}{6}, \frac{1}{12}, \frac{1}{24}, \ldots$

(v) $1^3, 2^3, 3^3, 4^3, \ldots$ (vi) $\frac{1}{2}, \frac{2}{3}, \frac{3}{4}, \frac{4}{5}, \ldots$

(vii) $1, 4, 9, 16, \ldots$ (viii) $1, 2, 6, 24, 120, \ldots$

(ix) $1, \frac{2}{3}, \frac{3}{9}, \frac{4}{27}, \ldots$ (x) $4, 2, 0, -2, \ldots$

(xi) $1, -1, 1, -1, \ldots$ (xii) $1, -\frac{1}{2}, \frac{1}{4}, -\frac{1}{8}. \ldots$

10.2. There is a problem which may be given to a class, and that is to add up the integers (whole numbers) from 1 to 100. After a short time it may be done on the blackboard as follows.

First write the numbers down in their natural order:

$$1 + 2 + 3 + \ldots + 98 + 99 + 100.$$

Now write the numbers down again in the opposite order, so that we have:

$$1 + \quad 2 + \quad 3 + \ldots + \quad 98 + \quad 99 + 100$$
$$100 + \quad 99 + \quad 98 + \ldots + \quad 3 + \quad 2 + \quad 1$$

$$101 + 101 + 101 + \ldots + 101 + 101 + 101.$$

The numbers in each column have been added together, and, since there are 100 terms in the top line, the total is $100 \times 101 = 10\ 100$. But this is twice the sum required, therefore the sum of the integers from 1 to 100 is 5 050.

10.3. If the terms of a sequence are considered as a sum, for instance:

$$1 + 2 + 3 + \ldots + 98 + 99 + 100,$$

or
$$1 + \tfrac{1}{2} + \tfrac{1}{4} + \tfrac{1}{8} + \ldots,$$

the expression is called a **series**. A series may end after a finite number of terms, in which case it is called a **finite series**; or it may be considered not to end, and it is then called an **infinite series**.

Arithmetical progressions

10.4. The method of §10.2, for finding the sum of a series, may only be applied to a certain type, which is usually called an arithmetical progression (often abbreviated to A.P.). For example:

$$1 + 3 + 5 + \ldots + 99,$$
$$7 + 11 + 15 + \ldots + 79,$$
$$3 - 2 - 7 - \ldots - 42,$$
$$1\tfrac{1}{8} + 1\tfrac{1}{4} + 1\tfrac{3}{8} + \ldots + 3\tfrac{1}{2},$$
$$-2 - 4 - 6 - \ldots - 16,$$

are arithmetical progressions. In such a series, any term may be obtained from the previous term by adding a certain number, called the **common difference**. Thus the common differences in the above progressions are 2, 4, -5, $\tfrac{1}{8}$, -2.

Example 1. *Find the third, tenth, twenty-first and "n"th terms of the A.P. with first term 6 and common difference 5.*

Position of term	1st	2nd	3rd	4th	10th	21st	nth
Value	6	$6+5$	$6+2\times5$	$6+3\times5$	$6+9\times5$	$6+20\times5$	$6+(n-1)\times5$

Note that to find the nth term $n - 1$ common differences are added to the first term.

The third, tenth, twenty-first, and nth terms are 16, 51, 106, and $5n + 1$.

Example 2. *Find the sum of the first twenty terms of the A.P.* $-4 - 1 + 2 + \ldots$.

To find the twentieth term, add 19 times the common difference to the first term: $-4 + 19 \times 3 = 53$.

Write S_{20} for the sum of the first twenty terms, then using the method of §10.2,

$$S_{20} = -4 - 1 + 2 + \ldots + 53.$$

Again, $\quad S_{20} = 53 + 50 + 47 + \ldots - 4.$

Adding, $\quad 2S_{20} = 49 + 49 + 49 + \ldots + 49,$

$$= 20 \times 49.$$

$$\therefore \ S_{20} = 490.$$

Therefore the sum of the first twenty terms of the A.P. is ·490.

Exercise 10a

1. Which of the following series are arithmetical progressions? Write down the common differences of those that are.

    (i) $7 + 8\frac{1}{2} + 10 + 11\frac{1}{2}$,   (ii) $-2 - 5 - 8 - 11$,

    (iii) $1 + 1\cdot1 + 1\cdot2 + 1\cdot3$,

    (iv) $1 + 1\cdot1 + 1\cdot11 + 1\cdot111$,

    (v) $\frac{1}{2} + \frac{5}{6} + \frac{7}{6} + \frac{3}{2}$,   (vi) $1^2 + 2^2 + 3^2 + 4^2$,

  (vii) $n + 2n + 3n + 4n$,   (viii) $1 + \frac{1}{2} + \frac{1}{3} + \frac{1}{4}$,

    (ix) $1\frac{1}{8} + 2\frac{1}{4} + 3\frac{3}{8} + 4\frac{1}{2}$,   (x) $19 + 12 + 5 - 2 - 9$,

    (xi) $1 - 2 + 3 - 4 + 5$,   (xii) $1 + 0\cdot8 + 0\cdot6 + 0\cdot4$.

2. Write down the terms indicated in each of the following A.P.s:

    (i) $3 + 11 + \ldots$, 10th, 19th,

   (ii) $8 + 5 + \ldots$, 15th, 31st,

  (iii) $\frac{1}{4} + \frac{7}{8} + \ldots$, 12th, nth,

   (iv) $50 + 48 + \ldots$, 100th, nth,

   (v) $7 + 6\frac{1}{2} + \ldots$, 42nd, nth,

   (vi) $3 + 7 + \ldots$, 200th, $(n + 1)$th.

3. Find the number of terms in the following A.P.s:

    (i) $2 + 4 + 6 + \ldots + 46$,

    (ii) $50 + 47 + 44 + \ldots + 14$,

    (iii) $2 \cdot 7 + 3 \cdot 2 + \ldots + 17 \cdot 7$,

    (iv) $6\frac{1}{4} + 7\frac{1}{2} + \ldots + 31\frac{1}{4}$,

    (v) $407 + 401 + \ldots - 133$,

    (vi) $2 - 9 - \ldots - 130$,

   (vii) $2 + 4 + \ldots + 4n$,

  (viii) $x + 2x + \ldots + nx$,

    (ix) $a + (a + d) + \ldots + \{a + (n - 1)d\}$,

    (x) $a + (a + d) + \ldots + l$.

4. Find the sums of the following A.P.s:

    (i) $1 + 3 + 5 + \ldots + 101$,

    (ii) $2 + 7 + 12 + \ldots + 77$,

   (iii) $- 10 - 7 - 4 - \ldots + 50$,

    (iv) $71 + 67 + 63 + \ldots - 53$,

    (v) $2 \cdot 01 + 2 \cdot 02 + 2 \cdot 03 + \ldots + 3 \cdot 00$,

   (vi) $1 + 1\frac{1}{6} + 1\frac{1}{3} + \ldots + 4\frac{1}{2}$,

  (vii) $x + 3x + 5x + \ldots + 21x$,

 (viii) $a + (a + 1) + \ldots + (a + n - 1)$,

   (ix) $a + (a + d) + \ldots + \{a + (n - 1)d\}$.

5. Find the sums of the following arithmetical progressions as far as the terms indicated:

    (i) $4 + 10 + \ldots$ 12th term,

   (ii) $15 + 13 + \ldots$ 20th term,

  (iii) $1 + 2 + \ldots$ 200th term,

  (iv) $20 + 13 + \ldots$ 16th term,

   (v) $6 + 10 + \ldots n$th term,

  (vi) $1\frac{1}{4} + 1 + \ldots n$th term.

6. The second term of an A.P. is 15, and the fifth is 21. Find common difference, the first term and the sum of the first ten terms.

7. The fourth term of an A.P. is 18, and the common difference is $- 5$. Find the first term and the sum of the first sixteen terms.

8. Find the difference between the sums of the first ten terms of the A.P.s whose first terms are 12 and 8, and whose common differences are respectively 2 and 3.

9. The first term of an A.P. is − 12, and the last term is 40. If the sum of the progression is 196, find the number of terms and the common difference.

10. Find the sum of the odd numbers between 100 and 200.

11. Find the sum of the even numbers, divisible by three, lying between 400 and 500.

12. The twenty-first term of an A.P. is $5\frac{1}{2}$, and the sum of the first twenty-one terms is $94\frac{1}{2}$. Find the first term, the common difference and the sum of the first thirty terms.

13. Show that the sum of the integers from 1 to n is $\frac{1}{2}n(n + 1)$.

14. The twenty-first term of an A.P. is 37 and the sum of the first twenty terms is 320. What is the sum of the first ten terms?

15. An advertisement for an appointment states that the post carries a salary of £1200 p.a. rising by annual increments of £80 to £1760 p.a. What is the total amount that a man would earn if he held the post for 20 years?

16. Show that the sum of the first n terms of the A.P. with first term a and common difference d is $\frac{1}{2}n\{2a + (n − 1)d\}$.

Geometrical progressions

10.5. Another series of common occurrence is the geometrical progression, for example:

$$1 + \tfrac{1}{2} + \tfrac{1}{4} + \tfrac{1}{8} + \ldots + \tfrac{1}{512},$$
$$3 + 6 + 12 + \ldots + 192,$$
$$\tfrac{16}{27} - \tfrac{8}{9} + \tfrac{4}{3} - \ldots + \tfrac{27}{4}.$$

In such a progression, the ratio of a term to the previous one is a constant, called the **common ratio**. Thus, the common ratios of the above progressions are respectively $\frac{1}{2}$, 2 and $-\frac{3}{2}$.

Qu. 2. Write down the third and fourth terms of the progressions which begin (i) $2 + 4 + \ldots$, (ii) $12 + 6 + \ldots$, (a) if they are A.P.s, (b) if they are G.P.s.

Example 3. *Find the third, tenth, twenty-first and "n"th terms of the G.P. which begins* $3 + 6 + \dots$.

Position of term	1st	2nd	3rd	4th	10th	21st	nth
Value	3	3×2	3×2^2	3×2^3	3×2^9	3×2^{20}	$3 \times 2^{n-1}$

Note that to find the nth term, the first term is multiplied by the $(n - 1)$th power of the common ratio.

The third, tenth, twenty-first, and nth terms are 12, 1536, 3 145 728, and $3 \times 2^{n-1}$.

Example 4. *Find the sum of the first eight terms of the geometrical progression* $2 + 6 + 18 + \dots$.

To find the eighth term, multiply the first term by the seventh power of the common ratio: 2×3^7.

Let S_8 be the sum of the first eight terms of the expression.

$$\therefore \ S_8 = 2 + 2 \times 3 + 2 \times 3^2 + \dots + 2 \times 3^7.$$

Now multiply both sides by the common ratio and write the terms obtained one place to the right, so that we have

$$S_8 = 2 + 2 \times 3 + 2 \times 3^2 + \dots + 2 \times 3^7.$$
$$3S_8 = \quad\ 2 \times 3 + 2 \times 3^2 + \dots + 2 \times 3^7 + 2 \times 3^8.$$

Substracting the top line from the lower,

$$2S_8 = -2 + 2 \times 3^8.$$
$$\therefore \ S_8 = 3^8 - 1.$$

Therefore the sum of the first eight terms is 6560.

Exercise 10b

1. Which of the following series are geometrical progressions? Write down the common ratios of those that are.

    (i) $3 + 9 + 27 + 81$,   (ii) $1 + \frac{1}{4} + \frac{1}{16} + \frac{1}{64}$,

    (iii) $-1 + 2 - 4 + 8$,   (iv) $1 - 1 + 1 - 1$,

(v) $1 + 1\frac{1}{2} + 1\frac{1}{4} + 1\frac{1}{8}$,   (vi) $a + a^2 + a^3 + a^4$,

(vii) $1 + 1\cdot1 + 1\cdot21 + 1\cdot331$,

(viii) $\frac{1}{2} + \frac{1}{6} + \frac{1}{12} + \frac{1}{36}$,   (ix) $2 + 4 - 8 - 16$,

(x) $\frac{3}{4} + \frac{9}{2} + 27$.

2. Write down the terms indicated in each of the following geometrical progressions. Do not simplify your answers.

(i) $5 + 10 + \ldots$, 11th, 20th;

(ii) $10 + 25 + \ldots$, 7th, 19th;

(iii) $\frac{2}{3} + \frac{3}{4} + \ldots$, 12th, nth;

(iv) $3 - 2 + \ldots$, 8th, nth;

(v) $\frac{2}{7} - \frac{3}{7} + \ldots$, 9th, nth;

(vi) $3 + 1\frac{1}{2} + \ldots$, 19th, $2n$th.

3. Find the number of terms in the following geometrical progressions:

(i) $2 + 4 + 8 + \ldots + 512$,

(ii) $81 + 27 + 9 + \ldots + \frac{1}{27}$,

(iii) $0\cdot03 + 0\cdot06 + 0\cdot12 + \ldots + 1\cdot92$,

(iv) $\frac{8}{81} - \frac{4}{27} + \frac{2}{9} - \ldots - 1\frac{11}{16}$,

(v) $5 + 10 + 20 + \ldots 5 \times 2^n$,

(vi) $a + ar + ar^2 + \ldots + ar^{n-1}$.

4. Find the sums of the geometrical progressions in No. 3. Simplify, but do not evaluate, your answers.

5. Find the sums of the following geometrical progressions as far as the terms indicated. Simplify, but do not evaluate, your answers.

(i) $4 + 12 + 36 + \ldots$, 12th term;

(ii) $15 + 5 + 1\frac{2}{3} + \ldots$, 20th term;

(iii) $1 - 2 + 4 - \ldots$, 50th term;

(iv) $24 - 12 + 6 - \ldots$, 17th term;

(v) $1\cdot1 + 1\cdot21 + 1\cdot331 + \ldots$, 23rd term;

(vi) $\frac{1}{2} + \frac{1}{4} + \frac{1}{8} + \ldots$, 13th term;

(vii) $3 + 6 + 12 + \ldots$, nth term;

(viii) $1 - \frac{1}{3} + \frac{1}{9} - \ldots$, nth term.

6. The third term of a geometrical progression is 10, and the sixth is 80. Find the common ratio, the first term and the sum of the first six terms.

7. The third term of a geometrical progression is 2, and the fifth is 18. Find two possible values of the common ratio, and the second term in each case.

8. The three numbers, $n - 2, n, n + 3$, are consecutive terms of a geometrical progression. Find n, and the term after $n + 3$.

9. A man starts saving on 1st April. He saves 1p the first day, 2p the second, 4p the third, and so on, doubling the amount every day. If he managed to keep on saving under this system until the end of the month (30 days), how much would he have saved? Give your answer in pounds, correct to three significant figures.

10. The first term of a G.P. is 16 and the fifth term is 9. What is the value of the seventh term?

11. Show that the sum of the series $4 + 12 + 36 + 108 + \ldots$ to 20 terms is greater than 3×10^9.

12. The numbers $n - 4$, $n + 2$, $3n + 1$ are in geometrical progression. Find the two possible values of the common ratio.

13. What is the common ratio of the G.P.

$$(\sqrt{2} - 1) + (3 - 2\sqrt{2}) + \ldots?$$

Find the third term of the progression.

14. Find the ratio of the sum of the first 10 terms of the series $\log x + \log x^2 + \log x^4 + \log x^8 + \ldots$ to the first term.

Formulae for the sums of A.P.s and G.P.s

10.6. The methods of Examples 2 and 4 will now be applied to general A.P.s and G.P.s to obtain formulae for their sums.

(i) If the first term of an A.P. is a, and the nth term is l, we may find the sum S_n of the first n terms.

We have

$$S_n = a + (a + d) + \ldots + (l - d) + l$$

$$\text{(where there are } n \text{ terms)},$$

and again,

$$S_n = l + (l - d) + \ldots + (a + d) + a.$$

Adding,

$$2S_n = (a + l) + (a + l) + \ldots + (a + l) + (a + l).$$

Now there are n terms on the right-hand side,

$$\therefore \ 2S_n = n(a + l).$$

$$\therefore \ S_n = \frac{n(a + l)}{2}.$$

(ii) If the first term of an A.P. is a, and the common difference is d, the nth term is $a + (n - 1)d$. Substituting $l = a + (n - 1)d$ in the formula above,

$$S_n = \frac{n}{2}\{a + a + (n - 1)d\}.$$

$$\therefore \ S_n = \frac{n}{2}\{2a + (n - 1)d\}.$$

(iii) If the first term of a G.P. is a and the common ratio is r, we may find the sum S_n of the first n terms.

The nth term is ar^{n-1}, therefore

$$S_n = a + ar + ar^2 + \ldots + ar^{n-1}.$$

$$\therefore \ rS_n = \quad ar + ar^2 + \ldots + ar^{n-1} + ar^n.$$

Subtracting,

$$S_n - rS_n = a - ar^n.$$

$$\therefore \ S_n(1 - r) = a(1 - r^n).$$

$$\therefore \ S_n = a\left(\frac{1 - r^n}{1 - r}\right).$$

An alternative formula for the sum of a G.P. is obtained by multiplying numerator and denominator by -1:

$$S_n = a\left(\frac{r^n - 1}{r - 1}\right).$$

This is more convenient if r is greater than 1.

Example 5. *In an arithmetical progression, the thirteenth term is 27, and the seventh term is three times the second term. Find the first term, the common difference and the sum of the first ten terms.*

[We have two unknowns (the first term and the common difference). We have two pieces of information:

(i) the thirteenth term is 27,

(ii) the seventh term is three times the second term.

Thus we can form two equations which will enable us to find the two unknowns.]

Let the first term be a, and let the common difference be d.

Then the thirteenth term is $a + 12d$, therefore

$$a + 12d = 27.$$

The seventh term is $a + 6d$, and the second term is $a + d$, therefore

$$a + 6d = 3(a + d).$$
$$\therefore 3d = 2a.$$

Substituting in the first equation,

$$a + 8a = 27,$$
$$\therefore a = 3,$$
and so
$$d = 2.$$

Therefore the first term is 3, and the common difference is 2.

To find the sum of the first ten terms: we know that

$$S_n = \frac{n}{2}\{2a + (n-1)d\},$$
$$\therefore S_{10} = \frac{10}{2}(6 + 9 \times 2),$$
$$= 5 \times 24.$$

Therefore the sum of the first ten terms is 120.

Example 6. *In a geometrical progression, the sum of the second and third terms is* 6, *and the sum of the third and fourth terms is* — 12. *Find the first term and the common ratio.*

[As in the last example, we have two unknowns (the first term and the common ratio). We have two pieces of information:

  (i) the sum of the second and third terms is 6,

  (ii) the sum of the third and fourth terms is — 12.

We may therefore write down two equations and these will enable us to find the two unknowns.]

Let the first term be a, and let the common ratio be r.

Then the second term is ar, and the third term is ar^2, therefore

$$ar + ar^2 = 6.$$

The third term is ar^2, and the fourth term is ar^3, therefore

$$ar^2 + ar^3 = -12.$$

Factorizing the left-hand sides of the equations,

$$ar(1 + r) = 6,$$
$$ar^2(1 + r) = -12.$$

We may eliminate a by dividing:

$$\frac{ar(1 + r)}{ar^2(1 + r)} = -\frac{6}{12}.$$

$$\therefore \frac{1}{r} = -\frac{1}{2}.$$

$$\therefore r = -2.$$

Substituting $r = -2$ in $ar(1 + r) = 6$,

$$a(-2)(-1) = 6.$$

$$\therefore a = 3.$$

Therefore the first term is 3, and the common ratio is — 2.

Example 7.　*The sum of a number of consecutive terms of an arithmetical progression is* $-19\frac{1}{2}$, *the first term is* $16\frac{1}{2}$, *and the common difference is* -3. *Find the number of terms.*

With the notation of §10.6,

$$S_n = \frac{n}{2}\{2a + (n-1)d\}.$$

Substituting $S_n = -19\frac{1}{2}$, $a = 16\frac{1}{2}$, $d = -3$:

$$-\frac{39}{2} = \frac{n}{2}\{33 - 3(n-1)\}.$$

$$\therefore \quad -39 = n(36 - 3n).$$

$$\therefore \quad 3n^2 - 36n - 39 = 0.$$

Dividing through by 3,

$$n^2 - 12n - 13 = 0,$$

$$\therefore \quad (n - 13)(n + 1) = 0.$$

$$\therefore \quad n = 13 \quad \text{or} \quad -1.$$

Therefore the number of terms is 13.

Example 8.　*What is the smallest number of terms of the geometrical progression,* $8 + 24 + 72 + \ldots$, *that will give a total greater than* 6 000 000?

With the notation of §10.6,

$$S_n = a\left(\frac{r^n - 1}{r - 1}\right).$$

Substituting $a = 8$ and $r = 3$,

$$S_n = 8\left(\frac{3^n - 1}{3 - 1}\right) = 4(3^n - 1).$$

Now if we solve the equation

$$4(3^n - 1) = 6\ 000\ 000,$$

the first integer greater than the value of n found from this will be the number of terms required.

To solve the equation:

$$3^n - 1 = 1\ 500\ 000.$$

$$\therefore 3^n = 1\ 500\ 001.*$$

Taking logarithms of both sides,

$$n \log_{10} 3 = \log_{10} 1\ 500\ 001,$$

$$\therefore n = \frac{\log_{10} 1\ 500\ 001}{\log_{10} 3}$$

$$\simeq \frac{6 \cdot 1761\dagger}{0 \cdot 4771}$$

6·176	0·7907
0·4771	$\bar{1}$·6786
12·94	1·1121

$$= 12 \cdot 94, \quad \text{by four-figure tables.}$$

Therefore the number of terms required to make a total exceeding 6 000 000 is 13.

Arithmetic and geometric means

10.7. If three numbers a, b, c are in arithmetical progression, b is called the **arithmetic mean** of a and c. The common difference of the progression is given by $b - a$ or $c - b$. Therefore

$$b - a = c - b.$$

$$\therefore 2b = a + c.$$

Therefore the arithmetic mean of a and c is

$$\frac{a + c}{2}.$$

This is the ordinary "average" of a and c.

If three numbers a, b, c are in geometrical progression,

---

* With four-figure tables, we cannot distinguish between 1 500 000 and 1 500 001.

† The sign $\simeq$ means " is approximately equal to."

b is called the **geometric mean** of a and c. The common ratio is given by b/a or c/b. Therefore

$$\frac{b}{a} = \frac{c}{b}.$$

$$\therefore\ b^2 = ac.$$

Therefore the geometric mean of a and c is

$$\sqrt{(ac)}.$$

If a rectangle is drawn with sides a and c, b is the side of a square whose area is equal to that of the rectangle.

Qu. 3. Find (i) the arithmetic mean, (ii) the geometric mean of 4 and 64.

Qu. 4. The reciprocal of the harmonic mean of two numbers is the arithmetic mean of their reciprocals. Find the harmonic mean of 5 and 20. Also find the arithmetic and geometric means of 5 and 20.

Qu. 5. Find an expression for the harmonic mean of a and c.

Exercise 10c

1. Find the sum of the even numbers up to and including 100.
2. How many terms of the series $2 - 6 + 18 - 54 + \dots$ are needed to make a total of $\frac{1}{2}(1 - 3^8)$?
3. The fifth term of an A.P. is 17 and the third term is 11. Find the sum of the first seven terms.
4. The fourth term of a G.P. is -6 and the seventh term is 48. Write down the first three terms of the progression.
5. Find the sum of the first eight terms of the G.P. $5 + 15 + \dots.$
6. What is the difference between the sums to ten terms of the A.P. and G.P. whose first terms are $-2 + 4 \dots$?
7. The sum of the second and fourth terms of an arithmetical progression is 15; and the sum of the fifth and sixth terms is 25. Find the first term and the common difference.
8. The second term of an arithmetical progression is three times the seventh; and the ninth term is 1. Find the first term, the common difference, and which is the first term less than 0.

9. In a geometrical progression, the sum of the second and third terms is 9; and the seventh term is eight times the fourth. Find the first term, the common ratio, and the fifth term.

10. The fourth term of an arithmetical progression is 15, and the sum of the first five terms is 55. Find the first term and the common difference, and write down the first five terms.

11. The sum of the first three terms of an arithmetical progression is 3, and the sum of the first five terms is 20. Find the first five terms of the progression.

12. The sum of the first two terms of a geometrical progression is 3, and the sum of the second and third terms is -6. Find the first term and the common ratio.

13. How many terms of the arithmetical progression

$$15 + 13 + 11 + \ldots,$$

are required to make a total of -36?

14. Which is the first term of the geometrical progression $5 + 10 + 20 + \ldots$, to exceed 400 000?

15. Find how many terms of the G.P. $1 + 3 + 9 + \ldots$ are required to make a total of more than a million.

16. The sum of the first six terms of an arithmetical progression is 21, and the seventh term is three times the sum of the third and fourth. Find the first term and the common difference.

17. In an arithmetical progression, the sum of the first five terms is 30, and the third term is equal to the sum of the first two. Write down the first five terms of the progression.

18. Find the difference between the sums of the first ten terms of the geometrical and arithmetical progressions, which begin, $6 + 12 + \ldots$.

19. The sum of the first n terms of a certain series is $n^2 + 5n$, for all integral values of n. Find the first three terms and prove that the series is an arithmetical progression.

20. The second, fourth, and eighth terms of an A.P. are in geometrical progression, and the sum of the third and fifth terms is 20. Find the first four terms of the progression.

21. A man pays a premium of £100 at the beginning of every year to an Insurance Company on the understanding that

at the end of fifteen years he can receive back the premiums which he has paid with $2\frac{1}{2}\%$ Compound Interest. What should he receive? ($\log_{10} 1\cdot025 = 0\cdot010\ 723\ 9$.)

22. A man earned in a certain year £200 from a certain source and his annual earnings from this time continued to increase at the rate of 5%. Find to the nearest £ the whole amount he received from this source in this year and the next seven years. ($\log_{10} 1\cdot05 = 0\cdot021\ 189\ 3$.)

Proof by induction

10.8. It sometimes happens that a result is found by some means which does not provide a proof. For example, consider the following table:

n	1	2	3	4	5
Sum of the integers up to n	1	3	6	10	15
n^3	1	8	27	64	125
Sum of the cubes of the integers up to n	1	9	36	100	225

Here the terms in the fourth row are the squares of the corresponding terms in the second row. Thus it is natural to suppose that

$$1^3 + 2^3 + \ldots + n^3 = (1 + 2 + \ldots + n)^2.$$

Now $1 + 2 + \ldots + n$ is an arithmetical progression whose sum is $\frac{1}{2}n(n + 1)$. Therefore we should suppose that

$$1^3 + 2^3 + \ldots + n^3 = \tfrac{1}{4}n^2(n + 1)^2.$$

In proof by induction, it is shown that if the result holds for some particular value of n, say k, then it also holds for $n = k + 1$. It is then verified that the result does hold for some value of n, usually 1 or 2.

Example 9. *Prove by induction that*

$$1^3 + 2^3 + \ldots + n^3 = \tfrac{1}{4}n^2(n + 1)^2.$$

Suppose the result holds for a particular value of n, say k; that is,

$$1^3 + 2^3 + \ldots + k^3 = \tfrac{1}{4}k^2(k + 1)^2.$$

Then, adding the next term of the series, $(k + 1)^3$, to both sides, we obtain

$$1^3 + 2^3 + \ldots + k^3 + (k + 1)^3 = \tfrac{1}{4}k^2(k + 1)^2 + (k + 1)^3,$$
$$= (k + 1)^2\left(\frac{k^2}{4} + k + 1\right),$$
$$= (k + 1)^2\left(\frac{k^2 + 4k + 4}{4}\right),$$
$$\therefore \ 1^3 + 2^3 + \ldots + (k + 1)^3 = \tfrac{1}{4}(k + 1)^2(k + 2)^2.$$

Now this is the formula with $n = k + 1$. Therefore if the result holds for $n = k$, then it also holds for $n = k + 1$; but if $n = 1$,

L.H.S. $= 1^3 = 1$, and R.H.S. $= \tfrac{1}{4} \times 1^2 \times 2^2 = 1$.

Therefore, since the result is true for $n = 1$, it follows, by what has been shown above, that it must also be true for $n = 2$. From this it follows that the result is true for $n = 3$, and so on, for all positive integral values of n.

Exercise 10d

Prove the following results by induction:

1. $1 + 2 + \ldots + n = \tfrac{1}{2}n(n + 1)$.
2. $1^2 + 2^2 + \ldots + n^2 = \tfrac{1}{6}n(n + 1)(2n + 1)$.
3. $1.2 + 2.3 + \ldots + n(n + 1) = \tfrac{1}{3}n(n + 1)(n + 2)$.
4. $1.3 + 2.4 + \ldots + n(n + 2) = \tfrac{1}{6}n(n + 1)(2n + 7)$.
5. $3 + 8 + \ldots + (n^2 - 1) = \tfrac{1}{6}n(n - 1)(2n + 5)$.
6. $a + ar + \ldots + ar^{n-1} = a\left(\dfrac{1 - r^n}{1 - r}\right)$.
7. $\dfrac{1}{1.2} + \dfrac{1}{2.3} + \ldots + \dfrac{1}{n(n + 1)} = \dfrac{n}{n + 1}$.
8. $\dfrac{1}{1.3} + \dfrac{1}{2.4} + \ldots + \dfrac{1}{n(n + 2)} = \dfrac{3}{4} - \dfrac{2n + 3}{2(n + 1)(n + 2)}$.
9. $\dfrac{3}{4} + \dfrac{5}{36} + \ldots + \dfrac{2n - 1}{n^2(n - 1)^2} = 1 - \dfrac{1}{n^2}$.

10. $\dfrac{1}{1.2.3} + \dfrac{1}{2.3.4} + \ldots + \dfrac{1}{n(n + 1)(n + 2)}$
$$= \dfrac{1}{4} - \dfrac{1}{2(n + 1)(n + 2)}.$$

11. $\dfrac{\mathrm{d}}{\mathrm{d}x}(x^n) = nx^{n-1}$. [Use the formula for differentiating a product.]

12. $1^2 + 3^2 + 5^2 + \ldots + (2n - 1)^2 = \frac{1}{3}n(4n^2 - 1)$.

13. $1^3 + 3^3 + 5^3 + \ldots + (2n - 1)^3 = n^2(2n^2 - 1)$.

14. $4^2 + 7^2 + 10^2 + \ldots + (3n + 1)^2$
$$= \frac{1}{2}n(6n^2 + 15n + 11).$$

15. Show that $\dbinom{n}{r} + \dbinom{n}{r - 1} = \dbinom{n + 1}{r}$, and prove by induction that
$$(1 + x)^n = 1 + nx + \ldots + \dbinom{n}{r}x^r + \ldots + x^n,$$
where $\dbinom{n}{r} = \dfrac{n!}{(n - r)!r!}$.

Further series

10.9. Certain series can be summed by means of the results:
$$1 + 2 + \ldots + n = \tfrac{1}{2}n(n + 1),$$
$$1^2 + 2^2 + \ldots + n^2 = \tfrac{1}{6}n(n + 1)(2n + 1),$$
$$1^3 + 2^3 + \ldots + n^3 = \tfrac{1}{4}n^2(n + 1)^2,$$

which appear in the last section and exercise.

It should be noted that they may be used to sum the series to more or less than n terms.   For instance,

$$1^3 + 2^3 + \ldots + (2n + 1)^3 = \tfrac{1}{4}(2n + 1)^2\{(2n + 1) + 1\}^2,$$
$$= \tfrac{1}{4}(2n + 1)^2(2n + 2)^2,$$
$$= \tfrac{1}{4}(2n + 1)^2 4(n + 1)^2,$$
$$= (2n + 1)^2(n + 1)^2.$$

Qu. 6.   Find the sums of the following series:

(i) $1 + 2 + \ldots + 2n$,

(ii) $1^2 + 2^2 + \ldots + (n + 1)^2$,

(iii) $1^3 + 2^3 + \ldots + (n - 1)^3$,

(iv) $1 + 2 + \ldots + (2n - 1)$,

(v) $1^2 + 2^2 + \ldots + (2n)^2$,

(vi) $1^3 + 2^3 + \ldots + (2n - 1)^3$.

Example 10. *Find the sum of the series*

$$1^3 + 3^3 + 5^3 + \ldots + (2n + 1)^3.$$

This series can be thought of as

$$1^3 + 2^3 + 3^3 + 4^3 + 5^3 + \ldots + (2n + 1)^3$$

with the even terms missing.

We found above that

$$1^3 + 2^3 + 3^3 + 4^3 + 5^3 + \ldots + (2n + 1)^3$$
$$= (2n + 1)^2(n + 1)^2,$$

and so it remains to find the sum of the series

$$2^3 + 4^3 + 6^3 + \ldots + (2n)^3$$
$$= 2^3 . 1^3 + 2^3 . 2^3 + 2^3 . 3^3 + \ldots + 2^3 . n^3,$$
$$= 8(1^3 + 2^3 + 3^3 + \ldots + n^3),$$
$$= 8 . \tfrac{1}{4}n^2(n + 1)^2 = 2n^2(n + 1)^2.$$

Therefore $1^3 + 3^3 + 5^3 + \ldots + (2n + 1)^3$

$$= (2n + 1)^2(n + 1)^2 - 2n^2(n + 1)^2,$$
$$= (n + 1)^2\{(2n + 1)^2 - 2n^2\},$$
$$= (n + 1)^2(4n^2 + 4n + 1 - 2n^2).$$

Therefore the sum is $(n + 1)^2(2n^2 + 4n + 1)$.

Example 11. *Find the sum of n terms of the series*
$2.3 + 3.4 + 4.5 + \ldots$.

The mth term of this series is $(m + 1)(m + 2)$, or $m^2 + 3m + 2$. Therefore we require the sum of:

$$1^2 + 3 \times 1 + 2$$
$$+ 2^2 + 3 \times 2 + 2$$
$$+ 3^2 + 3 \times 3 + 2$$
$$+ \ldots\ldots\ldots\ldots$$
$$+ n^2 + 3 \times n + 2.$$

Now the sums of the three columns are

$$1^2 + 2^2 + 3^2 + \ldots + n^2 = \tfrac{1}{6}n(n+1)(2n+1),$$
$$3(1 + 2 + 3 + \ldots + n) = \tfrac{3}{2}n(n+1),$$
$$(2 + 2 + 2 + \ldots + 2) = 2n.$$

Therefore the sum of the series is

$$\tfrac{1}{6}n(n+1)(2n+1) + \tfrac{3}{2}n(n+1) + 2n$$

$$= \frac{n}{6}\{(n+1)(2n+1) + 9(n+1) + 12\},$$

$$= \frac{n}{6}(2n^2 + 3n + 1 + 9n + 9 + 12),$$

$$= \frac{n}{6}(2n^2 + 12n + 22),$$

$$= \frac{n}{3}(n^2 + 6n + 11).$$

Therefore the sum of the first n terms of the series $2.3 + 3.4 + 4.5 + \ldots$, is $\tfrac{1}{3}n(n^2 + 6n + 11)$.

The Σ notation

10.10. It is useful to have a short way of writing expressions like

$$1^2 + 2^2 + \ldots + n^2.$$

This is done by writing

$$\sum m^2,$$

which means, "the sum of all the terms like m^2". For extra precision, however, numbers are placed below and above the Σ, to show where the series begins and ends. Thus

$$\sum_{1}^{n} m^2 = 1^2 + 2^2 + \ldots + n^2$$

and

$$\sum_{2}^{5} m(m+2) = 2.4 + 3.5 + 4.6 + 5.7.$$

Exercise 10e

1. Write in full:

(i) $\displaystyle\sum_{1}^{4} m^3,$ (ii) $\displaystyle\sum_{2}^{n} m^2,$ (iii) $\displaystyle\sum_{1}^{n}(m^2 + m),$

(iv) $\displaystyle\sum_{1}^{3} \frac{1}{m(m+1)},$ (v) $\displaystyle\sum_{2}^{5} 2^m,$ (vi) $\displaystyle\sum_{1}^{4}(-1)^m m^2,$

(vii) $\displaystyle\sum_{1}^{n} m^m,$ (viii) $\displaystyle\sum_{3}^{6} \frac{(-1)^m}{m},$ (ix) $\displaystyle\sum_{n}^{n+2} m(m-1),$

(x) $\displaystyle\sum_{n-2}^{n} \frac{m}{m+1}.$

2. Write in the Σ notation:

(i) $1 + 2 + 3 + \ldots + n,$

(ii) $1^4 + 2^4 + \ldots + n^4 + (n+1)^4,$

(iii) $1 + \frac{1}{2} + \frac{1}{3} + \frac{1}{4} + \frac{1}{5},$ (iv) $3^2 + 3^3 + 3^4 + 3^5,$

(v) $2.7 + 3.8 + 4.9 + 5.10 + 6.11,$

(vi) $1 + \frac{2}{3} + \frac{3}{8} + \frac{4}{27} + \frac{5}{81},$

(vii) $\dfrac{1.3}{4} + \dfrac{2.5}{6} + \dfrac{3.7}{8} + \dfrac{4.9}{10} + \dfrac{5.11}{12},$

(viii) $-1 + 2 - 3 + 4 - 5 + 6,$

(ix) $1 - 2 + 4 - 8 + 16 - 32,$

(x) $1.3 - 9.5 + 3.7 - 4.9 + 5.11.$

3. Use the results quoted at the beginning of §10.0 to find the sums of the following series:

(i) $1 + 2 + 3 + \ldots + (2n + 1),$

(ii) $1^2 + 2^2 + 3^2 + \ldots + (n-1)^2,$

(iii) $1^3 + 2^3 + 3^3 + \ldots + (2n)^3,$

(iv) $3 + 5 + 7 + \ldots + (2n+1),$

(v) $2 + 5 + 8 + 11 + \ldots,$ to n terms,

(vi) $5 + 9 + 13 + 17 + \ldots$ to n terms,

(vii) $2 + 5 + 10 + \ldots + (n^2 + 1),$

(viii) $1.2 + 2.3 + 3.4 + 4.5 + \ldots,$ to n terms,

(ix) $1.3 + 2.4 + 3.5 + 4.6 + \ldots,$ to n terms,

(x) $2^2 + 4^2 + 6^2 + \ldots + (2n)^2,$

(xi) $1^2 + 3^2 + 5^2 + \ldots + (2n - 1)^2$,
(xii) $2 + 10 + 30 + \ldots + (n^3 + n)$,
(xiii) $2 + 12 + 36 + \ldots + (n^3 + n^2)$.

Infinite geometrical progressions

10.11. Consider the geometrical progression

$$1 + \tfrac{1}{2} + \tfrac{1}{4} + \tfrac{1}{8} + \ldots + \frac{1}{2^{n-1}}.$$

The sum of these n terms, obtained by the formula of §10.6, is given by

$$S_n = \frac{1 - (\tfrac{1}{2})^n}{1 - \tfrac{1}{2}} = 2(1 - (\tfrac{1}{2})^n).$$

Now as n increases, $(\tfrac{1}{2})^n$ approaches zero; and $(\tfrac{1}{2})^n$ can be made as close to zero as we like, if n is large enough. Therefore the sum of n terms approaches 2, as closely as we please, as n increases.

This is what is meant by writing that the infinite series

$$1 + \tfrac{1}{2} + \tfrac{1}{4} + \ldots + \frac{1}{2^{m-1}} + \ldots = 2.$$

The limit 2 is called its **sum to infinity**.

In general, the sum of the geometrical progression

$$a + ar + ar^2 + \ldots + ar^{n-1} = a\left(\frac{1 - r^n}{1 - r}\right).$$

If r lies between -1 and $+1$, assuming that r^n approaches zero as n increases, the sum to infinity of the series

$$a + ar + ar^2 + \ldots + ar^{m-1} + \ldots = \frac{a}{1 - r}.$$

Example 12. *Express as fractions in their lowest terms:*
(i) $0 \cdot 0\dot{7}$, (ii) $0 \cdot 4\dot{5}$.

(i) $0 \cdot 0\dot{7}$ means $0 \cdot 0777\ldots$, which may be written

$$\frac{7}{100} + \frac{7}{1000} + \frac{7}{10\,000} + \ldots .$$

This is a geometrical progression, and in the notation of §10.6, $a = \frac{7}{100}$ and $r = \frac{1}{10}$. Therefore

$$S_n = \frac{7}{100}\left(\frac{1 - (\frac{1}{10})^n}{1 - \frac{1}{10}}\right).$$

Therefore the sum to infinity, S_∞, is given by

$$S_\infty = \frac{7}{100}\left(\frac{1}{\frac{9}{10}}\right) = \frac{7}{100} \times \frac{10}{9} = \frac{7}{90}.$$

$$\therefore \, 0 \cdot 0\dot{7} = \frac{7}{90}.$$

(ii) $0 \cdot 4\dot{5}$ means $0 \cdot 454 \, 545 \dots$, which may be written

$$\frac{45}{100} + \frac{45}{10\,000} + \frac{45}{1\,000\,000} + \dots.$$

In this geometrical progression, $a = \frac{45}{100}$, and $r = \frac{1}{100}$.

$$\therefore \, S_n = \frac{45}{100}\left(\frac{1 - (\frac{1}{100})^n}{1 - \frac{1}{100}}\right).$$

$$\therefore \, S_\infty = \frac{45}{100}\left(\frac{1}{\frac{99}{100}}\right) = \frac{45}{100} \times \frac{100}{99} = \frac{5}{11}.$$

$$\therefore \, 0 \cdot 4\dot{5} = \frac{5}{11}.$$

Exercise 10f

1. Write down the sums of the first n terms of the following
series, and deduce their sums to infinity:

(i) $1 + \frac{1}{3} + \frac{1}{9} + \frac{1}{27} + \dots,$

(ii) $12 + 6 + 3 + 1\frac{1}{2} + \dots,$

(iii) $\frac{3}{10} + \frac{3}{100} + \frac{3}{1000} + \frac{3}{10\,000} + \dots,$

(iv) $\frac{13}{100} + \frac{13}{10\,000} + \frac{13}{1\,000\,000} + \dots,$

(v) $0 \cdot 5 + 0 \cdot 05 + 0 \cdot 005 + \dots,$

(vi) $0 \cdot 54 + 0 \cdot 0054 + 0 \cdot 000\ 054 + \ldots,$
(vii) $1 - \frac{1}{2} + \frac{1}{4} - \frac{1}{8} + \ldots,$
(viii) $54 - 18 + 6 - 2 + \ldots$

2. Express the following recurring decimals as fractions in their lowest terms:

(i) $0 \cdot \dot{8},$ (ii) $0 \cdot \dot{1}\dot{2},$ (iii) $3 \cdot \dot{2},$
(iv) $2 \cdot \dot{6}\dot{9},$ (v) $1 \cdot 00\dot{4},$ (vi) $2 \cdot 9\dot{6}\dot{0}.$

3. If the sum to infinity of a G.P. is three times the first term, what is the common ratio?

4. The sum to infinity of a G.P. is 4 and the second term is 1. Find the first, third, and fourth terms.

5. The second term of a G.P. is 24 and its sum to infinity is 100. Find the two possible values of the common ratio and the corresponding first terms.

Exercise 10g (Miscellaneous)

1. Find the sum of the integers between 1 and 100 which are divisible by 3.

2. How many terms of the geometrical progression

$$\tfrac{1}{16} + \tfrac{1}{8} + \tfrac{1}{4} + \ldots$$

are needed to make a total of $2^{16} - \frac{1}{16}$?

3. Prove by induction that
$$1 \cdot 4 + 2 \cdot 5 + \ldots + n(n + 3) = \tfrac{1}{3}n(n + 1)(n + 5).$$

4. Show that the sums to infinity of the geometrical progressions $3 + \frac{9}{4} + \frac{27}{16} + \ldots, 4 + \frac{8}{3} + \frac{16}{9} + \ldots$ are equal.

5. How many terms of the arithmetical progression $2 + 3\frac{1}{4} + 4\frac{1}{2} + \ldots$ are needed to make a total of 204?

6. An arithmetical progression has thirteen terms whose sum is 143. The third term is 5. Find the first term.

7. The sum of n terms of a certain series is $3n^2 + 10n$ for all values of n. Find the nth term and show that the series is an arithmetical progression.

8. Find the sum of the series $2 + 6 + \ldots + (n^2 - n)$.

9. Show that the sum of the first n odd numbers is a perfect square. Show also, that $57^2 - 13^2$ is the sum of certain consecutive odd numbers, and find them.

10. What is the sum of the integers from 1 to 100, inclusive, which are not divisible by 6?

11. Find the sum of the first n terms of the geometrical progression $5 + 15 + 45 + \ldots$. What is the smallest number of terms which will give a total of more than 10^8?

12. The sum to infinity of a geometrical progression with a positive common ratio is 9 and the sum of the first two terms is 5. Find the first four terms of the progression.

13. Show that, if $\log a$, $\log b$, $\log c$ are consecutive terms of an arithmetical progression, then a, b, c are in geometrical progression.

14. The eighth term of an arithmetical progression is twice the third term, and the sum of the first eight terms is 39. Find the first three terms of the progression, and show that its sum to n terms is $\frac{3}{8}n(n+5)$.

15. Find the number n such that the sum of the integers from 1 to $n - 1$ is equal to the sum of the integers from $n + 1$ to 49.

16. Show that there are two possible geometrical progressions in each of which the first term is 8, and the sum of the first three terms is 14. Find the second term and the sum of the first seven terms in each progression.

17. Prove by induction that
$$\tfrac{1}{2} + \tfrac{1}{6} + \ldots + \frac{1}{n(n-1)} = 1 - \frac{1}{n}.$$

18. Find the sum of the series $3 + 6 + 11 + \ldots + (n^2 + 2)$.

19. If a and b are the first and last terms of an arithmetical progression of $r + 2$ terms, find the second and the $(r + 1)$th terms.

20. The sum of n terms of a certain series is $4^n - 1$ for all values of n. Find the first three terms and the nth term, and show that the series is a geometrical progression.

21. A child wishes to build up a triangular pile of toy bricks so as to have 1 brick in the top row, 2 in the second, 3 in the third and so on. If he has 100 bricks, how many rows can he complete and how many bricks has he left?

22. Show that the sum of the odd numbers from 1 to 55 inclusive is equal to the sum of the odd numbers from 91 to 105 inclusive.

23. The second, fifth, and eleventh terms of an arithmetical progression are in geometrical progression, and the seventh

term is 4. Find the first term and the common difference. What is the common ratio of the geometrical progression?

24. A chess board has 64 squares. Show that ten thousand million men each prepared to bring a million pounds could not bring sufficient money to put 1p on the first square, 2p on the second, 4p on the third, 8p on the fourth, and so on for the 64 squares.

25. Prove that

$$\log a + \log ax + \log ax^2 + \ldots \text{ to } n \text{ terms}$$
$$= n \log a + \tfrac{1}{2}n(n-1) \log x.$$

26. Given the series

$$1 + 2x + 3x^2 + 4x^3 + \ldots,$$

  (i) find the sum of the first n terms when $x = 1$,

  (ii) find, by multiplying by $1 - x$, the sum of the first n terms when x is not equal to unity.

CHAPTER 11

THE BINOMIAL THEOREM

Pascal's triangle

11.1. It is well known that

$$(a + b)^2 = a^2 + 2ab + b^2,$$

and it is the object of this chapter to show how higher powers of $a + b$ can be expanded with little difficulty.

Most readers will not be able to write down similar expressions for $(a + b)^3$ and $(a + b)^4$ without doing some work on paper, and so the long multiplication is given below. The reason for printing the coefficients in heavy type will appear later.

$$
\begin{array}{l}
1a^2 + 2ab\ + 1b^2 \\
\qquad\qquad a + b
\end{array}
\qquad
\begin{array}{l}
1a^3 + 3a^2b + 3ab^2\ + 1b^3 \\
\qquad\qquad\qquad\qquad a + b
\end{array}
$$

$$
\begin{array}{l}
1a^3 + 2a^2b + 1ab^2 \\
\quad\ \ 1a^2b + 2ab^2 + 1b^3
\end{array}
\qquad
\begin{array}{l}
1a^4 + 3a^3b + 3a^2b^2 + 1ab^3 \\
\quad\ \ 1a^3b + 3a^2b^2 + 3ab^3 + 1b^4
\end{array}
$$

$$
1a^3 + 3a^2b + 3ab^2 + 1b^3
\qquad
1a^4 + 4a^3b + 6a^2b^2 + 4ab^3 + 1b^4
$$

The results so far obtained are summarized below.

$$(a + b)^2 = \qquad\ \ 1a^2 +\ 2ab\ +\ 1b^2.$$
$$(a + b)^3 = \quad 1a^3 +\ 3a^2b +\ 3ab^2 +\ 1b^3.$$
$$(a + b)^4 = 1a^4 + 4a^3b + 6a^2b^2 + 4ab^3 + 1b^4.$$

It is clearer, however, if the coefficients are written alone.

$$
\begin{array}{ccccccc}
 & & 1 & & 2 & & 1 \\
 & 1 & & 3 & & 3 & & 1 \\
1 & & 4 & & 6 & & 4 & & 1
\end{array}
$$

237

The reader may be able to guess the next line; and, more important, he may be able to see how the table can be continued, obtaining each line from the previous one.

To show the construction of the table of coefficients, the last three lines of the long multiplications are written, leaving out the letters.

1	2	1		1	3	3	1	
	1	2	1		1	3	3	1
1	3	3	1	1	4	6	4	1

Thus it may be seen that every coefficient in the table is obtained from the two on either side of it in the row above. In this way the next line can be obtained:

For completeness, it may be observed that

$$(a + b)^0 = 1, \quad \text{and} \quad (a + b)^1 = 1a + 1b$$

Therefore the table of coefficients may be written in a triangle (known as Pascal's triangle, after the French mathematician and philosopher Blaise Pascal, 1623–1662) as follows:

When an expression is written as a series of terms, it is said to be **expanded**, and the series is called its **expansion**. Thus the expansion of $(a + b)^3$ is $a^3 + 3a^2b + 3ab^2 + b^3$.

Certain points should be noted about the expansion of $(a + b)^n$. They should be verified for the cases $n = 2, 3, 4$, in the expansions obtained so far.

(i) Reading from either end of each row, the *coefficients* are the same.

(ii) There are $(n + 1)$ terms.

(iii) Each term is of degree n.

(iv) The coefficients are obtained from the row in Pascal's triangle beginning $1, n$.

Example 1. *Expand $(a + b)^6$ in descending powers of a.*

There will be 7 terms, involving

$$a^6, \quad a^5b, \quad a^4b^2, \quad a^3b^3, \quad a^2b^4, \quad ab^5, \quad b^6,$$

each of which is of degree 6. Their coefficients, obtained from Pascal's triangle, are respectively

$$1, \quad 6, \quad 15, \quad 20, \quad 15, \quad 6, \quad 1.$$

Therefore the expansion of $(a + b)^6$ in descending powers of a is

$$a^6 + 6a^5b + 15a^4b^2 + 20a^3b^3 + 15a^2b^4 + 6ab^5 + b^6.$$

Example 2. *Expand $(2x + 3y)^3$ in descending powers of x.*

Here $a = 2x$ and $b = 3y$, and so there will be four terms involving

$$(2x)^3, \quad (2x)^2(3y), \quad (2x)(3y)^2, \quad (3y)^3.$$

Their coefficients, obtained from Pascal's triangle are respectively

$$1, \quad 3, \quad 3, \quad 1.$$

Therefore the expansion of $(2x + 3y)^3$, in descending powers of x is

$$(2x)^3 + 3(2x)^2(3y) + 3(2x)(3y)^2 + (3y)^3,$$

which simplifies to

$$8x^3 + 36x^2y + 54xy^2 + 27y^3.$$

Example 3. *Obtain the expansion of* $(2x - \frac{1}{2})^4$, *in descending powers of* x.

Here $a = 2x$ and $b = -\frac{1}{2}$, therefore the five terms of the expansion will involve

$$(2x)^4, \quad (2x)^3(-\tfrac{1}{2}), \quad (2x)^2(-\tfrac{1}{2})^2, \quad (2x)(-\tfrac{1}{2})^3, \quad (-\tfrac{1}{2})^4;$$

and their coefficients will be respectively

$$1, \qquad 4, \qquad 6, \qquad 4, \qquad 1.$$

$\therefore (2x - \frac{1}{2})^4$

$$= (2x)^4 + 4(2x)^3(-\tfrac{1}{2}) + 6(2x)^2(-\tfrac{1}{2})^2 + 4(2x)(-\tfrac{1}{2})^3 + (-\tfrac{1}{2})^4,$$

$$= 16x^4 + 4(8x^3)(-\tfrac{1}{2}) + 6(4x^2)(\tfrac{1}{4}) + 4(2x)(-\tfrac{1}{8}) + \tfrac{1}{16},$$

Therefore the expansion of $(2x - \frac{1}{2})^4$, in descending powers of x is

$$16x^4 - 16x^3 + 6x^2 - x + \tfrac{1}{16}.$$

Note that terms are alternately $+$ and $-$, according to the even or odd degree of $(-\frac{1}{2})$.

Example 4. *Use Pascal's triangle to obtain the value of* $(1 \cdot 002)^5$, *correct to six places of decimals.*

$1 \cdot 002$ may be written $(1 + 0 \cdot 002)$, so that the expansion of $(a + b)^5$ may be used, with $a = 1$ and $b = 0 \cdot 002$.

The terms in the expansion will involve

$$1, \quad (0 \cdot 002), \quad (0 \cdot 002)^2, \quad (0 \cdot 002)^3, \quad (0 \cdot 002)^4, \quad (0 \cdot 002)^5;$$

and their coefficients will be

$$1, \qquad 5, \qquad 10, \qquad 10, \qquad 5, \qquad 1,$$

respectively. Now the last three terms will make no difference to the answer, correct to six places of decimals. Therefore

$$(1 \cdot 002)^5 \simeq 1 + 5(0 \cdot 002) + 10(0 \cdot 002)^2,$$

$$= 1 + 0 \cdot 010 + 0 \cdot 000\ 040,$$

and so $(1\cdot002)^5 = 1\cdot010\ 040$, correct to six places of decimals.

Exercise 11a

1. Expand:

 (i) $(a + b)^5$,   (ii) $(x + y)^3$,   (iii) $(x + 2y)^4$,

 (iv) $(1 - z)^4$,   (v) $(2x + 3y)^4$,   (vi) $(4z + 1)^3$,

 (vii) $(a - b)^6$,   (viii) $(a - 2b)^3$,   (ix) $(3x - y)^4$,

 (x) $(2x + \frac{1}{3})^3$,   (xi) $\left(x - \dfrac{1}{x}\right)^5$,   (xii) $\left(\dfrac{x}{2} + \dfrac{2}{x}\right)^4$,

(xiii) $(a + b)^7$,   (xiv) $(a^2 - b^2)^5$,

 (xv) $(a - b)^3(a + b)^3$.

2. Without using tables, find the values of the following expressions:

 (i) $(1 + \sqrt{2})^3 + (1 - \sqrt{2})^3$;

 (ii) $(2 + \sqrt{3})^4 + (2 - \sqrt{3})^4$;

 (iii) $(1 + \sqrt{2})^3 - (1 - \sqrt{2})^3$, taking $\sqrt{2} = 1\cdot414$;

 (iv) $(2 + \sqrt{6})^4 - (2 - \sqrt{6})^4$, taking $\sqrt{6} = 2\cdot449$;

 (v) $(\sqrt{2} + \sqrt{3})^4 + (\sqrt{2} - \sqrt{3})^4$;

 (vi) $(\sqrt{6} + \sqrt{2})^3 - (\sqrt{6} - \sqrt{2})^3$, taking $\sqrt{2} = 1\cdot414$.

3. Write down the expansion of $(2 + x)^5$ in ascending powers of x. Taking the first three terms of the expansion, put $x = 0\cdot001$, and find the value of $(2\cdot001)^5$, correct to five places of decimals.

4. Write down the expansion of $(1 + \frac{1}{4}w)^4$. Taking the first three terms of the expansion, put $x = 0\cdot1$, and find the value of $(1\cdot025)^4$, correct to three places of decimals.

5. Expand $(2 - x)^6$ in ascending powers of x. Taking $x = 0\cdot002$, and using the first three terms of the expansion, find the value of $(1\cdot998)^6$ as accurately as you can. Examine the fourth term of the expansion to find to how many places of decimals your answer is correct.

6. Write down the first four terms of the expansion of $(1 + \frac{1}{8}x)^8$. Use as many of these terms as you need, to find the value of $(1\cdot0125)^8$, correct to three places of decimals.

7. Use the expansion of $(2 + x)^6$, to find the value of $(2\frac{1}{100})^6$, correct to three places of decimals.

8. Expand $(1 - x)^5$, and hence find, correct to three places of decimals, the value of $(\frac{49}{50})^5$.

Leading to the binomial theorem

11.2. In the last section it was shown how $(a + b)^n$ could be expanded, for a known value of n, by using Pascal's triangle. If n is large, this may involve a considerable amount of addition, and when (as is often the case) only the first few terms are required, it is much quicker to use a formula that will be obtained in the next section.

The last section began with the expansions of $(a + b)^2$ and $(a + b)^3$. Now, consider the expansions of $(a + b)(c + d)$ and $(a + b)(c + d)(e + f)$.

It is easily seen that

$$(a + b)(c + d) = ac + ad + bc + bd.$$

To obtain the expansion of $(a + b)(c + d)(e + f)$, each term of $ac + ad + bc + bd$ is multiplied by e and f, giving

$$ace + ade + bce + bde + acf + adf + bcf + bdf.$$

Note that each term contains one factor from each bracket, and that the expansion consists of the sum of all such combinations.

Now the expansion of $(a + b)(c + d)(c + f)(g + h)$ would be obtained by multiplying each term of the expansion by g and by h. So, continuing this method of expansion, it follows that, if the product of n factors is expanded, each term contains one factor from each bracket, and that the expansion consists of the sum of all such combinations.

The expansion of $(a + b)^5$ will be obtained by an argument making use of this fact.

$$(a + b)^5 = (a + b)(a + b)(a + b)(a + b)(a + b).$$

(i) Choosing an a from each bracket we obtain a^5.

(ii) The term in a^4 is obtained by choosing a b from one bracket, and a's from the other four. This can be done in ${}^5C_1$ ways, giving ${}^5C_1 a^4 b$.

(iii) The term in a^3 is obtained by choosing b's from two brackets, and a's from the other three. This can be done in ${}^5C_2$ ways, giving ${}^5C_2 a^3 b^2$.

(iv) Similarly, the terms in a^2 and a are ${}^5C_3 a^2 b^3$ and ${}^5C_4 a b^4$.

(v) Choosing a b from each bracket we obtain b^5.

$\therefore (a + b)^5$
$$= a^5 + {}^5C_1 a^4 b + {}^5C_2 a^3 b^2 + {}^5C_3 a^2 b^3 + {}^5C_4 a b^4 + b^5.$$

The binomial theorem

11.3. *If n is a positive integer,*

$$(a + b)^n = a^n + {}^nC_1 a^{n-1} b + \ldots + {}^nC_r a^{n-r} b^r + \ldots + b^n,$$

where $\qquad {}^nC_r = \dfrac{n!}{(n - r)! r!}.$

The expansion of $(a + b)^n$ is obtained as follows.

$(a + b)^n = (a + b)(a + b) \ldots (a + b),$ to n factors.

(i) Choosing an a from each bracket we obtain a^n.

(ii) The term in a^{n-1} is obtained by choosing a b from one bracket, and a's from the other $n - 1$. This can be done in ${}^nC_1$ ways, giving ${}^nC_1 a^{n-1} b$.

(iii) The term in a^{n-2} is obtained by choosing a b from two brackets, and a's from the other $n - 2$. This can be done in ${}^nC_2$ ways, giving ${}^nC_2 a^{n-2} b^2$.

(iv) The term in a^{n-r} is obtained by choosing a b from r brackets, and a's from the other $n - r$. This can be done in ${}^nC_r$ ways, giving ${}^nC_r a^{n-r} b^r$.

(v) Choosing a b from each bracket we obtain b^n.

This proves the theorem.

When only the first few terms of an expansion are required, the theorem is used in the form:

$$(a + b)^n = a^n + na^{n-1}b + \frac{n(n - 1)}{2!}a^{n-2}b^2 +$$

$$+ \frac{n(n - 1)(n - 2)}{3!}a^{n-3}b^3 + \ldots + b^n.$$

This follows immediately, since

$$^nC_1 = n, \quad ^nC_2 = \frac{n!}{(n - 2)!2!} = \frac{n(n - 1)}{2!}, \quad \text{and}$$

$$^nC_3 = \frac{n!}{(n - 3)!3!} = \frac{n(n - 1)(n - 2)}{3!}.$$

In case the name of the theorem is not understood, it may be helpful to remark that an expression with one term is called a mononomial, one which has two terms is a binomial, and one with three terms is a trinomial. Thus the theorem about the expansion of a power of two terms is called the binomial theorem.

Example 5. *Find the coefficient of x^{10} in the expansion of $(2x - 3)^{14}$.*

The term in $(2x)^{10}(- 3)^4$ is the only one needed, and by the binomial theorem it is

$$^{14}C_4(2x)^{10}(- 3)^4.$$

Therefore the coefficient of x^{10} is

$$\frac{14!}{10!4!} 2^{10} \times 3^4.$$

It is important to note that we could equally well have written the term as

$$^{14}C_{10}(2x)^{10}(- 3)^4,$$

because $^{14}C_{10} = {}^{14}C_4$. This is clear if they are written in factorial notation:

$$^{14}C_{10} = \frac{14!}{4!\,10!}, \qquad {}^{14}C_4 = \frac{14!}{10!\,4!}.$$

Alternatively, $^{14}C_{10}$ is the number of ways of choosing ten objects from fourteen unlike objects; but if ten are chosen, four are left, and so it must also be the number of ways of choosing four objects from fourteen unlike objects, which is $^{14}C_4$.

Qu. 1. Show that $^nC_{n-r} = {}^nC_r$.

It is useful to note in Example 5 that the numbers whose factorials appear in the coefficient

$$\frac{14!}{10!\,4!}$$

are all indices. 14 is the index of $2x - 3$, 10 is the index of $2x$ and 4 is the index of -3. That this is always the case should be clear if the term in $a^{n-r}b^r$ in the expansion of $(a + b)^n$ is written with factorial notation:

$$\frac{n!}{(n-r)!\,r!}a^{n-r}b^r.$$

Example 6. *Obtain the first four terms of the expansion of $(1 + \frac{1}{2}x)^{10}$ in ascending powers of x. Hence find the value of $(1 \cdot 005)^{10}$, correct to four decimal places.*

Using the second form of the binomial theorem,

$$(1 + \tfrac{1}{2}x)^{10} = 1 + 10\left(\frac{x}{2}\right) + \frac{10 \cdot 9}{2 \cdot 1}\left(\frac{x}{2}\right)^2 + \frac{10 \cdot 9 \cdot 8}{3 \cdot 2 \cdot 1}\left(\frac{x}{2}\right)^3 + \cdots,$$

$$= 1 + 5x + \frac{45}{4}x^2 + 15x^3 + \cdots .$$

Now $\frac{1}{2}x = 0 \cdot 005$, if $x = 0 \cdot 01$; so substituting this value of x,

$(1 \cdot 005)^{10} \simeq 1 + 5(0 \cdot 01) + 11 \cdot 25(0 \cdot 01)^2 + 15(0 \cdot 01)^3,$

$\qquad = 1 + 0 \cdot 05 + 0 \cdot 001\ 125 + 0 \cdot 000\ 015.$

Therefore $(1 \cdot 005)^{10} = 1 \cdot 0511$, correct to four places of decimals.

Example 7. *Obtain the expansion of* $(1 + x - 2x^2)^8$, *as far as the term in* x^3.

$(1 + x - 2x^2)^8$ may be written $\{1 + (x - 2x^2)\}^8$, which may then be expanded by the binomial theorem.

$$\{1 + (x - 2x^2)\}^8 = 1 + 8(x - 2x^2) + \frac{8 \times 7}{2!}(x - 2x^2)^2 +$$

$$+ \frac{8 \times 7 \times 6}{3!}(x - 2x^2)^3 + \ldots,$$

$$= 1 + 8(x - 2x^2) + 28(x^2 - 4x^3 + 4x^4) +$$

$$+ 56(x^3 + \text{other terms}) + \ldots,$$

$$= 1 + 8x - 16x^2 + 28x^2 - 112x^3 +$$

$$+ 56x^3 + \text{terms in } x^4 \text{ and higher powers.}$$

$$\therefore \ (1 + x - 2x^2)^8 = 1 + 8x + 12x^2 - 56x^3,$$

as far as the term in x^3.

Exercise 11b

1. Write down the terms indicated, in the expansions of the following, and simplify your answers.

   (i) $(x + 2)^8$, term in x^5;   (ii) $(3u - 2)^5$, term in u^3;
   (iii) $(2t - \frac{1}{2})^{12}$, term in t^7;   (iv) $(2x + y)^{11}$, term in x^3.

2. Write down, and simplify, the terms indicated, in the expansions of the following in ascending powers of x.

   (i) $(1 + x)^9$, 4th term;   (ii) $\left(2 - \frac{x}{2}\right)^{12}$, 4th term;
   (iii) $(3 + x)^7$, 5th term;   (iv) $(x + 1)^{20}$, 3rd term.

3. Write down, and simplify, the coefficients of the terms indicated, in the expansions of the following.

   (i) $(\frac{1}{2}t + \frac{1}{2})^{10}$, term in t^4;   (ii) $(4 + \frac{3}{4}x)^6$, term in x^3;
   (iii) $(2x - 3)^7$, term in x^5;   (iv) $(3 + \frac{1}{3}y)^{11}$, term in y^5.

4. Write down the coefficients of the terms indicated, in the expansions of the following in ascending powers of x.

(i) $(1 + x)^{16}$, 3rd term; (ii) $(2 - x)^{20}$, 18th term;
(iii) $(3 + 2x)^6$, 4th term; (iv) $(2 + \frac{3}{2}x)^8$, 5th term.

5. Write down the terms involving (i) $x^4\left(\dfrac{1}{x}\right)^2$, (ii) $x^3\left(\dfrac{1}{x}\right)^3$, in the expansion of $\left(x + \dfrac{1}{x}\right)^6$.

6. Write down the constant terms in the expansions of

(i) $\left(x - \dfrac{1}{x}\right)^8$, (ii) $\left(2x^2 - \dfrac{1}{2x}\right)^6$.

7. Find the coefficients of the terms indicated in the expansions of the following:

(i) $\left(x + \dfrac{1}{x}\right)^6$, term in x^4; (ii) $\left(2x + \dfrac{1}{x}\right)^7$, term in $\dfrac{1}{x^5}$;

(iii) $\left(x - \dfrac{2}{x}\right)^8$, term in x^6.

8. Find the ratio of the term in x^5 to the term in x^6, in the expansion of $(2x + 3)^{20}$.

9. Find the ratio of the term in x^7 to the term in x^8 in the expansion of $(3x + \frac{2}{3})^{17}$.

10. Find the ratio of the term in a^r to the term in a^{r+1} in the expansion of $(a + b)^n$.

11. Write down the first four terms of the expansions of the following, in ascending powers of x.

(i) $(1 + x)^{10}$, (ii) $(1 + \frac{1}{2}x)^9$, (iii) $(1 - x)^{11}$,
(iv) $(x + 1)^{12}$, (v) $(2 + \frac{1}{2}x)^8$, (vi) $(2 - \frac{1}{2}x)^7$.

12. Use the binomial theorem to find the values of:

(i) $(1 \cdot 01)^{10}$, correct to three places of decimals;
(ii) $(2 \cdot 001)^{10}$, correct to six significant figures;
(iii) $(0 \cdot 997)^{12}$, correct to three places of decimals;
(iv) $(1 \cdot 998)^8$, correct to two places of decimals.

13. Expand the following as far as the terms in x^3:

(i) $(1 + x + x^2)^3$, (ii) $(1 + 2x - x^2)^6$,
(iii) $(1 - x - x^2)^4$, (iv) $(2 + x + x^2)^5$,
(v) $(1 - x + x^2)^8$, (vi) $(2 + x - 2x^2)^7$,
(vii) $(3 - 2x + x^2)^4$, (viii) $(3 + x + x^3)^4$.

Convergent series

11.4. The series

$$1 + x + x^2 + \ldots + x^{n-1}$$

is a geometrical progression, with common ratio x, and may be summed by the method of §10.6. In this way

$$1 + x + x^2 + \ldots + x^{n-1} = \frac{1 - x^n}{1 - x}.$$

If x lies between -1 and $+1$,* we will assume that x^n approaches zero as n increases, which makes the right-hand side of the identity approach $1/(1-x)$.

Thus when we write

$$1 + x + x^2 + \ldots + x^r + \ldots = \frac{1}{1 - x},$$

we mean that the left-hand side can be made to differ as little as we please from the right-hand side, providing enough terms are taken. It must not be forgotten, however, that we have taken x to lie between -1 and $+1$.

A series of terms, whose sum approaches a finite value as the number of terms is increased indefinitely is called a **convergent** series, and the finite value is called its **sum to infinity**.

Thus $1 + x + x^2 + \ldots + x^r + \ldots$ is a convergent series, provided x lies between -1 and $+1$, and its sum is

$$\frac{1}{1 - x}.$$

To emphasize the necessity for the condition

$$-1 < x < +1,$$

---

* "x lies between -1 and $+1$" is written, "$-1 < x < 1$": the sign "$<$" means, "is less than".

the behaviour of the series for other values of x is examined below.

(i) If $x = 1$, $1 + x + x^2 + \ldots + x^{n-1} = n$. Therefore as n increases, the value of the series increases indefinitely.

(ii) If $x = -1$,
$$1 + x + x^2 + \ldots + x^{n-1}$$
$$= 1 - 1 + 1 - \ldots + (-1)^{n-1},$$
which is equal to 1 or 0, according to whether n is odd or even.

(iii) If x is greater than 1, x^n is greater than 1, and can be made as large as we like, if n is sufficiently large. Therefore the sum of the series, $(1 - x^n)/(1 - x)$, can be made as large as we like.

(iv) When x is less than -1, $1 - x$ is positive and x^n is numerically greater than 1. If n is even, x^n is positive, therefore $1 - x^n$ is negative and so the sum $(1 - x^n)/(1 - x)$ is negative. If n is odd, x^n is negative, therefore $1 - x^n$ is positive and so the sum is positive. Hence the sum is alternately positive and negative.

It is beyond the scope of this book to give tests to discover whether any particular series is convergent, but this section has been written to draw the reader's attention to the fact that series are not always convergent.

The binomial theorem for any index

11.5. It has been shown that
$$(a + b)^n = a^n + na^{n-1}b + \frac{n(n-1)}{2!}a^{n-2}b^2 + \ldots + b^n,$$
where n is a positive integer.

Now it will be *assumed* that
$$(1 + x)^n = 1 + nx + \frac{n(n-1)}{2!}x^2 + \frac{n(n-1)(n-2)}{3!}x^3 + \ldots,$$

(the series being continued indefinitely), for *any* value of n provided $-1 < x < +1$. The proof is beyond the scope of this book.

It should be remembered that, if n is a positive integer, there will only be a finite number of terms. (See §11.3.)

For the sake of those who go on to read other books, it should be added that the index, n, is often called the **exponent**.

Historical note. Pascal's triangle was given by a Chinese author of the early fourteenth century, but Pascal made considerable use of it in connection with problems on probability, and it became associated with his name. From it he obtained the theorem for positive integral exponents. The series for fractional and negative indices was given by Newton in 1676.

Example 8. *Use the binomial theorem to expand* $\dfrac{1}{1-x}$ *in ascending powers of x, as far as the term in x^3.*

(This example has been chosen because the result has already been established in §11.4.)

Since $\dfrac{1}{1-x}$ may be written $(1-x)^{-1}$, the binomial theorem may be used. Thus

$$(1-x)^{-1} = 1 + (-1)(-x) + \frac{(-1)(-2)}{2!}(-x)^2 +$$
$$+ \frac{(-1)(-2)(-3)}{3!}(-x)^3 + \ldots,$$
$$\therefore \frac{1}{1-x} = 1 + x + x^2 + x^3 + \ldots,$$

provided $-1 < x < 1$.

Example 9. *Obtain the first five terms of the expansion of* $\sqrt{(1+2x)}$ *in ascending powers of x. State the range of values of x for which the expansion is valid.*

Since $\sqrt{(1 + 2x)} = (1 + 2x)^{\frac{1}{2}}$, the binomial theorem may be used.

$$(1 + 2x)^{\frac{1}{2}} = 1 + \tfrac{1}{2}(2x) + \frac{(\tfrac{1}{2})(-\tfrac{1}{2})}{2!}(2x)^2 +$$

$$+ \frac{(\tfrac{1}{2})(-\tfrac{1}{2})(-\tfrac{3}{2})}{3!}(2x)^3 +$$

$$+ \frac{(\tfrac{1}{2})(-\tfrac{1}{2})(-\tfrac{3}{2})(-\tfrac{5}{2})}{4!}(2x)^4 + \ldots,$$

$$\therefore \sqrt{(1 + 2x)} = 1 + x - \tfrac{1}{2}x^2 + \tfrac{1}{2}x^3 - \tfrac{5}{8}x^4 + \ldots .$$

For the expansion to be valid, $-1 < 2x < +1$, therefore $-\tfrac{1}{2} < x < +\tfrac{1}{2}$.

Example 10. *Expand* $\dfrac{1}{(2 + x)^2}$ *in ascending powers of* x, *as far as the term in* x^3, *and state for what values of* x *the expansion is valid.*

First it may be observed that $\dfrac{1}{(2 + x)^2} = (2 + x)^{-2}$.

However, the binomial theorem has been stated for $(1 + x)^n$. Therefore a factor must be taken out, in order to leave the bracket in this form.

$$(2 + x)^{-2} = \{2(1 + \tfrac{1}{2}x)\}^{-2} = 2^{-2}(1 + \tfrac{1}{2}x)^{-2} = \tfrac{1}{4}(1 + \tfrac{1}{2}x)^{-2},$$

and this may now be expanded.

$$\left[\text{Alternatively}: \frac{1}{(2 + x)^2} = \frac{1}{2^2(1 + \tfrac{1}{2}x)^2} = \tfrac{1}{4}(1 + \tfrac{1}{2}x)^{-2}.\right]$$

$$\tfrac{1}{4}(1 + \tfrac{1}{2}x)^{-2} = \tfrac{1}{4}\left\{1 + (-2)\left(\frac{x}{2}\right) + \frac{(-2)(-3)}{2!}\left(\frac{x}{2}\right)^2 +\right.$$

$$\left. + \frac{(-2)(-3)(-4)}{3!}\left(\frac{x}{2}\right)^3 + \ldots\right\},$$

$$\therefore \frac{1}{(2 + x)^2} = \tfrac{1}{4}(1 - x + \tfrac{3}{4}x^2 - \tfrac{1}{2}x^3 + \ldots).$$

For the expansion to be valid, $-1 < \tfrac{1}{2}x < +1$, therefore $-2 < x < +2$.

Exercise 11c

1. Expand the following in ascending powers of x, as far as the terms in x^3; and state the ranges of values of x for which the expansions are valid.

(i) $(1 + x)^{-2}$, (ii) $(1 + x)^{\frac{1}{3}}$, (iii) $(1 + x)^{\frac{2}{3}}$,

(iv) $(1 - 2x)^{\frac{1}{2}}$, (v) $\left(1 + \dfrac{x}{2}\right)^{-3}$, (vi) $(1 - 3x)^{-\frac{1}{2}}$,

(vii) $\dfrac{1}{1 + 3x}$, (viii) $\sqrt{(1 - x^2)}$, (ix) $\sqrt[3]{(1 - x)}$,

(x) $\dfrac{1}{\sqrt{(1 + 2x)}}$, (xi) $\dfrac{1}{\left(1 + \dfrac{x}{2}\right)^2}$, (xii) $\sqrt{(1 - 2x)^3}$,

(xiii) $\dfrac{1}{2 + x}$, (xiv) $\sqrt{(2 - x)}$, (xv) $\sqrt[3]{(3 + x)}$,

(xvi) $\dfrac{1}{\sqrt{(2 + x^2)}}$, (xvii) $\dfrac{1}{(3 - x)^2}$ (xviii) $\dfrac{3}{\sqrt[3]{(3 - x^3)}}$.

2. Use the binomial theorem to find the values of the following:

(i) $\sqrt{(1 \cdot 001)}$, correct to six places of decimals.

(ii) $\dfrac{1}{(1 \cdot 02)^2}$, correct to four places of decimals.

(iii) $\sqrt{(0 \cdot 998)}$, correct to six places of decimals.

(iv) $\sqrt[3]{(1 \cdot 03)}$, correct to four places of decimals.

(v) $\dfrac{1}{\sqrt{(0 \cdot 98)}}$, correct to four places of decimals.

3. Find the first four terms of the expansions of the following in ascending powers of x:

(i) $\dfrac{1 + x}{1 - x}$, (ii) $\dfrac{x + 2}{(1 + x)^2}$, (iii) $\dfrac{1 - x}{\sqrt{(1 + x)}}$,

(iv) $\sqrt{\dfrac{1 + x}{1 - x}}$, [Multiply numerator and denominator by $\sqrt{(1 + x)}$.]

(v) $\dfrac{2x - 3}{x + 2}$, (vi) $\sqrt{\dfrac{(1 - x)^3}{1 + x}}$, (vii) $\dfrac{x + 3}{\sqrt[3]{(1 - 3x)}}$.

4. Find the first four terms of the expansion of $(1 - 8x)^{\frac{1}{2}}$ in ascending powers of x. Substitute $x = \frac{1}{100}$ and obtain the value of $\sqrt{23}$ correct to five significant figures.

5. Expand $(1 - x)^{\frac{1}{3}}$ in ascending powers of x as far as the fourth term. By taking the first two terms of the expansion and substituting $x = \frac{1}{1000}$, find the value of $\sqrt[3]{37}$, correct to six significant figures. [HINT: $27 \times 37 = 999$.]

6. Obtain the first four terms of the expansion of $(1 - 16x)^{\frac{1}{4}}$. Substitute $x = \dfrac{1}{10\ 000}$ and use the first two terms to find $\sqrt[4]{39}$. To how many significant figures is your answer accurate?

Exercise 11d (Miscellaneous)

1. Write down the sixth term of the expansion of $(3x + 2y)^{10}$ in ascending powers of x, and evaluate the term when $x = \frac{1}{2}$ and $y = \frac{1}{3}$.

2. (i) Expand $\left(2x + \dfrac{1}{2x}\right)^{5}$ in descending powers of x.

   (ii) Simplify $(\sqrt{2} + \sqrt{3})^4 - (\sqrt{2} - \sqrt{3})^4$.

3. Write down the expansion of $(a - b)^5$ and use the result to find the value of $(9\frac{1}{2})^5$ correct to the nearest 100.

4. Find the first three terms of the expansion of $(2 - \frac{1}{8}x)^{10}$ in ascending powers of x, and, by putting $x = \frac{1}{10}$, use your result to show that $1 \cdot 9875^{10} \simeq 961\frac{3}{4}$.

5. (i) Expand $(a + b)^{11}$ in descending powers of a as far as the fourth term.

   (ii) Find the middle term in the expansion of $(6x + \frac{1}{3}y)^{10}$.

   (iii) Find the constant term in the expansion of $\left(x^2 + \dfrac{2}{x}\right)^{9}$.

6. Expand $(x + 2)^5$ and $(x - 2)^4$. Obtain the coefficient of x^8 in the product of the expansions.

7. Obtain the expansion of $(x - 2)^2(1 - x)^6$ in ascending powers of x as far as the term in x^4.

8. (i) Expand $(2 + 3x)^4$ and simplify the coefficients.

   (ii) Obtain the first four terms in the expansion of $(1 + 2x + 3x^2)^6$ in ascending powers of x.

9. Find the first four terms in the expansions of

   (i) $(1 - x + 2x^2)^5$,   (ii) $(1 + x)^{-4}$,

in ascending powers of x.

10. (i) Write down the expansion of $(1 + x)^{-3}$ as far as the term in x^4, simplifying each term.

(ii) Write down the first four terms of the expansion of $(2 + \frac{1}{4}x)^{10}$ in ascending powers of x. Hence find the value of $2 \cdot 025^{10}$, correct to the nearest whole number.

11. (i) Find the middle term of the expansion of $(2x + 3)^8$, and the value of this term when $x = 1\frac{1}{2}$.

(ii) Find the first four terms in the expansion of $(1 - 2x)^{-2}$.

12. (i) Find the value of the fifth term in the expansion of $(\sqrt{2} + \sqrt{3})^8$.

(ii) Give the expansion of $(1 + x)^{\frac{1}{3}}$ up to and including the term in x^2. Hence, by putting $x = \frac{1}{8}$, calculate the cube root of 9, giving your answer correct to three decimal places.

13. Obtain the first four terms of the expansion of $(1 + 8x)^{\frac{1}{2}}$ in ascending powers of x. By putting $x = \frac{1}{100}$, obtain the value of $\sqrt{3}$, correct to five places of decimals.

14. Find the first three terms of the expansion of $(1 + 24x)^{\frac{1}{3}}$ in ascending powers of x. Obtain the value of $\sqrt[3]{2}$ correct to five places of decimals by putting $x = \frac{1}{1000}$.

15. Using the binomial theorem, expand $(1 + x)^{\frac{1}{3}}$ in ascending powers of x up to the term in x^4, simplifying each term. By putting $x = -\frac{1}{8}$ and using the terms up to x^2 only, show that $\sqrt[3]{7} \simeq 1 \cdot 913$.

16. Write down the first four terms of the expansions of $(1 + 5x)^{\frac{1}{5}}$, simplifying each term. Use the substitution $x = -\frac{1}{32}$ to show that $\sqrt[5]{27} \simeq 1 \cdot 933$.

17. If x is so small that its fourth and higher powers may be neglected, show that

$$\sqrt[4]{(1 + x)} + \sqrt[4]{(1 - x)} = a - bx^2,$$

and find the numbers a and b.

Hence by putting $x = \frac{1}{16}$ show that the sum of the fourth roots of 17 and of 15 is $3 \cdot 9985$ approximately.

18. Find the first four terms of the expansion of $\dfrac{x+3}{(1+x)^2}$ in ascending powers of x.

19. Show that, if x is small enough for its cube and higher powers to be neglected,

$$\sqrt{\frac{1-x}{1+x}} = 1 - x + \frac{x^2}{2}.$$

By putting $x = \frac{1}{8}$, show that $\sqrt{7} \simeq 2\frac{83}{128}$.

CHAPTER 12

THE GENERAL ANGLE AND PYTHAGORAS'
THEOREM

12.1. Consider a wheel which is free to rotate about a fixed axis, and suppose that one spoke is marked with a thin line of paint. If the wheel starts from rest and makes one revolution, the marked spoke turns through 360°, and if the wheel makes another revolution the spoke turns through 360° again. Thus we may say that the wheel has turned through a total of 720°, and by using angles greater than 360° the number of revolutions may be specified, as well as the position of the marked spoke.

Now on the x-axis of a graph the positive direction is usually taken to the right and the negative direction is opposite to this. Similarly, if the wheel mentioned above was rotating anti-clockwise, we could take that sense to be positive, and then a clockwise rotation would be considered negative. With axes OX, OY (see Fig. 12.1) angles measured from OX in an anti-clockwise sense are positive, and those measured in a clockwise sense are negative.

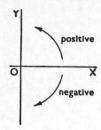

Fig. 12.1

Trigonometrical ratios of angles of any magnitude are required in connection with oscillating bodies and rotation about an axis, and in physics they arise in connection with such topics as alternating currents. But as the reader may only have had the six ratios defined for a limited range of angles, we will now give a general definition.

The axes divide the plane into four quadrants, and, as angles are measured in an anti-clockwise direction from OX, the quadrants are numbered as in Fig. 12.2. For the

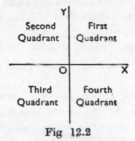

Fig 12.2

present, a point $P(x, y)$ and its coordinates will be given a suffix corresponding to the quadrant it lies in.

For an acute angle θ_1 (see Fig. 12.3),

$$\sin \theta_1 = \frac{y_1}{r}, \qquad \cos \theta_1 = \frac{x_1}{r}, \qquad \tan \theta_1 = \frac{y_1}{x_1}.$$

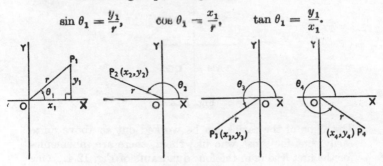

Fig. 12.3

Now $$\frac{\sin \theta_1}{\cos \theta_1} = \frac{y_1/r}{x_1/r} = \frac{y_1}{x_1} = \tan \theta_1,$$

so for an angle θ of *any* magnitude we shall define the six trigonometrical ratios as follows:

$$\sin \theta = \frac{y}{r}, \qquad \cos \theta = \frac{x}{r}, \qquad \tan \theta = \frac{\sin \theta}{\cos \theta},$$

$$\operatorname{cosec} \theta = \frac{1}{\sin \theta}, \qquad \sec \theta = \frac{1}{\cos \theta}, \qquad \cot \theta = \frac{1}{\tan \theta}.$$

For an angle θ_2 in the second quadrant (see Fig. 12.3), y_2 and r * are positive (abbreviated +ve) but x_2 is negative (abbreviated −ve), therefore

$$\sin \theta_2 \text{ is } +\text{ve}, \quad \cos \theta_2 \text{ is } -\text{ve}, \quad \tan \theta_2 \text{ is } -\text{ve}.$$

In the third quadrant r is positive, and x_3 and y_3 are both negative, hence

$$\sin \theta_3 \text{ is } -\text{ve}, \quad \cos \theta_3 \text{ is } -\text{ve}, \quad \tan \theta_3 \text{ is } +\text{ve}.$$

For an angle θ_4 in the fourth quadrant, r and x_4 are positive, and y_4 is negative, hence

$$\sin \theta_4 \text{ is } -\text{ve}, \quad \cos \theta_4 \text{ is } +\text{ve}, \quad \tan \theta_4 \text{ is } -\text{ve}.$$

These results can be summarized by writing which ratios are positive in each quadrant:

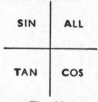

Fig. 12.4

The signs of the ratios can be worked out as above quite easily, but for those who like them, there are mnemonics for the first letters in the four quadrants of Fig. 12.4. One

---

* For the meaning to be ascribed to a negative value of r see §19.5.

such is All Silly Tom Cats. The signs of cosec θ, sec θ, cot θ are, of course, the same as their reciprocals.

A useful point to note is that angles for which OP is equally inclined to the positive or negative x-axis, have trigonometrical ratios of the same magnitude, their signs being determined as above. Thus the ratios of 150°, 210°, 330° are *numerically* the same as the ratios of 30°, since in each case the acute angle between OP and the x-axis is 30°:

$$\sin 150° = + \sin 30°, \qquad \sin 210° = - \sin 30°,$$
$$\cos 150° = - \cos 30°, \qquad \cos 210° = - \cos 30°,$$
$$\tan 150° = - \tan 30°, \qquad \tan 210° = + \tan 30°,$$

$$\sin 330° = - \sin 30°,$$
$$\cos 330° = + \cos 30°,$$
$$\tan 330° = - \tan 30°.$$

Qu. 1. Express in terms of the trigonometrical ratios of acute angles:

(i) sin 170°,	(ii) tan 300°,
(iii) cos 200°,	(iv) sin (− 50°),
(v) cos (− 20°),	(vi) sin 325°,
(vii) tan (− 140°),	(viii) cos 164°,
(ix) cosec 230°,	(x) tan 143°,
(xi) cos (− 130°),	(xii) sin 250°,
(xiii) tan (− 50°),	(xiv) cot 200°,
(xv) cos 293°,	(xvi) sin (− 230°),
(xvii) sec 142°,	(xviii) cot 156°,
(xix) cosec (− 53°),	(xx) sec (− 172°).

Graphs of sin θ, cos θ, tan θ

12.2. It is instructive to draw the graphs of sin θ, cos θ, and tan θ. Fig. 12.5 shows how the graph of sin θ may be drawn from the definition. Construct a circle of unit radius, then sin $\theta = y$. Dotted lines show this for $\theta = 30°$, 60°, 90°, and the rest of the figure is drawn similarly.

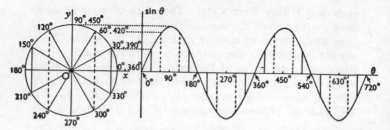

Fig. 12.5

It will be seen that the graph of sin θ repeats itself at intervals of 360°. (That this is so should be clear from the way it was drawn, because points on the graph separated by 360° correspond to the same point on the circle.) If a function repeats itself at regular intervals, like sin θ, it is called a **periodic** function, and the interval is called its **period**.

The graph of cos θ may be drawn in a similar way to that

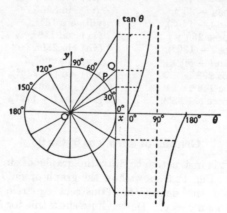

Fig. 12.6

of sin θ. In this case, since $\cos \theta = x/r$, the values of x are used instead of y.

The graph of tan θ may also be drawn from a unit circle, but in this case a tangent is drawn at the point $(1, 0)$. (See Fig. 12.6.) If P is any point on the circle, and OP meets the tangent at Q, then the y-coordinate of Q is equal to tan θ.

Qu. 2. Complete the graph of tan θ up to $\theta = 720°$.

Qu. 3. What are the periods of $\cos \theta$ and $\tan \theta$?

Trigonometrical ratios of 30°, 45°, 60°

12.3. The trigonometrical ratios of 30°, 45°, and 60° are frequently needed, and they may be obtained from two figures. Fig. 12.7 represents an equilateral triangle with an altitude constructed. The sides of the triangle are 2 units, and so, by Pythagoras' theorem, the altitude is $\sqrt{3}$ units. The ratios of 30° and 60° may now be read off. Fig. 12.8 represents a right-angled isosceles triangle with two sides of unit length. By Pythagoras' theorem the hypotenuse is $\sqrt{2}$ units, and so the ratios of 45° may be read off.

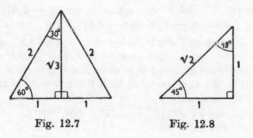

Fig. 12.7   Fig. 12.8

Qu. 4. Write down the values of (i) sin 30°, (ii) cos 30°, (iii) cos 45°, (iv) tan 30°, (v) sec 60°, (vi) cosec 60°, (vii) tan 45°, (viii) cosec 45°.

12.4. Equations in algebra have only a finite number of solutions, but in many cases trigonometrical equations have an unlimited number. For instance, the equation $\sin \theta = 0$ is satisfied by $\theta = 0°$, $\pm 180°$, $\pm 360°$, $\pm 540°$ and so on, indefinitely. In this book it will be specified for what range of values the roots are required.

Example 1. *Solve the equation* $\sin \theta = -\frac{1}{2}$ *for values of* θ *from* $0°$ *to* $360°$ *inclusive.*

The acute angle whose sine is $\frac{1}{2}$ is $30°$ and Fig. 12.9

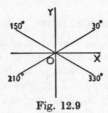

Fig. 12.9

indicates the angles between $0°$ and $360°$ whose sines are $\pm \frac{1}{2}$. But $\sin \theta$ is negative only in the third and fourth quadrants. Therefore the roots of the equation in the required range are $210°$ and $330°$.

Example 2. *Solve the equation* $\cos 2\theta = 0 \cdot 6428$, *for values of* θ *between* $-180°$ *and* $+180°$.

[Note that since θ must lie between $-180°$ and $+180°$, 2θ may lie between $-360°$ and $+360°$.]

From tables it is found that the acute angle whose cosine is $0 \cdot 6428$ is $50°$, and since $\cos 2\theta$ is positive only in the first and fourth quadrants

$$2\theta = -310°, \quad -50°, \quad 50°, \quad 310°.$$
$$\therefore \ \theta = -155°, \quad -25°, \quad 25°, \quad 155°.$$

Example 3. *Solve the equation** $2 \sin^2 \theta = \sin \theta$, *for values of* θ *from* $0°$ *to* $360°$ *inclusive.*

* In order to avoid brackets $(\sin \theta)^2$ is written $\sin^2 \theta$.

[This equation is a quadratic equation for $\sin \theta$, and may be solved by factorization.]

$$2 \sin^2 \theta - \sin \theta = 0.$$

$$\therefore \ \sin \theta (2 \sin \theta - 1) = 0$$

$$\therefore \ \sin \theta = 0 \quad \text{or} \quad \sin \theta = \tfrac{1}{2}.$$

If $\sin \theta = 0$, $\theta = 0°, 180°, 360°$. If $\sin \theta = \tfrac{1}{2}$, $\theta = 30°$, $150°$.

Therefore the roots of the equation, from $0°$ to $360°$ inclusive are $0°, 30°, 150°, 180°$, and $360°$.

(Note that if we had divided both sides of the equation by $\sin \theta$, giving $2 \sin \theta = 1$, we should have lost some of the roots.)

Example 4. *Solve the equation* $\tan \theta = 2 \sin \theta$, *for values of θ from $0°$ to $360°$ inclusive.*

[Equations are often solved by factorization, so look for a common factor.]

Remembering that $\tan \theta = \dfrac{\sin \theta}{\cos \theta}$ we may write

$$\frac{\sin \theta}{\cos \theta} = 2 \sin \theta.$$

$$\therefore \ 2 \sin \theta \cos \theta = \sin \theta.$$

$$\therefore \ 2 \sin \theta \cos \theta - \sin \theta = 0.$$

$$\therefore \ \sin \theta (2 \cos \theta - 1) = 0.$$

$$\therefore \ \sin \theta = 0 \quad \text{or} \quad \cos \theta = \tfrac{1}{2}.$$

If $\sin \theta = 0$, $\theta = 0°, 180°, 360°$. If $\cos \theta = \tfrac{1}{2}$, $\theta = 60°$, $300°$.

Therefore the required values of θ are $0°, 60°, 180°, 300°$, and $360°$.

Exercise 12a

1. Write down the values of the following, leaving surds in your answers:

  (i) $\cos 270°$,
  (ii) $\sin 540°$,
  (iii) $\cos(-180°)$,
  (iv) $\tan 135°$,
  (v) $\sin 150°$,
  (vi) $\cos 210°$,
  (vii) $\tan 120°$,
  (viii) $\cos(-30°)$,
  (ix) $\sin(-120°)$,
  (x) $\sin 405°$,
  (xi) $\cos(-135°)$,
  (xii) $\sin 225°$,
  (xiii) $\tan(-60°)$,
  (xiv) $\sin(-270°)$,
  (xv) $\tan 210°$.

2. Sketch the graph of $\sin \theta$, for values of θ from $-360°$ to $360°$.

3. Sketch the graph of $\cos \theta$, for values of θ from $0°$ to $720°$, and state its period.

4. Draw the graph of $\tan \theta$, for values of θ from $0°$ to $720°$. (This has been started in Fig. 12.6.) What is the period of $\tan \theta$?

5. Sketch the graphs of (i) $\cos 2\theta$, (ii) $\sin \frac{1}{2}\theta$, (iii) $\sin \frac{3}{2}\theta$, (iv) $\cos(\theta + 60°)$, (v) $\sin(\theta - 45°)$, for values of θ from $0°$ to $360°$, stating the period of each.

6. Find the values of θ from $-180°$ to $+180°$, inclusive, which satisfy the following equations:

  (i) $\cos \theta = -\frac{1}{2}$,
  (ii) $\tan \theta = 1$,
  (iii) $\operatorname{cosec} \theta = 2$,
  (iv) $\sin \theta = -0.7660$,
  (v) $\cos \theta = 0.6$,
  (vi) $\tan \theta = -\sqrt{3}$,
  (vii) $\sin \theta = \cos 150°$,
  (viii) $\cos \theta = \tan 135°$,
  (ix) $\sec \theta = \tan 110°$,
  (x) $\cos \theta = \sin 140°$,
  (xi) $\cos(\theta + 60°) = 0.5$,
  (xii) $\sin(\theta - 30°) = \sqrt{3}/2$,
  (xiii) $\tan(70° - \theta) = 2$,
  (xiv) $\cos(40° - \theta) = 0.2$.

7. Solve the following equations for values of θ from $0°$ to $360°$, inclusive:

  (i) $\sin^2 \theta = \frac{1}{4}$,
  (ii) $\tan^2 \theta = \frac{1}{3}$,
  (iii) $\sin 2\theta = \frac{1}{2}$,
  (iv) $\tan 2\theta = -1$,
  (v) $\cos 3\theta = \sqrt{3}/2$,
  (vi) $\sin 3\theta = -1$,
  (vii) $\sin^2 2\theta = 1$,
  (viii) $\sec 2\theta = 3$,
  (ix) $\tan^2 3\theta = 1$,
  (x) $4 \cos 2\theta = 1$,

(xi) $\sin (2\theta + 30°) = 0\cdot8$,   (xii) $\tan (3\theta - 45°) = \frac{1}{2}$,

(xiii) $\cos (\frac{1}{2}\theta - 40°) = 0\cdot9$,   (xiv) $\sec (2\theta - 60°) = -3$.

8. Solve the following equations for values of θ from $-180°$ to $+180°$, inclusive:

(i) $\tan^2 \theta + \tan \theta = 0$,   (ii) $2 \cos^2 \theta = \cos \theta$,

(iii) $3 \sin^2 \theta + \sin \theta = 0$,

(iv) $2 \sin^2 \theta - \sin \theta - 1 = 0$,

(v) $2 \cos^2 \theta + 3 \cos \theta + 1 = 0$,

(vi) $4 \cos^3 \theta = \cos \theta$,   (vii) $\tan \theta = \sin \theta$,

(viii) $\sec \theta = 2 \cos \theta$,   (ix) $\cot \theta = 5 \cos \theta$,

(x) $4 \sin^2 \theta = 3 \cos^2 \theta$,   (xi) $3 \cos \theta = 2 \cot \theta$,

(xii) $\tan \theta = 4 \cot \theta + 3$,

(xiii) $5 \sin \theta + 6 \operatorname{cosec} \theta = 17$,

(xiv) $3 \cos \theta + 2 \sec \theta + 7 = 0$.

9. Write down the maximum and minimum values of the following expressions, giving the smallest positive or zero value of θ for which they occur:

(i) $\sin \theta$,   (ii) $3 \cos \theta$,   (iii) $2 \cos \frac{1}{2}\theta$,

(iv) $-\frac{1}{2} \sin 2\theta$,   (v) $1 - 2 \sin \theta$,   (vi) $3 + 2 \cos 3\theta$,

(vii) $\dfrac{1}{2 + \sin \theta}$,   (viii) $\dfrac{1}{4 - 3 \cos \theta}$,   (ix) $\sec \frac{3}{4}\theta$,

(x) $\tan^2 \theta$,   (xi) $\dfrac{1}{1 + \operatorname{cosec} \theta}$,

(xii) $\dfrac{2}{3 - 2 \cot \theta}$,   (xiii) $\dfrac{\cos \theta}{\cos \theta + \sin \theta}$

10. State, with reasons, which of the following equations have no roots:

(i) $2 \sin \theta = 3$,   (ii) $\sin \theta + \cos \theta = 0$,

(iii) $\sin \theta + \cos \theta = 2$,   (iv) $3 \sin \theta + \operatorname{cosec} \theta = 0$,

(v) $4 \operatorname{cosec}^2 \theta - 1 = 0$,   (vi) $\operatorname{cosec} \theta = \sin \theta$,

(vii) $\sec \theta = \sin \theta$.

11. Sketch on the same axes, for values of θ from $-360°$ to $360°$, the graphs of (i) $\sin \theta$, $\operatorname{cosec} \theta$; (ii) $\cos \theta$, $\sec \theta$; (iii) $\tan \theta$, $\cot \theta$.

12.5. The reader who has drawn the graphs of $y = \sin \theta$ and $y = \cos \theta$ may have noticed that they are the same, except for the positions of the y-axes relative to the curves.

Figure 12.10 suggests that, for any angle α,

$$\cos \alpha = \sin (90° + \alpha),$$

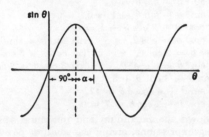

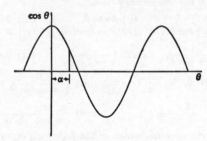

Fig. 12.10

and other relationships of this sort may be found from the graphs. Some people find the graphs help them to remember such relationships, but now it will be shown how they may be obtained from first principles.

For any value of θ, in the notation of §12.1 we have by definition

$$\sin \theta = \frac{y}{r}, \qquad \cos \theta = \frac{x}{r}.$$

Consider:

(i) ratios of $-\theta$. In Fig. 12.3, p. 257, the angle $-\theta$ is obtained by replacing (x, y) by $(x, -y)$,

$$\therefore \sin(-\theta) = -\frac{y}{r} = -\sin\theta, \quad \cos(-\theta) = \frac{x}{r} = \cos\theta.$$

(ii) ratios of $180° - \theta$. Replace (x, y) by $(-x, y)$, hence

$$\sin(180° - \theta) = \frac{y}{r} = \sin\theta, \cos(180° - \theta) = -\frac{x}{r} = -\cos\theta.$$

(iii) ratios of $180° + \theta$. Replace (x, y) by $(-x, -y)$, hence

$$\sin(180° + \theta) = -\frac{y}{r} = -\sin\theta,$$

$$\cos(180° + \theta) = -\frac{x}{r} = -\cos\theta.$$

[Note that in all these cases above, OP is inclined at an angle θ to the positive or negative x-axis, the ratios of these angles have the same magnitude as those of θ, and their signs are determined as on page 258 if θ is acute.]

(iv) ratios of $90° - \theta$. Replace (x, y) by (y, x), hence

$$\sin(90° - \theta) = \frac{x}{r} = \cos\theta, \quad \cos(90° - \theta) = \frac{y}{r} = \sin\theta.$$

(v) ratios of $90° + \theta$. Replace (x, y) by $(-y, x)$, hence

$$\sin(90° + \theta) = \frac{x}{r} = \cos\theta, \quad \cos(90° + \theta) = -\frac{y}{r} = -\sin\theta.$$

Qu. 5. Express the following in terms of the trigonometrical ratios of θ:

(i) $\tan(90° - \theta)$,	(ii) $\operatorname{cosec}(180° - \theta)$,
(iii) $\sec(90° + \theta)$,	(iv) $\cot(90° + \theta)$,
(v) $\sec(-\theta)$,	(vi) $\operatorname{cosec}(180° + \theta)$,
(vii) $\cos(270° - \theta)$,	viii) $\sin(360° + \theta)$,
(ix) $\tan(-\theta)$,	(x) $\sin(\theta - 90°)$,
(xi) $\cos(\theta - 180°)$,	(xii) $\sec(270° + \theta)$.

Pythagoras' theorem

12.6. The reader will be familiar with Pythagoras' theorem, and will have found that it is a very useful one. In trigonometry it retains its importance and provides relations between trigonometrical ratios.

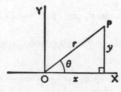

Fig. 12.11

In Fig. 12.11, the triangle is right-angled and so, by Pythagoras' theorem,

$$x^2 + y^2 = r^2.$$

But $\cos \theta = x/r$ and $\sin \theta = y/r$, so we divide by r^2 obtaining

$$\frac{x^2}{r^2} + \frac{y^2}{r^2} = 1.$$

$$\therefore \ \cos^2 \theta + \sin^2 \theta \equiv 1.$$

(If P is not in the first quadrant, OP^2 is still $x^2 + y^2$ by the distance formula of §1.8 and the proof continues as before.)

The $\equiv$ symbol is used to stress that the relationship is an identity, i.e. it holds for *all* values of θ.

Two similar identities can be deduced from this. Dividing through by $\cos^2 \theta$,

$$1 + \frac{\sin^2 \theta}{\cos^2 \theta} \equiv \frac{1}{\cos^2 \theta},$$

but $\tan \theta = \sin \theta / \cos \theta$ and $\sec \theta = 1/\cos \theta$, therefore

$$1 + \tan^2 \theta \equiv \sec^2 \theta.$$

Dividing the original identity by $\sin^2 \theta$,

$$\frac{\cos^2 \theta}{\sin^2 \theta} + 1 \equiv \frac{1}{\sin^2 \theta},$$

but $\cos \theta / \sin \theta = \cot \theta$ and $1/\sin \theta = \operatorname{cosec} \theta$, therefore

$$\cot^2 \theta + 1 \equiv \operatorname{cosec}^2 \theta.$$

Historical note. The equivalent of the identity

$$\cos^2 \theta + \sin^2 \theta = 1$$

is found in the *Syntaxis* written during the first century A.D. by Claudius Ptolemy. Instead of sines and cosines, he used chords. (If a chord subtends an angle 2θ at the centre of a circle, the ratio of the chord to the diameter of the circle is $\sin \theta$.)

Example 5. *Solve the equation* $1 + \cos \theta = 2 \sin^2 \theta$, *for values of* θ *between* $0°$ *and* $360°$.

[The square in the right-hand side indicates that the equation is a quadratic, and to solve it, we must write it in terms of either $\cos \theta$ or $\sin \theta$.] We know that

$$\cos^2 \theta + \sin^2 \theta = 1,$$

hence
$$\sin^2 \theta = 1 - \cos^2 \theta;$$

so substituting $2 - 2 \cos^2 \theta$ for $2 \sin^2 \theta$, we obtain

$$1 + \cos \theta = 2 - 2 \cos^2 \theta.$$

This quadratic for $\cos \theta$ is solved by factorization:

$$2 \cos^2 \theta + \cos \theta - 1 = 0,$$
$$\therefore \ (2 \cos \theta - 1)(\cos \theta + 1) = 0.$$
$$\therefore \ \cos \theta = \tfrac{1}{2} \ \text{ or } \ -1.$$

If $\cos \theta = \tfrac{1}{2}$, $\theta = 60°$, $300°$. If $\cos \theta = -1$, $\theta = 180°$.

Therefore the roots of the equation between $0°$ and $360°$ are $60°$, $180°$, and $300°$.

Example 6. *Simplify* $\dfrac{1}{\sqrt{(x^2 - a^2)}}$ *when* $x = a \operatorname{cosec} \theta$.

Substituting $x = a \operatorname{cosec} \theta$, we obtain

$$\frac{1}{\sqrt{(a^2 \operatorname{cosec}^2 \theta - a^2)}}.$$

But the $\operatorname{cosec}^2 \theta$ in the denominator suggests the use of the identity $\cot^2 \theta + 1 = \operatorname{cosec}^2 \theta$. With this the expression $(a^2 \operatorname{cosec}^2 \theta - a^2)$ may be simplified, giving

$$a^2 \operatorname{cosec}^2 \theta - a^2 = a^2(\cot^2 \theta + 1) - a^2 = a^2 \cot^2 \theta.$$

Thus the original expression becomes

$$\frac{1}{\sqrt{(a^2 \cot^2 \theta)}} = \frac{1}{a \cot \theta} = \frac{1}{a} \tan \theta.$$

Example 7. *Eliminate* θ *from the equations* $x = a \sin \theta$, $y = b \tan \theta$.

[Since $\sin \theta$ and $\tan \theta$ are the reciprocals of $\operatorname{cosec} \theta$ and $\cot \theta$ we use the identity $\operatorname{cosec}^2 \theta = \cot^2 \theta + 1$.]

$$\operatorname{cosec} \theta = \frac{a}{x} \quad \text{and} \quad \cot \theta = \frac{b}{y}.$$

Substituting into the identity

$$\operatorname{cosec}^2 \theta = \cot^2 \theta + 1,$$
$$\frac{a^2}{x^2} = \frac{b^2}{y^2} + 1.$$

Exercise 12b

1. If $s = \sin \theta$, simplify:

(i) $\sqrt{(1 - s^2)}$,   (ii) $\dfrac{s}{\sqrt{(1 - s^2)}}$,   (iii) $\dfrac{1 - s^2}{s}$.

2. If $c = \cos \theta$, simplify:

(i) $\sqrt{(1 - c^2)}$,   (ii) $\dfrac{\sqrt{(1 - c^2)}}{c}$,   (iii) $\dfrac{c}{1 - c^2}$.

3. If $t = \tan \theta$, simplify:

    (i) $\sqrt{(1 + t^2)}$,   (ii) $t(1 + t^2)$,   (iii) $\dfrac{t}{\sqrt{(1 + t^2)}}$.

4. If $c = \operatorname{cosec} \theta$, simplify:

    (i) $\sqrt{(c^2 - 1)}$,   (ii) $\dfrac{\sqrt{(c^2 - 1)}}{c}$,   (iii) $\dfrac{c}{c^2 - 1}$.

5. If $x = a \sin \theta$, simplify:

    (i) $a^2 - x^2$,   (ii) $\dfrac{1}{\sqrt{(a^2 - x^2)}}$,   (iii) $\dfrac{a^2 - x^2}{x}$.

6. If $y = b \cot \theta$, simplify:

    (i) $b^2 + y^2$,   (ii) $y\sqrt{(b^2 + y^2)}$,   (iii) $\dfrac{y}{b^2 + y^2}$.

7. If $z = a \sec \theta$, simplify:

    (i) $z^2 - a^2$,   (ii) $\dfrac{1}{\sqrt{(z^2 - a^2)}}$,   (iii) $\dfrac{\sqrt{(z^2 - a^2)}}{z}$.

In Nos. 8–13, solve the equations, giving values of θ from $0°$ to $360°$ inclusive.

8. $3 - 3 \cos \theta = 2 \sin^2 \theta$.   **9.** $\cos^2 \theta + \sin \theta + 1 = 0$.
10. $\sec^2 \theta = 3 \tan \theta - 1$.   **11.** $\operatorname{cosec}^2 \theta = 3 + \cot \theta$.
12. $3 \tan^2 \theta + 5 = 7 \sec \theta$.   **13.** $2 \cot^2 \theta + 8 = 7 \operatorname{cosec} \theta$.

14. If $\sin \theta = \frac{3}{5}$, find without using tables the values of (i) $\cos \theta$, (ii) $\tan \theta$.

15. If $\cos \theta = -\frac{8}{17}$, and θ is obtuse, find without using tables the values of (i) $\sin \theta$, (ii) $\cot \theta$.

16. If $\tan \theta = \frac{7}{24}$ and θ is reflex, find without using tables the values of (i) $\sec \theta$, (ii) $\sin \theta$.

Prove the following identities:

17. $\tan \theta + \cot \theta = \dfrac{1}{\sin \theta \cos \theta}$.
18. $\operatorname{cosec} \theta + \tan \theta \sec \theta = \operatorname{cosec} \theta \sec^2 \theta$.
19. $\sec^2 \theta - \operatorname{cosec}^2 \theta = \tan^2 \theta - \cot^2 \theta$.
20. $\cos^4 \theta - \sin^4 \theta = \cos^2 \theta - \sin^2 \theta$.
21. $(\sec \theta + \tan \theta)(\sec \theta - \tan \theta) = 1$.
22. $2 \cos^2 \theta - 1 = 1 - 2 \sin^2 \theta = \cos^2 \theta - \sin^2 \theta$.
23. $\sec^2 \theta + \operatorname{cosec}^2 \theta = \sec^2 \theta \operatorname{cosec}^2 \theta$.

24. $\sec^4 \theta - \operatorname{cosec}^4 \theta = \dfrac{\sin^2 \theta - \cos^2 \theta}{\cos^4 \theta \sin^4 \theta}.$

25. $\dfrac{1}{\tan^2 \theta + 1} + \dfrac{1}{\cot^2 \theta + 1} = 1.$

26. $(\sec^2 \theta - 1)(\operatorname{cosec}^2 \theta - 1) = 1.$

27. $\sqrt{(\sec^2 \theta - 1)} + \sqrt{(\operatorname{cosec}^2 \theta - 1)} = \sec \theta \operatorname{cosec} \theta.$

28. $\sqrt{(\sec^2 \theta - \tan^2 \theta)} + \sqrt{(\operatorname{cosec}^2 \theta - \cot^2 \theta)} = 2.$

29. $\dfrac{1 - \cos^2 \theta}{\sec^2 \theta - 1} = 1 - \sin^2 \theta.$

30. $\dfrac{\sec \theta - \operatorname{cosec} \theta}{\tan \theta - \cot \theta} = \dfrac{\tan \theta + \cot \theta}{\sec \theta + \operatorname{cosec} \theta}.$

31. $\dfrac{\cos \theta}{\sqrt{(1 + \tan^2 \theta)}} + \dfrac{\sin \theta}{\sqrt{(1 + \cot^2 \theta)}} = 1.$

32. $\dfrac{\operatorname{cosec}^2 \theta - 1}{\cos^2 \theta} + \dfrac{1}{1 - \sin^2 \theta} = \sec^2 \theta \operatorname{cosec}^2 \theta.$

33. $\dfrac{1 - \tan^2 \theta}{1 + \tan^2 \theta} = 1 - 2 \sin^2 \theta.$

34. $\dfrac{2 \tan \theta}{1 + \tan^2 \theta} = 2 \sin \theta \cos \theta.$

35. $(\cot \theta + \operatorname{cosec} \theta)^2 = \dfrac{1 + \cos \theta}{1 - \cos \theta}.$

36. If $\tan^2 \alpha - 2 \tan^2 \beta = 1$, show that $2 \cos^2 \alpha - \cos^2 \beta = 0$.

37. If $u = \dfrac{1 + \sin \theta}{\cos \theta}$, prove that $\dfrac{1}{u} = \dfrac{1 - \sin \theta}{\cos \theta}$ and deduce formulae for $\sin \theta$, $\cos \theta$, $\tan \theta$ in terms of u.

Eliminate θ from the following equations:

38. $x = a \cos \theta,\ y = b \sin \theta.$

39. $x = a \cot \theta,\ y = b \operatorname{cosec} \theta.$

40. $x = a \tan \theta,\ y = b \cos \theta.$

41. $x = 1 - \sin \theta,\ y = 1 + \cos \theta.$

42. $x = a \sec \theta,\ y = b + c \cos \theta.$

43. $x = a \operatorname{cosec} \theta,\ y = b \sec \theta.$

44. $x = 1 + \tan \theta,\ y = \cos \theta.$

45. $x = \sin \theta + \cos \theta,\ y = \sin \theta - \cos \theta.$

46. $x = \sec \theta + \tan \theta,\ y = \sec \theta - \tan \theta.$

47. $x = \sin \theta + \tan \theta,\ y = \sin \theta - \tan \theta.$

48. If $a \sin \theta = p - b \cos \theta,\ b \sin \theta = q + a \cos \theta,$ show that
$a^2 + b^2 = p^2 + q^2.$

Exercise 12c (Miscellaneous)

1. Express in terms of the ratios of acute angles:

   (i) $\cos 205°,$ (ii) $\tan 153°,$ (iii) $\sec 309°,$
   (iv) $\sin (-215°),$ (v) $\cot 406°,$ (vi) $\operatorname{cosec} 684°.$

2. Find the values of the following, leaving surds in your answers:

   (i) $\sin 270°,$ (ii) $\cos 150°,$ (iii) $\cot 210°,$
   (iv) $\cos 315°,$ (v) $\operatorname{cosec} 240°,$ (vi) $\sec 585°,$
   (vii) $\tan (-225°),$ (viii) $\sin (-690°),$ (ix) $\cos (-300°).$

3. Solve the following equations for values of θ from $0°$ to $360°$ inclusive:

   (i) $2 \sin \theta = 1,$ (ii) $\tan \theta + 1 = 0,$
  (iii) $\cos \theta = 0.8,$ (iv) $\tan 2\theta = 1,$
   (v) $\sec 2\theta = 4,$ (vi) $\sin \tfrac{1}{2}\theta = \tfrac{1}{2},$
  (vii) $3 \cos (\theta - 10°) = 1,$ (viii) $\sin (\theta + 30°) = 0.7,$
   (ix) $\cot \tfrac{1}{2}\theta = 0.9.$

4. Solve the following equations for values of θ from $-180°$ to $+180°$ inclusive:

   (i) $2 \sin^2 \theta + \sin \theta = 0,$ (ii) $3 \cos^2 \theta = 2 \sin \theta \cos \theta,$
  (iii) $2 \sin^2 \theta + 1 = 3 \sin \theta,$ (iv) $3 \cos^2 \theta = 7 \cos \theta + 6,$
   (v) $4 \sin \theta + \operatorname{cosec} \theta = 4,$ (vi) $10 \cos \theta + 1 = 2 \sec \theta,$
  (vii) $\tan \theta + 2 \cot \theta = 3,$
 (viii) $10 \sin \theta \cos \theta - 5 \sin \theta + 4 \cos \theta = 2.$

5. Find the maximum and minimum values of the following functions of θ. Give the smallest non-negative values of θ for which they occur.

   (i) $3 + 2 \sin \theta,$ (ii) $1 - 3 \cos \theta,$ (iii) $4 \sin \tfrac{2}{3}\theta,$
  (iv) $3 \sin^2 \tfrac{1}{2}\theta,$ (v) $\dfrac{1}{2 + 3 \cos \theta},$ (vi) $\dfrac{1}{3 - 2 \sin 2\theta}.$

6. Express in terms of the trigonometrical ratios of θ:

   (i) $\cot (90° - \theta),$ (ii) $\sin (90° + \theta),$
  (iii) $\cos (270° + \theta),$ (iv) $\tan (90° + \theta),$

(v) cosec $(360° - \theta)$,   (vi) sec $(180° - \theta)$,

(vii) sin $(\theta - 180°)$,   (viii) tan $(- \theta)$,

(ix) cos $(450° - \theta)$.

7. If $s = \sin \theta$ and $c = \cos \theta$, simplify:

(i) $\dfrac{1 - s^2}{1 - c^2}$,   (ii) $\dfrac{sc}{\sqrt{(1 - s^2)}}$,   (iii) $\dfrac{s}{c^2 - 1}$,

(iv) $\dfrac{c^4 - s^4}{c^2 - s^2}$,   (v) $\dfrac{s\sqrt{(1 - s^2)}}{c\sqrt{(1 - c^2)}}$,   (vi) $\dfrac{c}{s} + \dfrac{s}{c}$.

8. Solve the following equations for values of θ from $0°$ to $360°$ inclusive:

(i) $2 \cos^2 \theta + \sin \theta = 1$,   (ii) $5 \cos \theta = 2(1 + 2 \sin^2 \theta)$,

(iii) $2 \tan^2 \theta + \sec \theta = 1$,

(iv) $4 \cot^2 \theta + 39 = 24 \operatorname{cosec} \theta$,

(v) $5 \sec \theta - 2 \sec^2 \theta = \tan^2 \theta - 1$,

(vi) $\sec \theta + 3 = \cos \theta + \tan \theta (2 + \sin \theta)$.

(vii) $3 \sin^2 \theta - \sin \theta \cos \theta - 4 \cos^2 \theta = 0$.

9. Find, without using tables, the values of:

(i) $\sin \theta$, $\tan \theta$, if $\cos \theta = \frac{4}{5}$ and θ is acute.

(ii) $\sec \theta$, $\sin \theta$, if $\tan \theta = -\frac{5}{12}$ and θ is obtuse.

(iii) $\cos \theta$, $\cot \theta$, if $\sin \theta = \frac{15}{17}$ and θ is acute.

(iv) $\sin \theta$, $\sec \theta$, if $\cot \theta = \frac{20}{21}$ and θ is reflex.

Prove the following identities:

10. $\sec \theta + \operatorname{cosec} \theta \cot \theta = \sec \theta \operatorname{cosec}^2 \theta$.

11. $\sin^2 \theta (1 + \sec^2 \theta) = \sec^2 \theta - \cos^2 \theta$.

12. $\dfrac{1 - \cos \theta}{\sin \theta} = \dfrac{1}{\operatorname{cosec} \theta + \cot \theta}$.

13. $\dfrac{\tan \theta + \cot \theta}{\sec \theta + \operatorname{cosec} \theta} = \dfrac{1}{\sin \theta + \cos \theta}$.

14. $\sec^2 \theta = \dfrac{\operatorname{cosec} \theta}{\operatorname{cosec} \theta - \sin \theta}$.

15. $\dfrac{1 + \sin \theta}{1 - \sin \theta} = (\sec \theta + \tan \theta)^2$.

16. $\sec \theta - \sin \theta = \dfrac{\tan^2 \theta + \cos^2 \theta}{\sec \theta + \sin \theta}$.

17. $\dfrac{1 - \sin \theta + \cos \theta}{1 - \sin \theta} = \dfrac{1 + \sin \theta + \cos \theta}{\cos \theta}$.

Eliminate θ from the following pairs of equations:

18. $x = a \sec \theta, \ y = b \tan \theta.$
19. $x = 1 - \cos \theta, \ y = 1 + \sin \theta.$
20. $x = a \cot \theta, \ y = b \sin \theta.$
21. $x = a \sec \theta, \ y = b \cot \theta.$
22. $x = a \tan \theta, \ y = b \sin \theta.$
23. $x = \operatorname{cosec} \theta - \cot \theta, \ y = \operatorname{cosec} \theta + \cot \theta.$
24. $x = \sin \theta + \cos \theta, \ y = \tan \theta.$
25. $x = \cos \theta, \ y = \operatorname{cosec} \theta - \cot \theta.$

26. Plot the graph of $y = \sin x + \cos x$ for values of x from $-180°$ to $180°$ at intervals of $30°$. Find from your graph the maximum and minimum values of $\sin x + \cos x$, and the values of x for which they occur.

27. Plot the graph of $y = \sin x + 2 \cos x$ for values of x from $-180°$ to $180°$ at intervals of $30°$. Find from your graph the roots of the equation $\sin x + 2 \cos x = 1$ which lie in this range.

28. Plot the graphs of $y = \sin 2x$ and $y = \cos 3x$ on the same axes for values of x from $0°$ to $90°$. Find from your graph the root of the equation $\sin 2x = \cos 3x$ which lies in this range.

29. Solve the simultaneous equations

$$\sin (x + y) = \frac{1}{\sqrt{2}},$$
$$\cos 2x = -\tfrac{1}{2},$$

for values of x, y from $0°$ to $360°$ inclusive.

CHAPTER 13

COMPOUND ANGLES

13.1. Place a rectangular piece of cardboard PQRS in a vertical plane with two edges horizontal, and then turn it through an angle B. (See Fig. 13.1.) Take the diagonal PR as the unit of length and let angle RPQ be A.

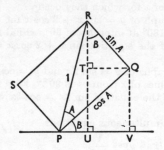

Fig. 13.1

What is the height of R above P?

One way to find this out is to drop a perpendicular RU from R to the horizontal through P, then from the triangle RPU, $RU = \sin (A + B)$.

Alternatively, since $RQ = \sin A$, $PQ = \cos A$ and angle QRU $= B$, the height of R above P can be found in two parts. First, the height of R above Q, $RT = \sin A \cos B$ (from triangle RTQ). Secondly, the height of Q above P, $QV = \cos A \sin B$ (from triangle PQV). Thus, equating the height of R above P obtained in the two ways,

$$\sin (A + B) = \sin A \cos B + \cos A \sin B.$$

276

How far to the right of P *is* R*?*

In triangle RPU, PU = cos (A + B).

Alternatively, the distance of Q to the right of P, PV = cos A cos B (from triangle PQV), and the distance of R to the left of Q, QT = sin A sin B (from triangle RTQ). So, equating the distance of R to the right of P obtained in these two ways,

$$\cos (A + B) = \cos A \cos B - \sin A \sin B.$$

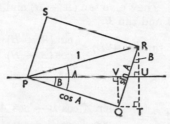

Fig 13.2

Consider now what happens if PQ is tilted through an angle B below the horizontal, as in Fig. 13.2. The height of R above P is now sin (A − B). R is a distance sin A cos B above Q, but Q is a distance cos A sin B below P, therefore

$$\sin (A - B) = \sin A \cos B - \cos A \sin B.$$

Further, R is a distance cos (A − B) to the right of P. Q is a distance cos A cos B to the right of P, but R is now a distance sin A sin B to the right of Q, therefore

$$\cos (A - B) = \cos A \cos B + \sin A \sin B.$$

The four identities just obtained have many applications apart from their use in trigonometry. They, or identities which will be derived from them, are needed in calculus, coordinate geometry and mechanics. Some applications are found in Chapters 16 and 19.

The outline of a good general proof of the identities (which may be taken on second reading) may be found in §13.5. For the present it will be assumed that they hold for all values of A and B.

Historical note. The equivalents of the identities for $\cos (A + B)$ and $\sin (A - B)$ were known to Ptolemy. The latter may be deduced from Ptolemy's theorem. (See Exercise 13e, No. 49.)

13.2. Two more identities will be deduced from the four just obtained. They give $\tan (A + B)$ and $\tan (A - B)$ in terms of $\tan A$ and $\tan B$.

$$\tan (A + B) = \frac{\sin (A + B)}{\cos (A + B)}.$$

Therefore, using the formulae for $\sin (A + B)$ and $\cos (A + B)$,

$$\tan (A + B) = \frac{\sin A \cos B + \cos A \sin B}{\cos A \cos B - \sin A \sin B}.$$

Dividing numerator and denominator of the right-hand side by $\cos A \cos B$,

$$\tan (A + B) = \frac{\dfrac{\sin A \cos B}{\cos A \cos B} + \dfrac{\cos A \sin B}{\cos A \cos B}}{\dfrac{\cos A \cos B}{\cos A \cos B} - \dfrac{\sin A \sin B}{\cos A \cos B}},$$

$$= \frac{\dfrac{\sin A}{\cos A} + \dfrac{\sin B}{\cos B}}{1 - \dfrac{\sin A}{\cos A} \cdot \dfrac{\sin B}{\cos B}},$$

$$\therefore \ \tan (A + B) = \frac{\tan A + \tan B}{1 - \tan A \tan B}.$$

Similarly, $\quad \tan (A - B) = \dfrac{\tan A - \tan B}{1 + \tan A \tan B}.$

For convenience, the six identities are printed together:

$$\cos (A + B) = \cos A \cos B - \sin A \sin B.$$
$$\cos (A - B) = \cos A \cos B + \sin A \sin B.$$
$$\sin (A + B) = \sin A \cos B + \cos A \sin B.$$
$$\sin (A - B) = \sin A \cos B - \cos A \sin B.$$
$$\tan (A + B) = \frac{\tan A + \tan B}{1 - \tan A \tan B}.$$
$$\tan (A - B) = \frac{\tan A - \tan B}{1 + \tan A \tan B}.$$

When memorizing these, note the following:

(i) the formulae for the ratios of $(A - B)$ are the same as those for $(A + B)$, except for the changes in signs.

(ii) the signs on the two sides of each of the sine formulae are the same, but in the cosine formulae they are different.

(iii) in the tangent formulae, the signs in the numerators are the same as in the corresponding sine formulae, and those in the denominators are the same as in the cosine formulae.

Example 1. *Find, without using tables, the value of* $\sin (120° + 45°)$, *leaving surds in the answer.*

Using the formula for $\sin (A + B)$,

$$\sin (120° + 45°) = \sin 120° \cos 45° + \cos 120° \sin 45°.$$

Reference to Figs. 12.7 and 12.8 on page 261 should remind the reader how to obtain the ratios of 30°, 45°, and 60°. Thus we have

$$\sin 120° = \sin 60° = \frac{\sqrt{3}}{2}, \quad \cos 120° = -\cos 60° = -\tfrac{1}{2};$$

$$\cos 45° = \sin 45° = \frac{1}{\sqrt{2}} = \frac{\sqrt{2}}{2}.$$

$$\therefore \sin (120° + 45°) = \frac{\sqrt{3}}{2} \cdot \frac{\sqrt{2}}{2} + \left(-\frac{1}{2}\right) \cdot \frac{\sqrt{2}}{2}.$$

$$\therefore \sin (120° + 45°) = \frac{\sqrt{2}}{4}(\sqrt{3} - 1).$$

Example 2. *If* $\sin A = \frac{3}{5}$ *and* $\cos B = \frac{15}{17}$, *where* A *is obtuse and* B *is acute, find, without using tables, the value of* $\sin (A + B)$.

$$\sin (A + B) = \sin A \cos B + \cos A \sin B.$$

So it is necessary to find the values of $\cos A$ and $\sin B$, and Fig. 13.3 and 13.4 indicate the method. In Fig. 13.3, the

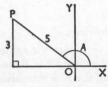

Fig. 13.3

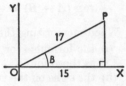

Fig. 13.4

third side of the right-angled triangle is 4 (by Pythagoras' theorem), hence the x-coordinate of P. is -4, therefore $\cos A = -\frac{4}{5}$. Similarly, in Fig. 13.4, the y-coordinate of P is 8, and therefore $\sin B = \frac{8}{17}$.

$$\therefore \sin (A + B) = \frac{3}{5} \cdot \frac{15}{17} + \left(-\frac{4}{5}\right) \cdot \frac{8}{17},$$

$$= \frac{45}{85} - \frac{32}{85},$$

$$\therefore \sin (A + B) = \frac{13}{85}.$$

Example 3. *If* $\sin (x + \alpha) = \cos (x - \beta)$, *find* $\tan x$ *in terms of* α *and* β.

Since $\qquad \sin (x + \alpha) = \cos (x - \beta),$

we have

$$\sin x \cos \alpha + \cos x \sin \alpha = \cos x \cos \beta + \sin x \sin \beta.$$

[Now $\tan x = \sin x/\cos x$, so collect terms in $\sin x$ on one side of the equation, and terms in $\cos x$ on the other.]

Thus

$$\sin x \cos \alpha - \sin x \sin \beta = \cos x \cos \beta - \cos x \sin \alpha.$$

$$\therefore \ \sin x (\cos \alpha - \sin \beta) = \cos x (\cos \beta - \sin \alpha).$$

$$\therefore \ \frac{\sin x}{\cos x} = \frac{\cos \beta - \sin \alpha}{\cos \alpha - \sin \beta}.$$

$$\therefore \ \tan x = \frac{\cos \beta - \sin \alpha}{\cos \alpha - \sin \beta}.$$

Exercise 13a

1. Without using tables, find the values of the following, leaving surds in your answers:

   (i) $\cos (45° - 30°)$,   (ii) $\sin (30° + 45°)$,

   (iii) $\sin (60° + 45°)$,   (iv) $\cos 105°$,

   (v) $\cos (120° + 45°)$,   (vi) $\sin 165°$,

   (vii) $\sin 15°$,   (viii) $\cos 75°$.

2. If $\sin A = \frac{3}{5}$ and $\sin B = \frac{5}{13}$, where A and B are acute angles, find, without using tables, the values of:

   (i) $\sin (A + B)$,   (ii) $\cos (A + B)$,   (iii) $\cot (A + B)$.

3. If $\sin A = \frac{4}{5}$ and $\cos B = \frac{12}{13}$, where A is *obtuse* and B is acute, find, without using tables, the values of:

   (i) $\sin (A - B)$,   (ii) $\tan (A - B)$,   (iii) $\tan (A + B)$.

4. If $\cos A = \frac{3}{5}$ and $\tan B = \frac{12}{5}$, where A and B are both reflex angles, find, without using tables, the values of:

   (i) $\sin (A - B)$,   (ii) $\tan (A - B)$,   (iii) $\cos (A + B)$.

5. If $\tan (x + 45°) = 2$, find, without using tables, the value of $\tan x$.

6. If $\tan (A + B) = \frac{1}{2}$ and $\tan A = 3$, find, without using tables, the value of $\tan B$.

7. If A and B are acute, $\tan A = \frac{1}{2}$ and $\tan B = \frac{1}{3}$, find without using tables the value of $A + B$.

8. If $\tan A = -\frac{1}{7}$ and $\tan B = \frac{3}{4}$, where A is obtuse and B is acute, find without using tables the value of $A - B$.

9. Express as single trigonometrical ratios:

   (i) $\frac{1}{2} \cos x - \frac{\sqrt{3}}{2} \sin x$,   (ii) $\frac{1}{\sqrt{2}} \sin x + \frac{1}{\sqrt{2}} \cos x$,

(iii) $\dfrac{\sqrt{3} + \tan x}{1 - \sqrt{3} \tan x}$,

(iv) $\cos 16° \sin 42° - \sin 16° \cos 42°$,

(v) $\dfrac{1}{\cos 24° \cos 15° - \sin 24° \sin 15°}$,

(vi) $\frac{1}{2} \cos 75° + \dfrac{\sqrt{3}}{2} \sin 75°$.

10. Find, without using tables, the values of:

(i) $\cos 75° \cos 15° + \sin 75° \sin 15°$,

(ii) $\sin 50° \cos 20° - \cos 50° \sin 20°$,

(iii) $\dfrac{\tan 10° + \tan 20°}{1 - \tan 10° \tan 20°}$,

(iv) $\cos 70° \cos 20° - \sin 70° \sin 20°$,

(v) $\dfrac{1}{\sqrt{2}} \cos 15° - \dfrac{1}{\sqrt{2}} \sin 15°$,

(vi) $\dfrac{\sqrt{3}}{2} \cos 15° - \frac{1}{2} \sin 15°$,

(vii) $\dfrac{1 - \tan 15°}{1 + \tan 15°}$,   (viii) $\cos 15° + \sin 15°$.

11. Find, without using tables, the value of $\tan A$, when $\tan (A - 45°) = \frac{1}{3}$.

12. Find, without using tables, the value of $\cot B$, when $\cot A = \frac{1}{4}$ and $\cot (A - B) = 8$.

13. From the following equations, find the values of $\tan x$:

(i) $\sin (x + 45°) = 2 \cos (x + 45°)$;

(ii) $2 \sin (x - 45°) = \cos (x + 45°)$;

(iii) $\tan (x - A) = \frac{2}{3}$, where $\tan A = 2$;

(iv) $\sin (x + 30°) = \cos (x + 30°)$.

14. If $\sin (x + \alpha) = 2 \cos (x - \alpha)$, prove that

$$\tan x = \frac{2 - \tan \alpha}{1 - 2 \tan \alpha}$$

15. If $\sin (x - \alpha) = \cos (x + \alpha)$, prove that $\tan x = 1$.

16. Solve, for values of x between $0°$ and $360°$, the equations:

(i) $2 \sin x = \cos (x + 60°)$,

(ii) $\cos (x + 45°) = \cos x$,

(iii) $\sin (x - 30°) = \frac{1}{2} \cos x$,
(iv) $3 \sin (x + 10°) = 4 \cos (x - 10°)$.

Prove the following identities:

17. $\sin (A + B) + \sin (A - B) = 2 \sin A \cos B$.

18. $\cos (A + B) - \cos (A - B) = - 2 \sin A \sin B$.

19. $\tan A + \tan B = \dfrac{\sin (A + B)}{\cos A \cos B}$.

20. $\cot A - \cot B = \dfrac{\sin (B - A)}{\sin A \sin B}$.

21. $\cot (A + B) = \dfrac{\cot A \cot B - 1}{\cot A + \cot B}$.

22. $\cot (A - B) = \dfrac{1 + \cot A \cot B}{\cot B - \cot A}$.

23. $\sec (A + B) = \dfrac{\sec A \sec B \operatorname{cosec} A \operatorname{cosec} B}{\operatorname{cosec} A \operatorname{cosec} B - \sec A \sec B}$.

24. $\dfrac{\cos (A - B) - \cos (A + B)}{\sin (A + B) + \sin (A - B)} = \tan B$.

25. $\dfrac{\cos (A + B) + \cos (A - B)}{\sin (A + B) - \sin (A - B)} = \cot B$.

26. $\tan (x + 45°) \tan (x - 45°) = - 1$.

27. $\sin (x + 60°) = \sin (120° - x)$.

28. $\sin (30° - x) = \cos (60° + x)$.

29. $\cos x + \cos (x + 120°) + \cos (x + 240°) = 0$.

30. $\sin (45° + A) + \sin (45° - A) = \sqrt{2} \cos A$.

31. $\cos B - \cos A \cos (A - B) = \sin A \sin (A - B)$.

32. $\sin B + \sin A \cos (A + B) + \cos A \sin (A - B)$
$$= 2 \sin A \cos A \cos B.$$

33. $\tan (A + B) - \tan A = \dfrac{\sin B}{\cos A \cos (A + B)}$.

34. $\tan (A + B + C)$
$$= \dfrac{\tan A + \tan B + \tan C - \tan A \tan B \tan C}{1 - \tan B \tan C - \tan C \tan A - \tan A \tan B}.$$

35. If A, B, C are angles of a triangle, show that
$$\tan A + \tan B + \tan C = \tan A \tan B \tan C.$$

[Use No. 34.]

36. If A, B, C are angles of a triangle, show that

$\tan \frac{1}{2}B \tan \frac{1}{2}C + \tan \frac{1}{2}C \tan \frac{1}{2}A + \tan \frac{1}{2}A \tan \frac{1}{2}B = 1$.

37. If $\sin (\alpha + \beta) = k \sin (\alpha - \beta)$, show that

$$\tan \alpha = \frac{k + 1}{k - 1} \tan \beta.$$

38. Prove that

$$\tfrac{1}{2}\{\tan (x + h) + \tan (x - h)\} - \tan x = \frac{\tan x \sin^2 h}{\cos^2 x - \sin^2 h}.$$

39. Prove that

$$\frac{\sin \alpha + \cos \alpha}{\sin (\theta - \alpha) \sin (\alpha - \beta)} + \frac{\sin \beta + \cos \beta}{\sin (\theta - \beta) \sin (\beta - \alpha)}$$

$$= \frac{\sin \theta + \cos \theta}{\sin (\theta - \alpha) \sin (\theta - \beta)}.$$

40. Prove that

$\sin^2 \alpha + \sin^2 \beta + 2 \sin \alpha \sin \beta \cos (\alpha + \beta) = \sin^2 (\alpha + \beta)$.

41. Find an expression for $\tan (\theta + 2\phi)$ in terms of $\tan \theta$ and $\tan \phi$. Check your result by putting $\phi = 45°$.

42. Show, without using tables, that if A, B, C are acute and $\tan A = \frac{1}{2}$, $\tan B = \frac{1}{5}$, $\tan C = \frac{1}{8}$, then $A + B + C = 45°$.

43. Write down the formulae for $\cos (A + B)$ and $\cos (A - B)$. Use them to find acute angles A and B such that

$\cos A \cos B = \frac{1}{3}$, $\sin A \sin B = \frac{1}{6}$.

The double angle formulae

13.3. The special cases of the identities on page 279, when $A = B$, are even more useful than the identities themselves. For convenience of reference, they are given together, below.

$$\cos 2A = \cos^2 A - \sin^2 A$$
$$= 2 \cos^2 A - 1,$$
$$= 1 - 2 \sin^2 A.$$
$$\sin 2A = 2 \sin A \cos A.$$
$$\tan 2A = \frac{2 \tan A}{1 - \tan^2 A}.$$

Further, it is useful to remember that

$$\cos^2 A = \tfrac{1}{2}(1 + \cos 2A), \quad \sin^2 A = \tfrac{1}{2}(1 - \cos 2A).$$

To prove the identities concerning $\cos 2A$, put $B = A$ in the identity

$$\cos (A + B) = \cos A \cos B - \sin A \sin B:$$
$$\cos 2A = \cos^2 A - \sin^2 A.$$

Now $\qquad\qquad \cos^2 A + \sin^2 A = 1,$

so substituting $\sin^2 A = 1 - \cos^2 A$, we obtain

$$\cos 2A = \cos^2 A - 1 + \cos^2 A.$$
$$\therefore \cos 2A = 2 \cos^2 A - 1.$$

If we had substituted $\cos^2 A = 1 - \sin^2 A$ in the identity $\cos 2A = \cos^2 A - \sin^2 A$, we should have obtained

$$\cos 2A = 1 - \sin^2 A - \sin^2 A.$$
$$\therefore \cos 2A = 1 - 2 \sin^2 A.$$

The expressions for $\cos^2 A$ and $\sin^2 A$ are obtained by changing the subjects in the formulae

$$\cos 2A = 2 \cos^2 A - 1 \quad \text{and} \quad \cos 2A = 1 - 2 \sin^2 A.$$

The identities for $\sin 2A$ and $\tan 2A$ are obtained immediately, when the substitution $B = A$ is made in the formulae for $\sin (A + B)$ and $\tan (A + B)$.

Example 4. *Solve the equation* $3 \cos 2\theta + \sin \theta = 1,$ *for values of θ from $0°$ to $360°$ inclusive.*

[The quadratic equation is liable to occur in various disguises. Here, $\sin \theta$ suggests that the equation may be a quadratic in $\sin \theta$, so we express $\cos 2\theta$ in terms of $\sin \theta$.]

We have $\qquad \cos 2\theta = 1 - 2 \sin^2 \theta,$

so, substituting in the equation

$$3 \cos 2\theta + \sin \theta = 1,$$

it follows that

$$3(1 - 2 \sin^2 \theta) + \sin \theta = 1.$$

This is a quadratic equation for $\sin \theta$, and it is solved by factorization.

$$3 - 6 \sin^2 \theta + \sin \theta = 1.$$
$$\therefore \ 6 \sin^2 \theta - \sin \theta - 2 = 0.$$
$$\therefore \ (3 \sin \theta - 2)(2 \sin \theta + 1) = 0.$$
$$\therefore \ \sin \theta = \tfrac{2}{3} \quad \text{or} \quad \sin \theta = -\tfrac{1}{2}.$$

If $\sin \theta = \tfrac{2}{3}$,
$$\theta = 41° \ 49' \quad \text{or} \quad 180° - 41° \ 49'.$$
If $\sin \theta = -\tfrac{1}{2}$,
$$\theta = 180° + 30° \quad \text{or} \quad 360° - 30°.$$

Therefore the values of θ between $0°$ and $360°$ which satisfy the equation are $41° \ 49'$, $138° \ 11'$, $210°$, and $330°$.

Example 5. *Prove that* $\sin 3A = 3 \sin A - 4 \sin^3 A$.

The left-hand side of the identity may be written as $\sin (A + 2A)$, so by using the formula for $\sin (A + B)$ we have:

$$\sin (A + 2A) = \sin A \cos 2A + \cos A \sin 2A.$$

But the right-hand side of the identity to be proved is in terms of $\sin A$, and this suggests that $\cos 2A$ should be expressed in terms of $\sin A$. (We have only one formula for $\sin 2A$, so it must be used.)

Therefore

$$\sin 3A = \sin A(1 - 2 \sin^2 A) + \cos A \,.\, 2 \sin A \cos A,$$
$$= \sin A - 2 \sin^3 A + 2 \sin A \cos^2 A.$$

Now $\cos^2 A$ must be expressed in terms of $\sin A$ by means of the identity $\cos^2 A = 1 - \sin^2 A$, therefore

$$\sin 3A = \sin A - 2 \sin^3 A + 2 \sin A(1 - \sin^2 A),$$
$$= \sin A - 2 \sin^3 A + 2 \sin A - 2 \sin^3 A.$$
$$\therefore \ \mathbf{sin \ 3A = 3 \sin A - 4 \sin^3 A.}$$

A formula for cos 3A in terms of cos A may be obtained from the expansion of cos $(2A + A)$. The proof is left as an exercise.

$$\cos 3A = 4 \cos^3 A - 3 \cos A.$$

Exercise 13b (Oral)

Express more simply:

1. 2 sin 17° cos 17°.

2. $\dfrac{2 \tan 30°}{1 - \tan^2 30°}$.

3. 2 cos² 42° − 1.

4. 2 sin $\frac{1}{2}\theta$ cos $\frac{1}{2}\theta$.

5. 1 − 2 sin² 22½°.

6. $\dfrac{2 \tan \frac{1}{2}\theta}{1 - \tan^2 \frac{1}{2}\theta}$.

7. cos² 15° − sin² 15°.

8. 2 sin 2A cos ²A.

9. 2 cos² $\frac{1}{2}\theta$ − 1.

10. 1 − 2 sin² 3θ.

11. $\dfrac{\tan 2\theta}{1 - \tan^2 2\theta}$.

12. sin x cos x.

13. $\dfrac{1 - \tan^2 20°}{\tan 20°}$.

14. sec θ cosec θ.

15. 1 − 2 sin² $\frac{1}{2}\theta$.

Exercise 13c

1. Evaluate without using tables:

   (i) 2 sin 15° cos 15°,

   (ii) $\dfrac{2 \tan 22\frac{1}{2}°}{1 - \tan^2 22\frac{1}{2}°}$,

   (iii) 2 cos² 75° − 1,

   (iv) 1 − 2 sin² 67½°,

   (v) cos² 22½° − sin² 22½°,

   (vi) $\dfrac{1 - \tan^2 15°}{\tan 15°}$,

   (vii) $\dfrac{1 - 2 \cos^2 25°}{1 - 2 \sin^2 65°}$,

   (viii) sec 22½° cosec 22½°.

2. Find, without using tables, the values of sin 2θ and cos 2θ when

   (i) sin $\theta = \frac{3}{5}$,   (ii) cos $\theta = \frac{12}{13}$,   (iii) sin $\theta = -\dfrac{\sqrt{3}}{2}$.

3. Find, without using tables, the value of tan 2θ when

   (i) tan $\theta = \frac{4}{3}$,   (ii) tan $\theta = \frac{8}{15}$,   (iii) cos $\theta = -\frac{5}{13}$.

4. Find, without using tables, the values of $\cos x$ and $\sin x$ when $\cos 2x$ is

   (i) $\frac{1}{8}$,   (ii) $\frac{7}{25}$,   (iii) $-\frac{119}{169}$.

5. Find, without using tables, the values of $\tan \dfrac{\theta}{2}$ when $\tan \theta$ is

   (i) $\frac{3}{4}$,   (ii) $\frac{4}{3}$,   (iii) $-\frac{12}{5}$.

6. If $t = \tan 22\frac{1}{2}°$, use the formula for $\tan 2\theta$ to show that $t^2 + 2t - 1 = 0$. Deduce the value of $\tan 22\frac{1}{2}°$.

Solve the following equations for values of θ from $0°$ to $360°$ inclusive.

7. $\cos 2\theta + \cos \theta + 1 = 0$.   8. $\sin 2\theta = \sin \theta$.

9. $\cos 2\theta = \sin \theta$.   10. $3 \cos 2\theta - \sin \theta + 2 = 0$.

11. $\sin 2\theta \cos \theta + \sin^2 \theta = 1$.   12. $\sin \theta = 6 \sin 2\theta$.

13. $2 \sin \theta (5 \cos 2\theta + 1) = 3 \sin 2\theta$.

14. $3 \tan \theta = \tan 2\theta$.   15. $3 \cot 2\theta + \cot \theta = 1$.

16. $4 \tan \theta \tan 2\theta = 1$.

17. Eliminate θ from the equations:

     (i) $x = \cos \theta$, $y = \cos 2\theta$;   (ii) $x = 2 \sin \theta$, $y = 3 \cos 2\theta$;

     (iii) $x = \tan \theta$, $y = \tan 2\theta$;   (iv) $x = 2 \sec \theta$, $y = \cos 2\theta$.

Prove the following identities:

18. $\dfrac{\cos 2A}{\cos A + \sin A} = \cos A - \sin A$.

19. $\dfrac{\sin A}{\sin B} + \dfrac{\cos A}{\cos B} = \dfrac{2 \sin (A + B)}{\sin 2B}$.

20. $\dfrac{\cos A}{\sin B} - \dfrac{\sin A}{\cos B} = \dfrac{2 \cos (A + B)}{\sin 2B}$.

21. $\tan A + \cot A = 2 \operatorname{cosec} 2A$.

22. $\cot A - \tan A = 2 \cot 2A$.

23. $\dfrac{1}{\cos A + \sin A} + \dfrac{1}{\cos A - \sin A} = \tan 2A \operatorname{cosec} A$.

24. $\dfrac{\sin 2A}{1 + \cos 2A} = \tan A = \dfrac{1 - \cos 2A}{\sin 2A}$.

25. $\cos 3A = 4 \cos^3 A - 3 \cos A$.

26. $\cot 2A = \dfrac{\cot^2 A - 1}{2 \cot A}$.

27. $\dfrac{\sin 4A + \sin 2A}{\cos 4A + \cos 2A + 1} = \tan 2A$.

28. $\operatorname{cosec} 2x - \cot 2x = \tan x$.

29. $\operatorname{cosec} 2x + \cot 2x = \cot x$.

30. $\tan x = \sqrt{\dfrac{1 - \cos 2x}{1 + \cos 2x}}$.

31. $2 \cos (45° + \theta) \cos (45° - \theta) = \cos 2\theta$.

32. $2 \cos (45° - \theta) \sin (45° + \theta) = 1 + \sin 2\theta$.

33. $\dfrac{\sin 2\theta + \cos 2\theta + 1}{\sin 2\theta - \cos 2\theta + 1} = \cot \theta$.

34. $\sin 3A + \sin A = 2 \sin 2A \cos A$.

35. $\cos 3A + \cos A = 2 \cos 2A \cos A$.

36. $\sin 2\theta (2 \cos 2\theta + 1) = 2 \cos \theta \sin 3\theta$.

37. $\sin^4 \theta + \cos^4 \theta = \frac{1}{4}(\cos 4\theta + 3)$.

38. $\cos 4\theta = 1 - 8 \sin^2 \theta + 8 \sin^4 \theta$.

39. $(1 + \cos 2\theta)(2 \cos 2\theta - 1) = 2 \cos \theta \cos 3\theta$.

40. $4 \sin^3 A \cos 3A + 4 \cos^3 A \sin 3A = 3 \sin 4A$.

41. If $x = \cos \theta + \cos 2\theta$, $y = \sin \theta + \sin 2\theta$, show that
$$x^2 - y^2 = \cos 2\theta + 2 \cos 3\theta + \cos 4\theta$$
and that
$$2xy = \sin 2\theta + 2 \sin 3\theta + \sin 4\theta.$$

42. If $\cos 2A \cos 2B = \cos 2\theta$, prove that
$$\sin^2 A \cos^2 B + \cos^2 A \sin^2 B = \sin^2 \theta.$$

43. If $\tan^2 \alpha - 3 \tan^2 \beta = 2$, show that $\cos 2\beta - 3 \cos 2\alpha = 2$.

44. $\sin 2x = \dfrac{2 \tan x}{1 + \tan^2 x}$,

45. $\cos 2x = \dfrac{1 - \tan^2 x}{1 + \tan^2 x}$.

From above we obtain the results that if $t = \tan \frac{1}{2}\theta$,

$$\tan \theta = \frac{2t}{1 - t^2}, \quad \sin \theta = \frac{2t}{1 + t^2}, \quad \cos \theta = \frac{1 - t^2}{1 + t^2}.$$

46. If $t = \tan \frac{1}{2}\theta$, express in terms of t:

    (i) $1 + \sin \theta$,   (ii) $1 + \cos \theta$,

    (iii) $\sec^2 \frac{1}{2}\theta \, (3 \sin \theta + 4 \cos \theta - 1)$,

    (iv) $\sqrt{\left\{\dfrac{1 - \sin \theta}{1 + \sin \theta}\right\}}$,   (v) $\dfrac{3 \cos \theta + 1}{1 - 2 \sin \theta}$,

    (vi) $\cot \theta \cot \frac{1}{2}\theta$.

47. Use the substitution $t = \tan \frac{1}{2}\theta$ to solve the following equations, giving values of θ from $0°$ to $360°$ inclusive.

    (i) $2 \cos \theta + 3 \sin \theta - 2 = 0$,

    (ii) $7 \cos \theta + \sin \theta - 5 = 0$,

    (iii) $3 \cos \theta - 4 \sin \theta + 3 = 0$,

    (iv) $3 \cos \theta + 4 \sin \theta = 2$.

13.4 Two applications of the identities of §13.2 follow in the next examples.

Example 6. *Solve the equation* $3 \cos \theta + 4 \sin \theta = 2$, *for values of θ from $0°$ to $360°$, inclusive.*

The solution is obtained by dividing both sides of the equation by some number, so as to leave it in the form

$$\cos \alpha \cos \theta + \sin \alpha \sin \theta = \text{constant.}$$

Comparing this with

$$3 \cos \theta + \qquad 4 \sin \theta = 2,$$

it follows that

$$\frac{\cos \alpha}{3} = \frac{\sin \alpha}{4}, \qquad \therefore \tan \alpha = \tfrac{4}{3}.$$

From tables we find that $\alpha = 53° \, 08'$, and from Fig. 13.5

Fig. 13.5

it follows that $\sin \alpha = \frac{4}{5}$ and $\cos \alpha = \frac{3}{5}$. Therefore we divide the original equation by 5, giving

$$\frac{3}{5} \cos \theta + \frac{4}{5} \sin \theta = \frac{2}{5}.$$

$$\therefore \cos \theta \cos 53° \ 08' + \sin \theta \sin 53° \ 08' = 0·4.$$

$$\therefore \cos (\theta - 53° \ 08') = 0·4.$$

$$\therefore \theta - 53° \ 08' = 66° \ 25' \quad \text{or} \quad 293° \ 35'.$$

Therefore the roots of the equation in the range from 0° to 360° are 119° 33' and 346° 43'.

Qu. 1. What advantage is there in using the formula for $\cos (A - B)$, rather than that for $\sin (A + B)$ in Example 6?

Example 7. *Find the maximum and minimum values of* $2 \sin \theta - 5 \cos \theta$, *and the corresponding values of* θ *between* 0° *and* 360°.

This will be solved by writing

$$2 \sin \theta - 5 \cos \theta = k(\cos \alpha \sin \theta - \sin \alpha \cos \theta),$$

where k and α are to be found. Comparing the two forms of the expression,

$$\frac{\sin \alpha}{\cos \alpha} = \frac{5}{2}, \quad \text{i.e. } \tan \alpha = 2·5.$$

From tables it is found that $\alpha = 68° \ 12'$; and from Fig. 13.6,

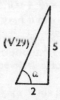

Fig. 13.6

it follows that $\cos \alpha = 2/\sqrt{29}$, and $\sin \alpha = 5/\sqrt{29}$. So we may write

$$2 \sin \theta - 5 \cos \theta = \sqrt{29}\left(\frac{2}{\sqrt{29}} \sin \theta - \frac{5}{\sqrt{29}} \cos \theta\right),$$
$$= \sqrt{29} \sin (\theta - 68° \ 12').$$

Now the greatest value of $\sin x$ is 1, and this occurs when $x = 90°$, and the least value of $\sin x$ is -1, when $x = 270°$. (Values of x less than $0°$ or greater than $360°$ have been ignored.)

Therefore $\sqrt{29} \sin (\theta - 68° \, 12')$ has a maximum value of $\sqrt{29}$ when $\theta - 68° \, 12' = 90°$; and it has a minimum value of $-\sqrt{29}$ when $\theta - 68° \, 12' = 270°$.

Therefore the maximum and minimum values of

$$2 \sin \theta - 5 \cos \theta$$

are $\sqrt{29}$ and $-\sqrt{29}$; and are given by $\theta = 158° \, 12'$ and $338° \, 12'$ respectively.

Exercise 13d

Solve the following equations for values of θ from $0°$ to $360°$ inclusive.

1. $\sqrt{3} \cos \theta + \sin \theta = 1$.
2. $5 \sin \theta - 12 \cos \theta = 6$.
3. $\sin \theta + \cos \theta = \frac{1}{2}$.
4. $\cos \theta - 7 \sin \theta = 2$.
5. $2 \sin \theta + 7 \cos \theta = 4$.
6. $3 \tan \theta - 2 \sec \theta = 4$.
7. $4 \cos \theta \sin \theta + 15 \cos 2\theta = 10$.
8. $\cos \theta + \sin \theta = \sec \theta$.
9. Prove that

$$\cos \theta - \sin \theta \equiv \sqrt{2} \cos (\theta + 45°) \equiv - \sqrt{2} \sin (\theta - 45°).$$

10. Show that $\sqrt{3} \cos \theta - \sin \theta$ may be written as

$$2 \cos (\theta + 30°) \quad \text{or} \quad 2 \sin (60° - \theta).$$

Find the maximum and minimum values of the expression, and state the values of θ between $0°$ and $360°$ for which they occur.

11. Show that $3 \cos \theta + 2 \sin \theta$ may be written in the form $\sqrt{13} \cos (\theta - \alpha)$, where $\alpha = \tan^{-1} \frac{2}{3}$. Hence find the maximum and minimum values of the function, giving the corresponding values of θ in the range from $-180°$ to $+180°$. ($\tan^{-1} \frac{2}{3}$ is an abbreviation for "the angle between $-90°$ and $+90°$ whose tangent is $\frac{2}{3}$".)

12. Show that $3 \cos \theta + 4 \sin \theta$ may be expressed in the form $R \cos (\theta - \alpha)$, where α is acute. Find the values of R and α.

13. By expressing $\cos \theta + 2 \sin \theta$ in the form $R \sin (\theta + \alpha)$, where α is acute, find the maximum and minimum values of the expression, giving the values of θ between $-180°$ and $180°$ for which they occur.

Find the maximum and minimum values of the following expressions, stating the values of θ, from $0°$ to $360°$ inclusive, for which they occur.

14. $\cos \theta + \sin \theta$. **15.** $4 \sin \theta - 3 \cos \theta$.
16. $\sqrt{3} \sin \theta + \cos \theta$. **17.** $8 \cos \theta - 15 \sin \theta$.
18. $\sin \theta - 6 \cos \theta$. **19.** $\cos (\theta + 60°) - \cos \theta$.
20. $3\sqrt{2} \cos (\theta + 45°) + 7 \sin \theta$.

21. $\dfrac{1}{\cos \theta - \sin \theta}.$ **22.** $(2 \cos \theta + 3 \sin \theta)^2$.

23. $\dfrac{1}{(\sin \theta - 2 \cos \theta)^2}.$

Method for a general proof of the addition formulae

13.5. (i) Note that the cosine formula (§15.2) holds for angles of any magnitude.

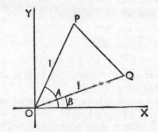

Fig. 13.7

(ii) In Fig. 13.7, OP and OQ, of unit length, make angles of A and B with the x-axis, and angle $QOP = A - B$ (all angles being measured in the positive sense.) Therefore, by the cosine formula,

$$PQ^2 = 2 - 2 \cos (A - B).$$

(iii) P is the point $(\cos A, \sin A)$ and Q is $(\cos B, \sin B)$, therefore by the distance formula,

$$PQ^2 = (\cos A - \cos B)^2 + (\sin A - \sin B)^2.$$

(iv) Equating the expressions for PQ^2 it follows that $\cos (A - B) = \cos A \cos B + \sin A \sin B$.

(v) To obtain $\cos (A + B)$, replace B by $- B$.

(vi) To obtain $\sin (A + B)$, replace A by $(90° - A)$ in $\cos (A - B)$.

(vii) To obtain $\sin (A - B)$, replace B by $- B$ in $\sin (A + B)$.

Exercise 13e (Miscellaneous)

1. If $\sin A = \frac{5}{13}$, $\sin B = \frac{8}{17}$, where A and B are acute, find without using tables the values of:

(i) $\cos (A + B)$,   (ii) $\sin (A - B)$,   (iii) $\tan (A + B)$.

2. If $\cos A = \frac{15}{17}$, $\sin B = \frac{20}{29}$, where A is reflex and B is obtuse, find without using tables the values of:

(i) $\sin (A + B)$,   (ii) $\cos (A - B)$,   (iii) $\cot (A - B)$.

3. Find, without using tables, the values of:

(i) $\cos 80° \cos 20° + \sin 80° \sin 20°$,

(ii) $\dfrac{\tan 15° + \tan 30°}{1 - \tan 15° \tan 30°}$,

(iii) $\sin 40° \cos 50° + \sin 50° \cos 40°$.

4. Find, without using tables, the values of $\sin x$ and $\cos x$ when $\cos 2x$ is (i) $\frac{1}{9}$, (ii) $\frac{49}{81}$.

5. Find, without using tables, the value of $\tan \theta$ when $\tan 2\theta$ is (i) $- \frac{20}{21}$, (ii) $\frac{36}{77}$.

6. If $\sin \theta = \frac{35}{37}$, where θ is acute, find without using tables the values of (i) $\sin 2\theta$, (ii) $\cos 2\theta$.

7. If $t = \tan 15°$, show that $t^2 + 2\sqrt{3}t - 1 = 0$, and deduce that $t = 2 - \sqrt{3}$.

Solve the following equations, giving values of θ from $0°$ to $360°$ inclusive.

8. $\sin \theta = 2 \cos (\theta + 60°)$.   9. $\sin (\theta - 45°) = 3 \cos \theta$.

10. $4 \cos (\theta - 60°) = 5 \cos (\theta - 30°)$.

11. $\cos 2\theta + 5 \cos \theta = 2.$ **12.** $2 \sin 2\theta = 3 \sin \theta.$

13. $\sin 2\theta \cos \theta + 3 \sin^2 \theta = 3.$

14. $\tan 2\theta + \tan \theta = 0.$ **15.** $4 \cos \theta - 3 \sin \theta = 1.$

16. $3 \cos \theta + 2 \sin \theta = 2 \cdot 5.$ **17.** $5 \cot \theta - 10 \operatorname{cosec} \theta = 12.$

Eliminate θ from the equations:

18. $x = 2 \cos 2\theta,\ y = 3 \cos \theta.$ **19.** $x = 2 \tan \theta,\ y = \tan 2\theta.$

20. $x = \cot \theta,\ y = \cot 2\theta.$ **21.** $x = 3 \cos \theta,\ y = \cos 3\theta.$

22. $x = \sin \theta,\ y = \cos 3\theta.$

If $t = \tan \frac{1}{2}\theta$, express in terms of t:

23. $3 \cos \theta + 4 \sin \theta + 5.$ **24.** $\sqrt{\left(\dfrac{1 + \sin \theta}{1 - \sin \theta}\right)}.$

25. $\sec^2 \frac{1}{2}\theta\ (20 \cos \theta - 21 \sin \theta + 29).$

Find the maximum and minimum values of the following, giving the values of θ between $0°$ and $360°$ for which they occur:

26. $5 \cos \theta - 12 \sin \theta.$ **27.** $12 \cos \theta + 35 \sin \theta.$

28. $48 \cos \theta - 55 \sin \theta.$

29. If θ_1 and θ_2 are roots (not in the same quadrant) of the equation $\tan 2\theta = c$, show that $\tan \theta_1 \tan \theta_2 = -1.$

30. If $x = a \cos \theta + b \cos 3\theta,\ y = a \sin \theta + b \sin 3\theta$, show that $x^2 + y^2 = (a + b)^2 - 4ab \sin^2 \theta.$

Prove the following identities:

31. $\sin (x + 120°) = \sin (60° - x).$

32. $\cos (x + 240°) = \sin (x - 30°).$

33. $\sin x + \sin (x + 240°) + \sin (x + 480°) = 0.$

34. $\sin B + \cos A \sin (A - B) = \sin A \cos (A - B).$

35. $\cot A - \cot (A + B) = \dfrac{\sin B}{\sin A \sin (A + B)}.$

36. $\tan 3A - \tan A = 2 \sin A \sec 3A.$

37. $\cot A - \cot 2A = \operatorname{cosec} 2A.$

38. $\dfrac{\cos A}{\sin B} + \dfrac{\sin A}{\cos B} = \dfrac{2 \cos (A - B)}{\sin 2B}.$

39. $\dfrac{1}{\cot A - \tan A} = \frac{1}{2} \tan 2A.$

40. $\cos B - \sin A \sin (A - B) - \cos A \cos (A + B)$
$$= \sin 2A \sin B.$$

41. $\cos 3A = -4 \cos A \sin (A + 30°) \sin (A - 30°).$

42. $\sin 2\theta\,(1 - 2\cos 2\theta) = -\,2\sin\theta\cos 3\theta$.

43. $(\cos 2\theta - 1)(1 + 2\cos 2\theta) = -\,2\sin\theta\sin 3\theta$.

44. $\cos 3\theta + \sin 3\theta = (\cos\theta - \sin\theta)(1 + 2\sin 2\theta)$.

45. $\tan 3A = \dfrac{3\tan A - \tan^3 A}{1 - 3\tan^2 A}$.

46. If $2A + B = 45°$, show that

$$\tan B = \frac{1 - 2\tan A - \tan^2 A}{1 + 2\tan A - \tan^2 A}.$$

47. In the acute-angled triangle OPQ, the altitude OR makes angles A and B with OP and OQ. Show by means of areas that if $OP = q$, $OQ = p$, $OR = r$,

$$pq\sin(A + B) = qr\sin A + pr\sin B.$$

Write down $\cos A$ and $\cos B$ in terms of p, q, r, and deduce the identity

$$\sin(A + B) = \sin A\cos B + \cos A\sin B.$$

48. With the data of No. 47, except that angle OQP is to be taken obtuse in this question, show that

$$\sin(A - B) = \sin A\cos B - \cos A\sin B.$$

49. Ptolemy's theorem states that in a cyclic quadrilateral PQRS, $QS.PR = PQ.RS + SP.QR$. Take PQ to be a diameter of the circle, angle $QPS = A$ and angle $QPR = B$, and show that

$$\sin A\cos B = \sin(A - B) + \cos A\sin B.$$

CHAPTER 14

THE FACTOR FORMULAE

14.1. Factors are very useful, in algebra, for solving equations and simplifying expressions, and when dealing with trigonometrical ratios, it is often convenient to be able to factorize a sum of two terms. On the other hand, it is sometimes useful to express a product as a sum or difference of two terms, and it is to this that we turn first.

In the last chapter it was shown that

$$\cos (A + B) = \cos A \cos B - \sin A \sin B,$$
$$\cos (A - B) = \cos A \cos B + \sin A \sin B.$$

Adding,

$$\cos (A + B) + \cos (A - B) = 2 \cos A \cos B,$$

and subtracting,

$$\cos (A + B) - \cos (A - B) = - 2 \sin A \sin B.$$

Now, keeping the formulae for $\cos (A + B)$ and $\cos (A - B)$ in mind, and working from them, work through the next exercise.

Exercise 14a (Oral)

Express as a sum or difference of two cosines:

1. $- 2 \sin x \sin y.$
2. $2 \cos x \cos y.$
3. $2 \cos 3\theta \cos \theta.$
4. $- 2 \sin (S + T) \sin (S - T).$
5. $2 \sin 5x \sin 3x.$
6. $2 \cos (x + y) \cos (x - y).$
7. $2 \cos \dfrac{A + B}{2} \cos \dfrac{A - B}{2}.$
8. $- 2 \sin \dfrac{B + C}{2} \sin \dfrac{B - C}{2}.$
9. $- 2 \sin (x + 45°) \sin (x - 45°).$
10. $2 \cos (2x + 30°) \cos (2x - 30°).$

297

14.2. Following the same method as in the last section, we have

$$\sin (A + B) = \sin A \cos B + \cos A \sin B,$$
$$\sin (A - B) = \sin A \cos B - \cos A \sin B.$$

Adding,

$$\sin (A + B) + \sin (A - B) = 2 \sin A \cos B,$$

and subtracting,

$$\sin (A + B) - \sin (A - B) = 2 \cos A \sin B.$$

Again, keeping the formulae for $\sin (A + B)$ and $\sin (A - B)$ in mind, work through the next exercise.

Exercise 14b (Oral)

Express as a sum or difference of two sines:

1. $2 \sin x \cos y.$

2. $2 \cos x \sin y.$

3. $2 \sin 3\theta \cos \theta.$

4. $2 \sin (S + T) \cos (S - T).$

5. $2 \cos 5x \sin 3x.$

6. $2 \cos (x + y) \sin (x - y).$

7. $- 2 \cos 4x \sin 2x.$

8. $2 \sin \dfrac{A + B}{2} \cos \dfrac{A - B}{2}.$

9. $2 \cos \dfrac{A + B}{2} \sin \dfrac{A - B}{2}.$

10. $2 \sin \dfrac{R - S}{2} \cos \dfrac{R + S}{2}.$

The factor formulae

14.3. We may now proceed to the question of factorizing a sum or difference of two cosines or sines. The last two sections have indicated the method, for it was shown that

$$\cos (A + B) + \cos (A - B) = 2 \cos A \cos B,$$
$$\cos (A + B) - \cos (A - B) = -2 \sin A \sin B,$$
$$\sin (A + B) + \sin (A - B) = 2 \sin A \cos B,$$
$$\sin (A + B) - \sin (A - B) = 2 \cos A \sin B.$$

Here, the right-hand sides of the identities are in factors, but it would be more convenient if the left-hand sides were in the form $\cos P + \cos Q$, etc. Therefore let

$$P = A + B \quad \text{and} \quad Q = A - B.$$

Adding, $P + Q = 2A. \quad \therefore A = \dfrac{P + Q}{2}.$

Subtracting,

$$P - Q = 2B. \quad \therefore B = \dfrac{P - Q}{2}.$$

Substituting into the four identities above,

$$\cos P + \cos Q = \quad 2 \cos \frac{P + Q}{2} \cos \frac{P - Q}{2},$$

$$\cos P - \cos Q = -\, 2 \sin \frac{P + Q}{2} \sin \frac{P - Q}{2},$$

$$\sin P + \sin Q = \quad 2 \sin \frac{P + Q}{2} \cos \frac{P - Q}{2},$$

$$\sin P - \sin Q = \quad 2 \cos \frac{P + Q}{2} \sin \frac{P - Q}{2}.$$

Remember how these identities were obtained: this will make it easier to remember them. Many people find it helpful to remember them in the form, "cos plus cos, equals two cos semi-sum, cos semi-diff."

Example 1. *Solve the equation* $\sin 3x + \sin x = 0$, *for values of x from* $- 180°$ *to* $+ 180°$, *inclusive.*

$$\sin 3x + \sin x = 0,$$

therefore, using the formula for $\sin P + \sin Q$,

$$2 \sin 2x \cos x = 0.$$

$$\therefore \sin 2x = 0 \quad \text{or} \quad \cos x = 0.$$

Now x may lie in the range from $- 180°$ to $180°$, therefore $2x$ lies in the range from $- 360°$ to $360°$.

If $\sin 2x = 0$,

$$2x = - 360°, \quad - 180°, \quad 0°, \quad 180°, \quad 360°.$$

$$\therefore x = - 180°, \quad - 90°, \quad 0°, \quad 90°, \quad 180°.$$

If $\cos x = 0$, $x = - 90°, 90°$.

Therefore the roots of the equation in the range from — 180° to 180° are

$$-180°, \quad -90°, \quad 0°, \quad 90° \quad \text{and} \quad 180°.$$

Example 2. *Solve the equation*

$$\cos(x + 30°) - \cos(x + 48°) = 0 \cdot 2,$$

for values of x from 0° to 360° inclusive.

[The difference of two cosines suggests using one of the last identities.]

$$\cos(x + 30°) - \cos(x + 48°) = 0 \cdot 2.$$

$$\therefore \; -2 \sin(x + 39°) \sin(-9°) = 0 \cdot 2.$$

But $\sin(-9°) = -\sin 9°$,

$$\therefore \; 2 \sin(x + 39°) \sin 9° = 0 \cdot 2.$$

$$\therefore \; \sin(x + 39°) = \frac{0 \cdot 2}{2 \sin 9°},$$

$$= 0 \cdot 1 \operatorname{cosec} 9°.$$

$$\therefore \; \sin(x + 39°) = 0 \cdot 6392.$$

$$\therefore \; x + 39° = 39° \, 44' \quad \text{or} \quad 140° \, 16'.$$

Therefore the roots of the equation in the range from 0° to 360° are 0° 44′ and 101° 16′.

Example 3. *Solve the equation*

$$\sin(x + 17°) \cos(x - 12°) = 0 \cdot 7,$$

for values of x from 0° to 360° inclusive.

[The product of sine and cosine suggest that the L.H.S. may be expressed as a sum of two sines.]

$$\sin(x + 17°) \cos(x - 12°) = 0 \cdot 7.$$

$$\therefore \; 2 \sin(x + 17°) \cos(x - 12°) = 1 \cdot 4.$$

$$\therefore \; \sin(2x + 5°) + \sin 29° = 1 \cdot 4.$$

$$\therefore \; \sin(2x + 5°) = 0 \cdot 9152.$$

$$\therefore \ 2x + 5° = 66° \ 14', \quad 180° - 66° \ 14',$$
$$360° + 66° \ 14', \quad 540° - 66° \ 14'.$$

$$\therefore \ 2x = 61° \ 14', \quad 108° \ 46', \quad 421° \ 14', \quad 468° \ 46'.$$

$$\therefore \ x = 30° \ 37', \quad 54° \ 23', \quad 210° \ 37', \quad 234° \ 23'.$$

Example 4. *Prove the identity*

$$\cos^2 A - \cos^2 B = \sin (A + B) \sin (B - A).$$

[A neat method is to use $\cos^2 A = \frac{1}{2}(1 + \cos 2A)$, $\cos^2 B = \frac{1}{2}(1 + \cos 2B).$]

$$\cos^2 A - \cos^2 B = \frac{1}{2}(\cos 2A - \cos 2B),$$
$$= \frac{1}{2}\{- 2 \sin (A + B) \sin (A - B)\}.$$
$$\therefore \ \cos^2 A - \cos^2 B = \sin (A + B) \sin (B - A).$$

Example 5. *Prove the identity*

$$\cos 2(\alpha + \beta) - \cos 2\alpha + \cos 2\beta - 1$$
$$= - 4 \sin (\alpha + \beta) \cos \alpha \sin \beta.$$

[Look to see how factors of the R.H.S. may be obtained in the L.H.S., thus $\sin (\alpha + \beta)$ is obtained from:]

$$\cos 2\beta - \cos 2\alpha = - 2 \sin (\alpha + \beta) \sin (\beta - \alpha).$$

[$\sin (\alpha + \beta)$ is also a factor of $\cos 2(\alpha + \beta) - 1$ since]

$$\cos 2(\alpha + \beta) - 1 = - 2 \sin^2 (\alpha + \beta).$$

$$\therefore \ \cos 2(\alpha + \beta) - \cos 2\alpha \cos 2\beta - 1$$
$$= - 2 \sin^2 (\alpha + \beta) - 2 \sin (\alpha + \beta) \sin (\beta - \alpha),$$
$$= - 2 \sin (\alpha + \beta)\{\sin (\alpha + \beta) + \sin (\beta - \alpha)\},$$
$$= - 2 \sin (\alpha + \beta)(2 \sin \beta \cos \alpha),$$
$$= - 4 \sin (\alpha + \beta) \cos \alpha \sin \beta.$$

This proves the identity.

Qu. 1. Why is the L.H.S. in Example 5 paired as above? Why not take it as $\cos 2(\alpha + \beta) - \cos 2\alpha$ and $\cos^2 \beta - 1$?

Exercise 14c (Oral)

Express the following in factors:

1. $\cos x + \cos y$.
2. $\sin 3x + \sin 5x$.
3. $\sin 2y - \sin 2z$.
4. $\cos 5x + \cos 7x$.
5. $\cos 2A - \cos A$.
6. $\sin 4x - \sin 2x$.
7. $\cos 3A - \cos 5A$.
8. $\sin 5\theta + \sin 7\theta$.
9. $\sin (x + 30°) + \sin (x - 30°)$.
10. $\cos (y + 10°) + \cos (y - 80°)$.
11. $\sin 3\theta - \sin 5\theta$.
12. $\cos (x + 30°) - \cos (x - 30°)$.
13. $\cos \dfrac{3x}{2} - \cos \dfrac{x}{2}$.
14. $\sin 2(x + 40°) + \sin 2(x - 40°)$.
15. $\cos (90° - x) + \cos y$.
16. $\sin A + \cos B$.
17. $\sin 3x + \sin 90°$.
18. $1 + \sin 2x$.
19. $\cos A - \sin B$.
20. $\frac{1}{2} + \cos 2\theta$.

Exercise 14d

Prove the following identities:

1. $\dfrac{\cos B + \cos C}{\sin B - \sin C} = \cot \dfrac{B - C}{2}$.
2. $\dfrac{\cos B - \cos C}{\sin B + \sin C} = - \tan \dfrac{B - C}{2}$.
3. $\dfrac{\sin B + \sin C}{\cos B + \cos C} = \tan \dfrac{B + C}{2}$.
4. $\dfrac{\sin B - \sin C}{\sin B + \sin C} = \cot \dfrac{B + C}{2} \tan \dfrac{B - C}{2}$.
5. $\sin x + \sin 2x + \sin 3x = \sin 2x \, (2 \cos x + 1)$.
6. $\cos x + \sin 2x - \cos 3x = \sin 2x \, (2 \sin x + 1)$.
7. $\cos 3\theta + \cos 5\theta + \cos 7\theta = \cos 5\theta \, (2 \cos 2\theta + 1)$.
8. $\cos \theta + 2 \cos 3\theta + \cos 5\theta = 4 \cos^2 \theta \cos 3\theta$.
9. $1 + 2 \cos 2\theta + \cos 4\theta = 4 \cos^2 \theta \cos 2\theta$.
10. $\sin \theta - 2 \sin 3\theta + \sin 5\theta = 2 \sin \theta \, (\cos 4\theta - \cos 2\theta)$.
11. $\cos \theta - 2 \cos 3\theta + \cos 5\theta = 2 \sin \theta \, (\sin 2\theta - \sin 4\theta)$.
12. $\sin x - \sin (x + 60°) + \sin (x + 120°) = 0$.
13. $\cos x + \cos (x + 120°) + \cos (x + 240°) = 0$.

Solve the following equations, for values of x from $0°$ to $360°$ inclusive:

14. $\cos x + \cos 5x = 0.$ **15.** $\cos 4x - \cos x = 0.$

16. $\sin 3x - \sin x = 0.$ **17.** $\sin 2x + \sin 3x = 0.$

18. $\sin (x + 10°) + \sin x = 0.$

19. $\cos (2x + 10°) + \cos (2x - 10°) = 0.$

20. $\cos (x + 20°) - \cos (x - 70°) = 0.$

21. $\sin (2x + 60°) + \sin (3x - 40°) = 0.$

22. $\cos \dfrac{3x}{2} - \cos \dfrac{x}{2} = 0.$

23. $\sin \dfrac{2x}{3} + \sin \dfrac{x}{3} = 0.$

24. $\sin 3x + \cos 2x = 0.$ (HINT: $\cos 2x = \sin (90° - 2x).$)

25. $\cos 3x + \sin x = 0.$

26. $\sin (x + 30°) + \sin (x - 30°) = \dfrac{\sqrt{3}}{2}.$

27. $\cos (x + 120°) + \cos x = 1.$

28. $\cos (2x + 45°) - \cos (2x - 45°) = 1.$

29. $\sin (x + 70°) - \sin (x + 10°) = \dfrac{\sqrt{3}}{2}.$

30. $2 \sin (x + 15°) \cos (x - 15°) = 1.$

31. $\cos (x + 60°) \cos (x - 60°) = \frac{1}{4}.$

32. $2 \cos (x + 20°) \sin (x - 5°) = 0 \cdot 1.$

33. $\sin (2x + 40°) \sin (2x + 10°) = 0 \cdot 2.$

34. If A, B, C are in arithmetical progression, show that

$$\tan B = \frac{\sin A + \sin C}{\cos A + \cos C}.$$

Prove the following identities:

35. $\dfrac{\sin (2A + B) + \sin B}{\cos (2A + B) + \cos B} = \tan (A + B).$

36. $\dfrac{\sin (2A - B) + \sin B}{\cos (2A - B) - \cos B} = - \cot (A - B).$

37. $\sin (3\theta + \phi) + \sin (\theta + 3\phi)$
$$= 2 \sin (\theta + \phi) (\cos 2\theta + \cos 2\phi).$$

38. $2 \sin (45° + A) \sin (45° - B)$
$$= \cos (A + B) + \sin (A - B).$$

39. $\dfrac{\sin A - \sin 2A + \sin 3A}{\cos A - \cos 2A + \cos 3A} = \tan 2A.$

40. $\cos (A - B)(\cos 2A + \cos 2B) - \cos (A + B)$
$$= \cos 2(A - B) \cos (A + B).$$

41. $\sin^2 A - \sin^2 B = \sin (A + B) \sin (A - B).$

42. $\sin 2(\alpha + \beta) - \sin 2\alpha - \sin 2\beta$
$$= - 4 \sin (\alpha + \beta) \sin \alpha \sin \beta.$$

43. $\sin 2(\alpha + \beta) + \sin 2\alpha + \sin 2\beta$
$$= 4 \sin (\alpha + \beta) \cos \alpha \cos \beta.$$

44. $\dfrac{\sin 2(\alpha + \beta) + \sin 2\alpha - \sin 2\beta}{\sin 2(\alpha + \beta) - \sin 2\alpha + \sin 2\beta} = \tan \alpha \cot \beta.$

45. $\cos 2(\alpha + \beta) + \cos 2\alpha - \cos 2\beta - 1$
$$= - 4 \sin (\alpha + \beta) \sin \alpha \cos \beta.$$

46. $\dfrac{\cos 2(\alpha + \beta) + \cos 2\alpha + \cos 2\beta + 1}{\cos 2(\alpha + \beta) - \cos 2\alpha - \cos 2\beta + 1} = - \cot \alpha \cot \beta.$

14.4 Example 6. *Solve the equation*

$$\cos 6x + \cos 4x + \cos 2x = 0,$$

for values of x from 0° *to* 180° *inclusive.*

[Remember that equations are very often solved by factorization, so look to see whether any of the three terms is a factor of the sum of the other pair. cos $4x$ is a factor of cos $6x$ + cos $2x$, so group cos $6x$ and cos $2x$ together.]

$$\cos 4x + \cos 6x + \cos 2x = 0.$$
$$\therefore \ \cos 4x + 2 \cos 4x \cos 2x = 0.$$
$$\therefore \ \ \cos 4x(1 + 2 \cos 2x) = 0.$$
$$\therefore \ \cos 4x = 0 \quad \text{or} \quad \cos 2x = - \tfrac{1}{2}.$$

If $\cos 4x = 0,$
$$4x = 90°, \quad 270°, \quad 450°, \quad 630°.$$
$$\therefore \ x = 22\tfrac{1}{2}°, \quad 67\tfrac{1}{2}°, \quad 112\tfrac{1}{2}°, \quad 157\tfrac{1}{2}°$$

If $\cos 2x = - \tfrac{1}{2},$
$$2x = 120°, \quad 240°.$$
$$\therefore \ x = 60°, \quad 120°.$$

Therefore the roots of the equation in the range $0°$ to $180°$ are $22\frac{1}{2}°$, $60°$, $67\frac{1}{2}°$, $112\frac{1}{2}°$, $120°$, $157\frac{1}{2}°$.

Example 7. *If A, B, C are the angles of a triangle, prove that*

$$\cos A + \cos B + \cos C - 1 = 4 \sin \frac{A}{2} \sin \frac{B}{2} \sin \frac{C}{2}.$$

Split the left-hand side into two pairs of terms. Now,

$$\cos A + \cos B = 2 \cos \frac{A+B}{2} \cos \frac{A-B}{2}.$$

But since $A + B = 180° - C$, $\dfrac{A+B}{2} = 90° - \dfrac{C}{2}$; therefore $\cos \dfrac{A+B}{2} = \sin \dfrac{C}{2}$. Seeing this factor $\sin \dfrac{C}{2}$ on the right-hand side, write

$$\cos C - 1 = -2 \sin^2 \frac{C}{2}.$$

Therefore

$$\cos A + \cos B + \cos C - 1 = 2 \sin \frac{C}{2} \cos \frac{A-B}{2} - 2 \sin^2 \frac{C}{2},$$

$$= 2 \sin \frac{C}{2} \left(\cos \frac{A-B}{2} - \sin \frac{C}{2} \right).$$

On the right-hand side of the identity to be proved, $\sin \dfrac{C}{2}$ is multiplied by a function of A and B, so in the last bracket we must express $\sin \dfrac{C}{2}$ in terms of A and B. This has been done above.

$$\therefore \cos A + \cos B + \cos C - 1$$

$$= 2 \sin \frac{C}{2} \left(\cos \frac{A-B}{2} - \cos \frac{A+B}{2} \right),$$

$$= -2 \left(\cos \frac{A+B}{2} - \cos \frac{A-B}{2} \right) \sin \frac{C}{2},$$

$$= -2\left(-2\sin\frac{A}{2}\sin\frac{B}{2}\right)\sin\frac{C}{2},$$

$$\therefore \cos A + \cos B + \cos C - 1 = 4\sin\frac{A}{2}\sin\frac{B}{2}\sin\frac{C}{2}.$$

Exercise 14e (Oral)

Given that A, B, C are angles of a triangle, express the following in terms of the other angles:

1. $\sin(B + C)$.   **2.** $\cos B$.   **3.** $\cos(C + A)$.

4. $\tan A$.   **5.** $\cot C$.   **6.** $\sin\dfrac{A}{2}$.

7. $\sin 2A$.   **8.** $\cos\dfrac{B + C}{2}$.   **9.** $\tan\dfrac{C}{2}$.

10. $\cos(2A + 2B)$.   **11.** $\cos\dfrac{B}{2}$.   **12.** $\sin 2(B + C)$.

13. $\sec\dfrac{C + A}{2}$.   **14.** $\cot\dfrac{A}{2}$.   **15.** $\sin 3A$.

Exercise 14f

Solve the following equations for values of θ from $0°$ to $180°$ inclusive:

1. $\cos\theta + \cos 3\theta + \cos 5\theta = 0$.
2. $\sin 2\theta + \sin 4\theta + \sin 6\theta = 0$.
3. $\sin\theta - 2\sin 2\theta + \sin 3\theta = 0$.
4. $\cos\frac{1}{2}\theta + 2\cos\frac{3}{2}\theta + \cos\frac{5}{2}\theta = 0$.
5. $\sin\theta + \cos 2\theta - \sin 3\theta = 0$.
6. $\sin\theta + \sin 2\theta + \sin 3\theta + \sin 4\theta = 0$.
7. $\cos\theta + \cos 3\theta + \cos 5\theta + \cos 7\theta = 0$.

Prove the following identities. A, B, C are to be taken as the angles of a triangle.

8. $\sin A + \sin(B - C) = 2\sin B \cos C$.
9. $\cos A - \cos(B - C) = -2\cos B \cos C$.
10. $\sin(A + B) + \sin(A + C) = 2\cos\dfrac{A}{2}\cos\dfrac{B - C}{2}$.

11. $\cos (A + B) + \cos (A + C) = -2 \sin \dfrac{A}{2} \cos \dfrac{B - C}{2}.$

12. $\cos \dfrac{A}{2} + \sin \dfrac{B - C}{2} = 2 \sin \dfrac{B}{2} \cos \dfrac{C}{2}.$

13. $\sin \dfrac{C}{2} - \cos \dfrac{A - B}{2} = -2 \sin \dfrac{A}{2} \sin \dfrac{B}{2}.$

14. $\sin A + \sin B + \sin C = 4 \cos \dfrac{A}{2} \cos \dfrac{B}{2} \cos \dfrac{C}{2}.$

15. $\sin 2A + \sin 2B + \sin 2C = 4 \sin A \sin B \sin C.$

16. $\cos 2A + \cos 2B + \cos 2C + 1 = -4 \cos A \cos B \cos C.$

17. $\tan A + \tan B + \tan C = \tan A \tan B \tan C.$

18. $\sin A + \sin B - \sin C = 4 \sin \dfrac{A}{2} \sin \dfrac{B}{2} \cos \dfrac{C}{2}.$

19. $\cos 2A + \cos 2B - \cos 2C = 1 - 4 \sin A \sin B \cos C.$

CHAPTER 15

SOLUTION OF TRIANGLES

15.1. It is supposed that the reader has already learned the sine and cosine formulae, but it is possible that he is not familiar with the proofs that follow, which make use of the definitions of sine and cosine in terms of coordinates.

The sine formula

In any triangle ABC,

$$\frac{a}{\sin A} = \frac{b}{\sin B} = \frac{c}{\sin C} = 2R,$$

where R is the radius of the circumcircle of the triangle.

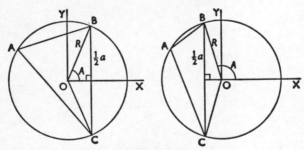

Fig. 15.1

Take axes with origin at the centre of the circle and OX perpendicular to BC. From geometry, in both cases, $\angle BOC = 2\angle BAC$ (reflex angle BOC when angle BAC is obtuse) and **OX** bisects angle **BOC** and the chord **BC**. Then from the definition of the sine of an angle,

$$\sin A = \frac{\frac{1}{2}a}{R}.$$

$$\therefore \frac{a}{\sin A} = 2R.$$

Similarly, $\qquad \dfrac{b}{\sin B} = \dfrac{c}{\sin C} = 2R,$

$$\therefore \frac{a}{\sin A} = \frac{b}{\sin B} = \frac{c}{\sin C} = 2R.$$

Example 1. *Solve the triangle in which $a = 4{\cdot}73$, $c = 3{\cdot}58$ and $C = 42° 12'$.*

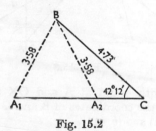

Fig. 15.2

Fig. 15.2 indicates that two triangles may be constructed with these data.

By the sine formula,

$$\frac{4{\cdot}73}{\sin A} = \frac{b}{\sin B} = \frac{3{\cdot}58}{\sin 42° 12'} = 2R.$$

$$\therefore A_1 = 62° 34' \text{ and } A_2 = 117° 26'$$

$3{\cdot}58$ $\sin 42° 12'$	$0{\cdot}5539$ $\bar{1}{\cdot}8272$
$2R$	$0{\cdot}7267$
$4{\cdot}73$ $2R$	$0{\cdot}6749$ $0{\cdot}7267$
$\sin 62° 34'$	$\bar{1}{\cdot}9482$

(i) If $A_1 = 62° 34'$, $B_1 = 75° 14'$

sin 75° 14'	$\bar{1}$·9854
$2R$	0·7267

$\therefore b_1 = 5\cdot15$, correct to 3 sig. fig.

5·153	0·7121

(ii) If $A_2 = 117° 26'$, $B_2 = 20° 22'$

sin 20° 22'	$\bar{1}$·5416
$2R$	0·7267

$\therefore b_2 = 1\cdot86$, correct to 3 sig. fig.

1·855	0·2683

The cosine formula

15.2. *In any triangle* ABC, $a^2 = b^2 + c^2 - 2bc \cos A$.

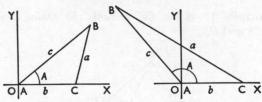

Fig. 15.3

Take axes with origin at A and OX along AC as in Fig. 15.3, then the coordinates of B and C are respectively $(c \cos A, c \sin A)$ and $(b, 0)$.

Using the distance formula,

$$a^2 = (b - c \cos A)^2 + (c \sin A)^2,$$
$$= b^2 - 2bc \cos A + c^2 \cos^2 A + c^2 \sin^2 A.$$

Now $\cos^2 A + \sin^2 A = 1$,

$$\therefore a^2 = b^2 + c^2 - 2bc \cos A.$$

NOTE.—The formula

$$a^2 = b^2 + c^2 - 2bc \cos A$$

remains valid for any magnitude of A. (See *Mathematical Gazette*, XXXVIII, p. 212; Note by A. Hurrell.)

Example 2. *Solve the triangle in which* $a = 14 \cdot 3$, $c = 17 \cdot 5$ *and* $B = 25° 36'$.

By the cosine formula,

$$b^2 = c^2 + a^2 - 2ca \cos B.$$

c^2	$306 \cdot 3$		2	$0 \cdot 3010$
a^2	$204 \cdot 5$		c	$1 \cdot 2430$
			a	$1 \cdot 1553$
$c^2 + a^2$	$510 \cdot 8$		$\cos B$	$\bar{1} \cdot 9551$
$2ca \cos B$	$451 \cdot 2$			
b^2	$59 \cdot 6$		$451 \cdot 2$	$2 \cdot 6544$

$\therefore b = 7 \cdot 72$, " correct " to three significant figures.

By the sine formula,

$$\frac{\sin A}{14 \cdot 3} = \frac{\sin 25° 36'}{7 \cdot 72}$$

$\sin 25° 36'$	$\bar{1} \cdot 6356$
$7 \cdot 72$	$0 \cdot 8876$

$$= \frac{\sin C}{17 \cdot 5} = \frac{1}{2R}.$$

$1/2R$	$\bar{2} \cdot 7480$
$14 \cdot 3$	$1 \cdot 1553$
$\sin 53° 10'$	$\bar{1} \cdot 9033$

$$\therefore A = 53° 10',$$

$$\therefore C = 101° 14'.$$

As a check C has also been calculated by the sine formula which
gives the value $101° 36'$.

$1/2R$	$\bar{2} \cdot 7480$
$17 \cdot 5$	$1 \cdot 2430$
$\sin 78° 24'$	$\bar{1} \cdot 9910$

The error has arisen in the value of b. Although square
tables and logarithms gave values correct to four figures,
b^2 contains only three. In fact, correct to four significant
figures and the nearest minute, $b^2 = 59 \cdot 37$, $b = 7 \cdot 705$,
$A = 53° 19'$, $C = 101° 05'$.

Since the cosine formula is unsuitable for work with
logarithms and at times leads to inaccurate answers, other
methods of solving triangles have been devised to take its
place. When two sides and the included angle are given,
the tangent formula is a good method.

The tangent formula

15.3. *In any triangle* ABC,

$$\tan \frac{B-C}{2} = \frac{b-c}{b+c} \cot \frac{A}{2}.$$

[Formulae involving only angles are usually easier to manipulate than those which involve sides and angles, so we express $(b-c)/(b+c)$ in terms of $\sin B$ and $\sin C$.]

By the sine formula,

$$\frac{a}{\sin A} = \frac{b}{\sin B} = \frac{c}{\sin C} = 2R.$$

$$\therefore \frac{b-c}{b+c} = \frac{2R \sin B - 2R \sin C}{2R \sin B + 2R \sin C},$$

$$= \frac{\sin B - \sin C}{\sin B + \sin C}.$$

Using the factor formulae of §14.3, it follows that

$$\frac{b-c}{b+c} = \frac{2 \cos \dfrac{B+C}{2} \sin \dfrac{B-C}{2}}{2 \sin \dfrac{B+C}{2} \cos \dfrac{B-C}{2}},$$

$$= \cot \frac{B+C}{2} \tan \frac{B-C}{2}.$$

Now A, B, C are angles of a triangle,

$$\therefore B + C = 180° - A, \quad \therefore \frac{B+C}{2} = 90° - \frac{A}{2}.$$

$$\therefore \cot \frac{B+C}{2} = \cot \left(90° - \frac{A}{2}\right) = \tan \frac{A}{2}.$$

$$\therefore \frac{b-c}{b+c} = \tan \frac{A}{2} \tan \frac{B-C}{2}.$$

$$\therefore \tan \frac{B-C}{2} = \frac{b-c}{b+c} \cot \frac{A}{2}.$$

Example 3. *Solve the triangle in which* $a = 14\cdot3$, $c = 17\cdot5$ *and* $B = 25° 36'$.

[In this example, the tangent rule may be used either for $\tan \frac{1}{2}(C - A)$ or $\tan \frac{1}{2}(A - C)$; but as c is greater than a, it follows that C is greater than A, so we use the formula for $\tan \frac{1}{2}(C - A)$ in order to avoid a negative angle.]

$$\tan \frac{C - A}{2} = \frac{c - a}{c + a} \cot \frac{B}{2},$$

$$= \frac{3 \cdot 2}{31 \cdot 8} \cot 12° \; 48'.$$

$3 \cdot 2$ $\cot 12°\ 48'$	$0 \cdot 5051$ $0 \cdot 6436$
$31 \cdot 8$	$1 \cdot 1487$ $1 \cdot 5024$
$\tan 23°\ 53'$	$\bar{1} \cdot 6463$

$$\therefore \; \frac{C - A}{2} = 23° \; 53'.$$

But * $$\frac{C + A}{2} = 77° \; 12'.$$

Adding, $C = 101° \; 05'.$

Subtracting, $A = 53° \; 19'.$

By the sine formula,

$$\frac{b}{\sin 25° \; 36'} = \frac{14 \cdot 3}{\sin 53° \; 19'}.$$

$14 \cdot 3$ $\sin 25°\ 36'$ $\operatorname{cosec} 53°\ 19'$	$1 \cdot 1553$ $\bar{1} \cdot 6356$ $0 \cdot 0958$
$7 \cdot 704$	$0 \cdot 8867$

$$\therefore b = 14 \cdot 3 \sin 25° \; 36' \operatorname{cosec} 53° \; 19'.$$

$$\therefore \; C = 101° \; 05', \quad A = 53° \; 19', \quad b = 7 \cdot 704.$$

The reader should compare these results with those obtained in Example 2, in which the same triangle was solved by the cosine and sine formulae.

Exercise 15a

Solve the following triangles:

1. (Sine formula, acute angled.)
   (i) $a = 12$, $B = 59°$, $C = 73°$;
   (ii) $A = 75° \; 37'$, $b = 5 \cdot 6$, $C = 48° \; 15'$;
   (iii) $A = 73° \; 10'$, $B = 61° \; 44'$, $c = 171$.

$$* \left(C + A = 180° - B, \quad \therefore \frac{C + A}{2} = 90° - \frac{B}{2} \right).$$

2. (Sine formula, obtuse angled.)
    (i) $A = 36°$, $b = 2·37$, $C = 49°$;
    (ii) $A = 123° 12'$, $a = 11·5$, $C = 37° 08'$;
    (iii) $a = 136$, $B = 104° 10'$, $C = 43° 05'$.

3. (Sine formula, ambiguous case.)
    (i) $b = 17·6$, $C = 48° 15'$, $c = 15·3$;
    (ii) $B = 129°$, $b = 7·89$, $c = 4·56$;
    (iii) $A = 28° 15'$, $a = 8·5$, $b = 14·8$;
    (iv) $A = 65° 14'$, $a = 146$, $c = 153$;
    (v) $A = 164° 11'$, $a = 5·64$, $b = 2·51$.

4. (Cosine formula, acute angled.)
    (i) $a = 5$, $b = 8$, $c = 7$;
    (ii) $a = 10$, $b = 12$, $c = 9$;
    (iii) $a = 17$, $b = 13$, $c = 18$.

5. (Cosine formula, acute angled.)
    (i) $A = 60°$, $b = 8$, $c = 15$;
    (ii) $a = 14$, $B = 53°$, $c = 12$;
    (iii) $a = 11$, $b = 9$, $C = 43° 12'$.

6. (Cosine formula, obtuse angled.)
    (i) $a = 8$, $b = 10$, $c = 15$;
    (ii) $a = 11$, $b = 31$, $c = 24$;
    (iii) $a = 27$, $b = 35$, $c = 46$.

7. (Cosine formula, obtuse angled.)
    (i) $a = 17$, $B = 120°$, $c = 63$;
    (ii) $A = 104° 15'$, $b = 10$, $c = 12$;
    (iii) $a = 31$, $b = 42$, $C = 104° 10'$.

8. (Tangent formula.)
    (i) $A = 54°$, $b = 1·46$, $c = 1·05$;
    (ii) $a = 1·72$, $b = 1·96$, $C = 40°$;
    (iii) $a = 10·2$, $B = 110°$, $c = 8·4$;
    (iv) $a = 3·4$, $b = 4·1$, $C = 120°$;
    (v) $A = 59° 18'$, $b = 12·6$, $c = 13·4$;
    (vi) $a = 156$, $b = 140$, $C = 34° 54'$.

9. Two points A and B on a straight coastline are 1 km apart, B being due East of A. If a ship is observed on bearings 167° and 205° from A and B respectively, what is its distance from the coastline?

10. A boat is sailing directly towards a cliff. The angle of elevation of a point on the top of the cliff and straight

ahead of the boat increases from 10° to 15° as the ship sails a distance of 50 m.   What is the height of the cliff?

11. A triangle is taken with sides 10, 11, 15 cm.   By how much does its largest angle differ from a right angle?

12. A ship rounds a headland by sailing first 4 n.m. on a course of 069° then 5 n.m. on a course of 295°. Calculate the distance and bearing of its new position from its original position.

13. A man travelling along a straight level road in the direction 053° observes a pylon on a bearing of 037°.   800 m further along the road the bearing of the pylon is 296°. Calculate the distance of the pylon from the road.

14. What is the length of the common chord of two circles, radii 7 cm and 9 cm, whose centres are 11 cm apart?

15. A ship A is 3·2 n.m. N 17° 30′ W of a lightship, and at the same time a ship B is 4·5 n.m. N 70° 42′ E of the lightship. Calculate the distance and bearing of A from B.

16. The diagonals of a parallelogram are 30 cm and 48 cm, and one side is 21 cm.   Calculate the length of the other side, and the acute angle of the parallelogram.

17. In the cyclic quadrilateral ABCD, AB = 7, BC = 8, CD = 8, DA = 15.   Calculate the angle ADC and the length of AC.

18. Show that the angle B in the triangle ABC in which $a = n^2 - 1$, $b = n^2 - n + 1$, $c = n^2 - 2n$ is 60°.

The half-angle formulae

15.4. If the three sides of a triangle are given, the following formulae are suitable for work with logarithms.

In any triangle ABC

$$\sin \tfrac{1}{2}A = \sqrt{\left\{\frac{(s-b)(s-c)}{bc}\right\}}, \quad \cos \tfrac{1}{2}A = \sqrt{\left\{\frac{s(s-a)}{bc}\right\}},$$

$$\tan \tfrac{1}{2}A = \sqrt{\left\{\frac{(s-b)(s-c)}{s(s-a)}\right\}},$$

where $s = \dfrac{a+b+c}{2}$.

From §13.3 we have
$$\sin^2 \tfrac{1}{2}A = \tfrac{1}{2}(1 - \cos A), \quad \cos^2 \tfrac{1}{2}A = \tfrac{1}{2}(1 + \cos A);$$

but by the cosine formula, $\cos A = (b^2 + c^2 - a^2)/(2bc)$, therefore

$$\sin^2 \tfrac{1}{2}A = \tfrac{1}{2}\left(1 - \frac{b^2 + c^2 - a^2}{2bc}\right),$$

$$= \tfrac{1}{2}\left(\frac{2bc - b^2 - c^2 + a^2}{2bc}\right),$$

$$= \frac{a^2 - (b-c)^2}{4bc},$$

$$= \frac{(a-b+c)(a+b-c)}{4bc}.$$

$$\cos^2 \tfrac{1}{2}A = \tfrac{1}{2}\left(1 + \frac{b^2 + c^2 - a^2}{2bc}\right),$$

$$= \tfrac{1}{2}\left(\frac{2bc + b^2 + c^2 - a^2}{2bc}\right),$$

$$= \frac{(b+c)^2 - a^2}{4bc},$$

$$= \frac{(b+c+a)(b+c-a)}{4bc}.$$

But since $s = \tfrac{1}{2}(a + b + c)$, it follows that

$$s - a = \frac{b+c-a}{2}, \ s - b = \frac{a-b+c}{2}, \ s - c = \frac{a+b-c}{2}.$$

$$\therefore \ \sin^2 \tfrac{1}{2}A = \frac{(s-b)(s-c)}{bc}, \quad \cos^2 \tfrac{1}{2}A = \frac{s(s-a)}{bc},$$

from which the formulae for $\sin \tfrac{1}{2}A$ and $\cos \tfrac{1}{2}A$ follow immediately.

The formula for $\tan \tfrac{1}{2}A$ follows at once, since

$$\tan \tfrac{1}{2}A = \frac{\sin \tfrac{1}{2}A}{\cos \tfrac{1}{2}A},$$

$$= \sqrt{\left\{\frac{(s-b)(s-c)}{bc}\right\}} \cdot \sqrt{\left\{\frac{bc}{s(s-a)}\right\}},$$

$$= \sqrt{\left\{\frac{(s-b)(s-c)}{s(s-a)}\right\}}.$$

15.5. At this point it is appropriate to obtain another formula for $\tan \frac{1}{2}A$.

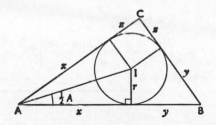

Fig. 15.4

In Fig. 15.4, I is the centre of the incircle of triangle ABC and the lengths of the tangents from A, B, C are x, y, z, respectively. Then angle $IAB = \frac{1}{2}A$, therefore

$$\tan \frac{1}{2}A = \frac{r}{x}.$$

Now, the circumference of triangle ABC is $2x + 2y + 2z$, but it is also equal to $a + b + c$, therefore

$$x + y + z = \frac{1}{2}(a + b + c);$$

but we write $s = \frac{1}{2}(a + b + c)$, and from the figure, $y + z = a$; therefore

$$x = s - a,$$

$$\therefore \tan \frac{1}{2}A = \frac{r}{s - a}.$$

We may easily obtain an expression for r, the radius of the inscribed circle, by equating the two formulae for $\tan \frac{1}{2}A$; thus

$$\frac{r}{s - a} = \sqrt{\left\{ \frac{(s - b)(s - c)}{s(s - a)} \right\}},$$

$$\therefore r = \sqrt{\left\{ \frac{(s - a)(s - b)(s - c)}{s} \right\}}.$$

Example 4. *Calculate the angles of the triangle* ABC, *in which* $a = 1\cdot73$, $b = 1\cdot58$, $c = 2\cdot50$.

We use $\qquad \tan \tfrac{1}{2}A = \dfrac{r}{s-a}$,

where $\qquad r = \sqrt{\left\{ \dfrac{(s-a)(s-b)(s-c)}{s} \right\}}$,

and $\qquad s = \tfrac{1}{2}(a+b+c)$.

a	$1\cdot73$	
b	$1\cdot58$	
c	$2\cdot50$	
$2s$	$5\cdot81$	
s	$2\cdot905$	
$s-a$	$1\cdot175$	$0\cdot0701$
$s-b$	$1\cdot325$	$0\cdot1222$
$s-c$	$0\cdot405$	$\bar{1}\cdot6075$
$\dfrac{(s-a)(s-b)(s-c)}{s}$		$\bar{1}\cdot7998$
		$0\cdot4631$
r^2		$\bar{1}\cdot3367$
r		$\bar{1}\cdot6684$
$s-a$		$0\cdot0701$
$\tan \tfrac{1}{2}A$		$\bar{1}\cdot5983$
r		$\bar{1}\cdot6684$
$s-b$		$0\cdot1222$
$\tan \tfrac{1}{2}B$		$\bar{1}\cdot5462$
r		$\bar{1}\cdot6684$
$s-c$		$\bar{1}\cdot6075$
$\tan \tfrac{1}{2}C$		$0\cdot0609$

$\tfrac{1}{2}A = 21° 38'$,
$\therefore \; A = 43° 16'$.

$\tfrac{1}{2}B = 19° 23'$,
$\therefore \; B = 38° 46'$.

$\tfrac{1}{2}C = 49° 00'$,
$\therefore \; C = 98° 00'$.

NOTE. Two checks should be made.

(i) $(s-a)+(s-b)+(s-c) = s$, before looking up any logarithims.

(ii) $A + B + C = 180°$. Any error in the half-angles will be doubled when A, B, C are found, so a total of precisely $180°$ must not be expected every time.

Exercise 15b

1. Calculate the length of the radius of the inscribed circle of triangle **ABC** when

   (i) $a = 6,$ $b = 8,$ $c = 10;$
   (ii) $a = 3,$ $b = 5,$ $c = 7;$
   (iii) $a = 12,$ $b = 15,$ $c = 17.$

2. Use the half-angle formulae to calculate the angles of the triangle **ABC** when

   (i) $a = 27,$ $b = 35,$ $c = 34;$
   (ii) $a = 65,$ $b = 61,$ $c = 56;$
   (iii) $a = 34·1,$ $b = 33·1,$ $c = 32·0;$
   (iv) $a = 71·6,$ $b = 52·4$ $c = 41·5;$
   (v) $a = 51·46,$ $b = 23·75,$ $c = 41·63.$

The area of a triangle

15.6. Fig. 15.5 illustrates the data from which it is sometimes necessary to calculate the area of a triangle.

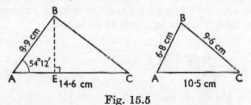

Fig. 15.5

In the first case the area can easily be found by dropping a perpendicular from **B** to meet AC at **E**. Then from the right-angled triangle **BAE**, $BE = 8·9 \sin 54° 12'$. Therefore the area of the triangle is $\frac{1}{2} \cdot 14·6 \cdot 8·9 \sin 54° 12' = 52·7$ cm². This method will now be used to obtain a general formula for the area of a triangle in terms of b, c, A.

Given a triangle ABC, take axes as shown in Fig. 15·6. For both cases, from the definition of the sine of an angle,

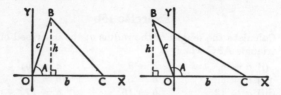

Fig. 15.6

$\sin A = h/c$, where h is the length of the altitude through B, therefore

$$h = c \sin A.$$

But the area of the triangle $\triangle = \frac{1}{2}bh$, therefore

$$\triangle = \tfrac{1}{2}\, bc \sin A.$$

The area of the second triangle in Fig. 15.5 could be found by calculating the angle A and applying the last formula, and that is how we shall obtain an expression for the area of a triangle in terms of its sides.

From §15.4,

$$\sin \tfrac{1}{2}A = \sqrt{\left\{\frac{(s-b)(s-c)}{bc}\right\}}, \cos \tfrac{1}{2}A = \sqrt{\left\{\frac{s(s-a)}{bc}\right\}}.$$

But $\triangle = \frac{1}{2}bc \sin A$, and $\sin A = 2 \sin \frac{1}{2}A \cos \frac{1}{2}A$.

$$\therefore \triangle = \tfrac{1}{2}bc.2\sqrt{\left\{\frac{(s-b)(s-c)}{bc}\right\}}\sqrt{\left\{\frac{s(s-a)}{bc}\right\}},$$

$$= bc\sqrt{\left\{\frac{(s-b)(s-c)s(s-a)}{b^2c^2}\right\}},$$

$$\therefore \triangle = \sqrt{\{s(s-a)(s-b)(s-c)\}},$$

$$\text{where } s = \tfrac{1}{2}(a + b + c).$$

NOTE. When applying this formula, use the check $(s - a) + (s - b) + (s - c) = s$, as in Example 4.

Historical note. A proof of this formula was given by Heron (or Hero) in the *Metrica* (sometime between 150 B.C. and A.D. 250), but the result may well have been known to Archimedes (287–212 B.C.). It is known as Heron's (or Hero's) formula.

Qu. 1. If a is greater than $b + c$, no triangle can be constructed. What effect does this have on the formula for $\triangle$?

There is a simple result connecting the area of a triangle $\triangle$, the radius of the inscribed circle r, and the semi-perimeter s.

$$\triangle = rs.$$

It may be proved using the formulae already obtained for $\triangle$ and r, but the geometrical proof is instructive.

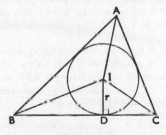

Fig. 15.7

In Fig. 15.7 I is the centre of the inscribed circle of triangle ABC. Thus the altitude ID of the triangle IBC is also a radius of the circle, therefore

$$\triangle IBC = \tfrac{1}{2}ar,$$

similarly, $\qquad \triangle ICA = \tfrac{1}{2}br,$

and $\qquad \triangle IAB = \tfrac{1}{2}cr.$

Adding, $\qquad \triangle ABC = \frac{1}{2}r(a + b + c)$,

but $s = \frac{1}{2}(a + b + c)$,

$$\therefore \ \triangle = rs.$$

Exercise 15c

1. Calculate the areas of the triangles in which:
   (i) $A = 60°$ $b = 3$, $c = 5$;
   (ii) $C = 110°$, $a = 14$, $b = 11$;
   (iii) $B = 90°$, $c = 8\cdot6$, $b = 11\cdot4$;
   (iv) $a = 8$, $b = 11$, $c = 13$;
   (v) $a = 12\cdot3$, $b = 14\cdot1$, $c = 13\cdot6$;
   (vi) $a = 17\cdot6$, $b = 16\cdot9$, $c = 16\cdot1$;
   (vii) $a = 209$, $b = 313$, $c = 390$.

2. A field is in the shape of a quadrilateral ABCD in which AB = 50 m, BC = 80 m, CD = 90 m, DA = 100 m, AC = 100 m. What is its area?

3. The area of a triangle ABC is 6 cm², AB = 3 cm and AC = 5 cm. Calculate two possible lengths of BC.

4. The area of a triangle is $4\sqrt{6}$ cm², the perimeter is 16 cm and one side is 7 cm. Calculate the lengths of the other sides.

5. Calculate the area of the cyclic quadrilateral ABCD in which AB = 2 cm, BC = 3 cm, CD = 4 cm, DA = 5 cm.

6. Calculate the area of the triangle ABC in which $a = 24\cdot6$, $b = 14\cdot8$, $c = 19\cdot5$. Calculate also the radius of the inscribed circle and the angle B.

7. Show that $\sin A = \dfrac{2}{bc}\sqrt{\{s(s - a)(s - b)(s - c)\}}$.

   Use this formula to calculate the angles of the triangle ABC in which $a = 17\cdot8$ cm, $b = 21\cdot4$ cm, $c = 19\cdot5$ cm.

8. r_1 is the radius of the escribed circle touching BC on the opposite side to A. Prove that $\triangle = r_1(s - a)$. (Compare method on p. 321.)

9. Prove that $\triangle = \frac{1}{2}a^2 \dfrac{\sin B \sin C}{\sin (B + C)}$.

10. Prove that $abc = 4R\triangle$.

11. Use the formulae $2bc \cos A = b^2 + c^2 - a^2$,

$$2bc \sin A = 4\triangle$$

to show that $\triangle = \sqrt{\{s(s-a)(s-b)(s-c)\}}$. (Square and add the two formulae.)

12. Prove that, in triangle ABC, $\cot C = \dfrac{a}{c} \operatorname{cosec} B - \cot B$.

(Drop a perpendicular from A to meet BC at D. Express AD and DC in terms of a, c, B.)

Use this result, and the similar one for $\cot A$, to calculate the remaining angles of the triangle in which $a = 14 \cdot 7$, $c = 17 \cdot 3$, $B = 64° \ 12'$.

13. Prove that $a \sin \frac{1}{2}(B - C) = (b - c) \cos \frac{1}{2}A$ and that $a \cos \frac{1}{2}(B - C) = (b + c) \sin \frac{1}{2}A$. (Use the method of §15.3.)

Use these two formulae to solve the triangle in which $b = 12 \cdot 5$, $c = 13 \cdot 8$, $A = 64°$.

14. Prove that $a = b \cos C + c \cos B$.

15. A line through the vertex A of a triangle ABC meets the base at K, making $AKC = x$. Prove that

$$\frac{BK}{KC} = \frac{\cot B - \cot x}{\cot C + \cot x}.$$

16. Prove that in a triangle ABC the length of the perpendicular from A to BC is $(bc/a) \sin A$, and that, if M is the mid-point of BC, then $\tan AMB = \dfrac{2bc}{b^2 - c^2} \sin A$.

17. In the triangle ABC, prove that, if the internal bisector of A meets BC at D, then $AD = \dfrac{2bc}{b + c} \cos \dfrac{A}{2}$.

18. If $2b = a + c$, prove that $\cot \frac{1}{2}C = 3 \tan \frac{1}{2}A$. (C.)

19. Solve the following triangles ABC by the most appropriate methods:

(i) $a = 23 \cdot 4$,	$b = 19 \cdot 6$,	$C = 67° \ 08'$;
(ii) $c = 11 \cdot 6$,	$A = 54° \ 12'$,	$B = 26° \ 23'$;
(iii) $a = 5$,	$b = 8$,	$c = 9$;
(iv) $a = 146$,	$b = 138$,	$c = 157$;
(v) $b = 16 \cdot 3$,	$c = 14 \cdot 5$,	$A = 78° \ 30'$;
(vi) $a = 4 \cdot 96$,	$b = 6 \cdot 01$,	$A = 31° \ 09'$;

(vii) $a = 9$,	$b = 11$,	$C = 60°$;
(viii) $c = 12.5$,	$a = 16.9$,	$B = 153° 18'$;
(ix) $a = 12$,	$b = 16$,	$B = 130°$;
(x) $a = 10.5$,	$c = 12.6$,	$B = 112°$.

Exercise 15d (Problems in three dimensions)

1. A cuboid is formed by joining the vertices AA′, BB′ CC′, DD′ of two rectangles ABCD and A′B′C′D′. AB = 4 cm, BC = 3 cm, CC′ = 2 cm. Calculate:
   (i) the length of AC′,
   (ii) the angle AC′ makes with the base ABCD,
   (iii) the angle between the plane ABC′D′ and the base.

2. A hall is 12 m long, 9 m wide and 8 m high. Calculate the length of a diagonal of the room and the angle it makes with the horizontal. Also find the inclination to the horizontal of a plane joining the bottom of one of the shorter walls to the top of the opposite wall.

3. A right pyramid has a square base of side 8 cm, and the length of a slant edge is 9 cm. Calculate:
   (i) the height of the pyramid,
   (ii) the inclination of a slant edge to the base,
   (iii) the base angle of a slant face.

4. A right pyramid of height 17 cm is on a square base of side 12 cm. Calculate:
   (i) the angle between a slant face and the base,
   (ii) the length of a slant edge,
   (iii) the angle between two adjacent slant faces.

5. A wireless mast is held vertically by four stays 20 m long, fixed to the mast at the same height and joined to the four corners of a square on level ground. If each stay is inclined at 63° to the horizontal, calculate the height of the top of each stay and the length of a side of the square.

6. A man travelling in a direction N 57° E along a road in level country sees the top of a spire due N at an angle of elevation of 18°. When he has gone 400 m along the road, he finds himself due E of the spire. Calculate:
   (i) the height of the spire,

(ii) its angle of elevation from the second point of observation.

7. A flagstaff is placed at one corner of a level rectangular playground 30 m wide and 40 m long. If the angle of elevation of the top of the flagstaff from the opposite corner of the playground is 13°, calculate the height of the pole, and the angles of elevation of the top from the other two corners of the playground.

8. A rectangular mirror measuring 24 cm by 18 cm with the longer edges horizontal is inclined at 30° to the vertical. Calculate:

   (i) the height of the top edge above the bottom edge,
   (ii) the angle which a diagonal makes with the horizontal.

9. A house on a hill may be reached from a level main road either by a straight road 500 m long inclined at 5° to the horizontal, or by a straight footpath 250 m long at right angles to the main road. Calculate:

   (i) the height of the house above the main road,
   (ii) the inclination of the footpath to the horizontal.

10. A rectangular roof slopes at 50° to the horizontal. What is the inclination to the horizontal of a line along the roof which makes 70° with the line of greatest slope?

11. The roofs round a rectangular courtyard slope at 45°. What is the inclination to the horizontal of the line in which two adjacent roofs meet?

12. CD is a vertical pole 9 m high with D on level ground, A and B are points 23 m apart on the ground, angle CAB = 34° and angle CBA = 61°. Calculate the angles of elevation of C from A and B.

13. A man walking along a path sees a tree in a direction making 16° with the path, and the angle of elevation of the top of the tree is 5°. After walking 100 m along the path he sees the tree in a direction making 25° with the path. Calculate the height of the tree.

14. From a point D on top of a cliff 80 m high, two boats A and B are observed on bearings of 202° and 140° respectively. If B is 750 m from A on a bearing of 110°, calculate the angles of depression of A and B from D.

15. The tops of two hills are observed from a point A. Their horizontal distances are 3000 m and 5000 m, and their angles of elevation are 2° and 5° respectively. If their bearings differ by 120°, calculate the angle of elevation of the top of the higher from the top of the lower hill.

16. The angle of elevation of the top T of a flagstaff AT on level ground is 26° 34′ from B, 40 m due N of A. A point C is 44 m on a bearing of 240° from B. Calculate the angle of elevation of T from C.

17. The cross-bar of a Rugby football goal is 5·58 m long and 3 m high. What angle does the cross-bar subtend at a point on the ground at distances of 12 m and 15 m from the feet of the posts?

18. The angles of elevation of a point on the top of a cliff from two buoys in the sea due S are α and β. $(\beta > \alpha.)$ Show that if the distance between the buoys is d, then the height of the cliff is $d/(\cot \alpha - \cot \beta)$.

19. The angles of elevation of the top T of a vertical post TC are observed to be α and β from points A and B due S and due E of the post. If the distance AB $= d$, show that the height of the post is $d/\sqrt{(\cot^2 \beta + \cot^2 \alpha)}$.

20. A level road running on a bearing of $a°$ passes by a hill. When an observer is on the road due S of the hill, the angle of elevation of the top is $\theta°$, and after he has travelled a distance d along the road, his bearing from the hill is S $\beta°$ E. Show that the height of the hill is

$$d \sin (a + \beta) \operatorname{cosec} \beta \tan \theta.$$

21. A plane slopes at θ to the horizontal. Show that a line along the plane making an angle ϕ with the line of greatest slope is inclined at an angle $\sin^{-1} (\cos \phi \sin \theta)$ to the horizontal.

22. A rectangular door a m wide, b m high, opens through an angle 2θ. Show that if a diagonal makes an angle 2ϕ with its original position, $\sin \phi = a \sin \theta/\sqrt{(a^2 + b^2)}$.

23. A, B, C are three points in order along a straight level road, and AB $=$ BC $= d$. If the angles of elevation of the top of a hill h above the level of the road are α, β, γ, show that $d = h(\frac{1}{2} \cot^2 \alpha - \cot^2 \beta + \frac{1}{2} \cot^2 \gamma)^{\frac{1}{2}}$.

CHAPTER 16

RADIANS

16.1. The fact that there are 90 degrees in a right angle has been familiar to the reader since the time when he began geometry; but he may not have realized that the number is an arbitrary one which has come down to us from the Babylonian civilization. Indeed, 100 degrees to the right angle were introduced after the French Revolution, but later dropped, and in 1938 a similar attempt was made by the Germans. The following example also illustrates the arbitrary nature of the number of degrees in a right angle.

Example 1. *An arc* AB *of a circle, centre* O, *subtends an angle of* $x°$ *at* O. *Find expressions in terms of* x *and the radius,* r, *for* (i) *the length of the arc* AB, (ii) *the area of the sector* OAB. (See Fig. 16.1.)

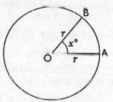

Fig. 16.1

(i) The length of an arc of a given circle is proportional to the angle it subtends at the centre. But an angle of 360° is subtended by an arc of length $2\pi r$, therefore an angle of $x°$ is subtended by an arc of length

$$\frac{x}{360} \times 2\pi r.$$

327

Therefore the length of arc AB is $\left(\dfrac{\pi}{180}\right)xr$.

(ii) The area of a sector of a given circle is proportional to the angle at the centre. But a sector containing an angle of 360° is the whole circle, which has an area of πr^2, therefore a sector containing an angle of $x°$ has an area of

$$\frac{x}{360} \times \pi r^2.$$

Therefore the area of the sector OAB is $\frac{1}{2}\left(\dfrac{\pi}{180}\right)xr^2$.

Thus, in both the length of an arc and the area of a sector, there appears a factor of $\pi/180$, which is due to the unit of measurement of the angle AOB. This suggests a new unit for measuring angles, which is called a **radian**, such that an

$$\text{angle in radians} = \frac{\pi}{180} \text{ (angle in degrees)}.$$

If we let θ radians equal x degrees, then, referring to Fig. 16.1,

the **length of arc AB = $r\theta$,**

and   the **area of sector OAB = $\frac{1}{2}r^2\theta$.**

If, then, we construct an angle of 1 radian, the arc AB will be of length r, and so *an arc of a circle equal to the radius subtends at the centre an angle of 1 radian*. Radians are sometimes termed circular measure, and are denoted by rad. It follows from the relation above, by putting the angle in degrees equal to 180, that

$$\pi \text{ rad} = 180°.$$

Hence 1 radian = 57·296 degrees and 1 degree = 0·017 453 radians, both correct to five significant figures.

The use of radians extends far beyond finding lengths of arcs and areas of sectors. In later sections it is shown how they have applications in mechanics and calculus.

Exercise 16a (Oral)

1. Convert to degrees:

   (i) $\dfrac{\pi}{2}$ rad,   (ii) $\dfrac{\pi}{4}$ rad,   (iii) $\dfrac{\pi}{3}$ rad,   (iv) $\dfrac{2\pi}{3}$ rad,

   (v) $\dfrac{\pi}{6}$ rad,   (vi) $\dfrac{3\pi}{2}$ rad,   (vii) $\dfrac{5\pi}{2}$ rad,   (viii) 4π rad,

   (ix) 5π rad,   (x) $\dfrac{4\pi}{3}$ rad,   (xi) $\dfrac{7\pi}{2}$ rad,   (xii) $\dfrac{3\pi}{4}$ rad.

2. Convert to radians, leaving π in your answer:

   (i) $360°$,   (ii) $90°$,   (iii) $45°$,   (iv) $15°$,

   (v) $60°$,   (vi) $120°$,   (vii) $300°$,   (viii) $270°$,

   (ix) $540°$,   (x) $30°$,   (xi) $150°$,   (xii) $450°$.

3. What is the length of an arc which subtends an angle of $0·8$ rad at the centre of a circle of radius 10 cm?

4. An arc of a circle subtends an angle of $1·2$ rad at any point on the remaining part of the circumference. Find the length of the arc, if the radius of the circle is 4 cm.

5. An arc of a circle subtends an angle of $0·5$ rad at the centre. Find the radius of the circle, if the length of the arc is 3 cm.

6. Find, in radians, the angle subtended at the centre of a circle of radius $2·5$ cm by an arc 2 cm long.

7. What is the area of a sector containing an angle of $1·5$ rad, in a circle of radius 2 cm?

8. The radius of a circle is 3 cm. What is the angle contained by a sector of area 18 cm^2?

9. An arc subtends an angle of 1 rad at the centre of a circle, and a sector of area 72 cm^2 is bounded by this arc and two radii. What is the radius of the circle?

10. The arc of a sector in a circle, radius 2 cm, is 4 cm long. What is the area of the sector?

Exercise 16b

1. Express in radians, leaving π in your answers:

   (i) $22\frac{1}{2}°$,   (ii) $1080°$,   (iii) $12'$,   (iv) $37°\ 30'$.

2. Express in degrees:

   (i) $\dfrac{2\pi}{5}$ rad,   (ii) $\dfrac{\pi}{36}$ rad,   (iii) $\dfrac{7\pi}{12}$ rad,   (iv) $\dfrac{7\pi}{2}$ rad.

3. Find the length of an arc of a circle, which subtends an angle of 31° at the centre, if the radius of the circle is 5 cm.

4. The chord AB of a circle subtends an angle of 60° at the centre. What is the ratio of chord AB to arc AB?

5. An arc of a circle, radius 2·5 cm, is 3 cm long. What is the angle subtended by the arc at the centre
   (i) in radians,
   (ii) in degrees and minutes?

6. A segment is cut off a circle of radius 5 cm by a chord AB, 6 cm long. What is the length of the minor arc AB?

7. Find from tables the values of:
   (i) sin 1 rad,   (ii) cos 1·27 rad,   (iii) tan 2 rad,
   (iv) sin 10 rad,   (v) tan − 0·7 rad,   (vi) sec 0·61 rad.

8. What is the area of a sector containing an angle of 1·4 rad in a circle whose radius is 2·4 cm?

9. A chord AB subtends an angle of 120° at O, the centre of a circle with radius 12 cm. Find the area of:
   (i) sector AOB,
   (ii) triangle AOB,
   (iii) the minor segment AB.

10. An arc AB of a circle with radius 6 cm subtends an angle of 40° at the centre. Find the area bounded by the diameter BC, CA and the arc AB.

11. Two equal circles of radius 5 cm are situated with their centres 6 cm apart. Calculate what area lies within both circles.

12. A chord PQ of a circle with radius r, subtends an angle θ at the centre. Show that the area of the minor segment PQ is $\frac{1}{2}r^2(\theta - \sin \theta)$, and write down the area of the major segment PQ in terms of r and θ.

13. AB is a diameter of a circle with radius a. C is a point on the circumference such that angle CAB = ϕ. Show that the perimeter of the area bounded by CA, AB, and the arc BC is $2a(\cos \phi + 1 + \phi)$; and find an expression for the area in terms of a and ϕ.

14. A circle of radius r is drawn with its centre on the circumference of another circle of radius r. Show that the area common to both circles is $2r^2\left(\dfrac{\pi}{3} - \dfrac{\sqrt{3}}{4}\right)$.

Graphical solution of equations

16.2.   *A chord* AB *subtends an angle* θ *at the centre of a circle of radius r.   For what value of* θ *does the chord bisect the sector* OAB?   (See Fig. 16.2.)

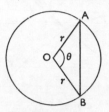

Fig. 16.2

The area of sector OAB = $\frac{1}{2}r^2\theta$, and the area of triangle OAB = $\frac{1}{2}r^2 \sin \theta$, therefore if AB bisects the sector,

$$\tfrac{1}{2} r^2\theta = 2(\tfrac{1}{2}r^2 \sin \theta).$$
$$\therefore \theta = 2 \sin \theta.$$

It must, of course, be remembered that θ is in radians. We solve this equation graphically.

First stage.   Rewrite the equation as

$$\sin \theta = \tfrac{1}{2}\theta,$$

(this simplifies the plotting slightly) and plot the graphs of sin θ and $\frac{1}{2}\theta$. It is sufficient, at this stage, to take values of θ at intervals of 10°, and a glance at Fig. 16.2 suggests that we may expect a root rather more than $\frac{1}{2}\pi$ rad. (Note that although θ is in radians, tables of sin θ are given in degrees, so it is advisable to choose a scale on the axis of θ from which angles can easily be read off in degrees.) The graph of $y = \sin \theta$ has been plotted directly from sine tables in Fig. 16.3 and radian tables give the values of $\frac{1}{2}\theta$ corresponding to 80° and 130°. From the point of intersection of the two graphs we see that the required root of the equation lies between 108° and 110°.

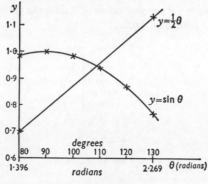

Fig. 16.3

Second stage. An enlargement of part of the first graph is now drawn. Angles have been taken at half degree intervals from 108° to 110°, as in Fig. 16.4, and the root may be read off as 108° 36′. Radian tables convert this to 1·895 rad. Therefore when $\theta = 1·895$ the chord bisects the sector.

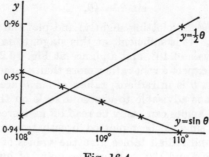

Fig. 16.4

Qu. 1. What graphs would you plot in order to solve the following equations?

(i) $\tan \theta = 2\theta$,

(ii) $\theta = 1 + \sin 2\theta$,

(iii) $2\theta = 1 - \cos \theta$,

(iv) $4 \sin^2 \frac{1}{2}\theta = \theta$,

(v) $\theta \tan \theta = 2$.

Exercise 16c

Solve the following equations graphically, taking values of θ between the limits indicated.

1. $\tan \theta = 1 + \theta$, from 0 to $\frac{1}{2}\pi$.

2. $\theta = \cos \theta$, from 0 to $\frac{1}{2}\pi$.

3. $\theta = 3 \sin 2\theta$, from 0 to $\frac{1}{2}\pi$.

4. $2\theta = 2 \sec \theta - 3$, from $-\frac{1}{2}\pi$ to $\frac{1}{2}\pi$.

5. $\theta = 2 \sin^2 \theta$, from 0 to π.

6. $2 \sin \theta + \cos \theta = 1 \cdot 5$, from 0 to π.

7. $3 \cos \theta - 4 \sin \theta = 2$, from $-\pi$ to $\frac{1}{2}\pi$.

8. An arc **AB** of a circle subtends an angle 2θ at the centre, and the tangents to the circle at **A** and **B** meet at **T**. For what value of θ is the length of arc **AB** equal to the length of **TA**?

9. A chord **AB**, which divides the area of a circle in the ratio $3:1$, subtends an angle θ at the centre. Find graphically the value of θ.

10. A belt runs round a circular pulley and a driving shaft of negligible radius. If the length of the belt is 10 times the radius of the pulley, what angle does the part of the pulley which is not in contact with the belt subtend at the centre?

Angular velocity

16.3. A man who buys an electric motor is usually interested in the rate at which it goes, and he may be told that it does 12 000 revolutions per minute (rev/min). On the other hand the drum of a barograph turns at the rate of 49 degrees per day. In either case the rate of turning, which is called average angular velocity, is given by

$$\text{average angular velocity} = \frac{\text{angle turned}}{\text{time taken}}.$$

Qu. 2. Find the average angular velocity of the second hand of a watch:

(i) in degrees per second (deg/sec),
(ii) in rev/min.

Qu. 3. Convert:

(i) 500 rev/min into deg/sec,
(ii) 1 rev/week into deg/h.

In many cases of turning, however, the angular velocity is not constant, so consider the average angular velocity in a small interval of time Δt. If the angle turned through in this time is $\Delta \theta$ radians,

$$\text{average angular velocity} = \frac{\Delta \theta}{\Delta t} \text{ rad/sec.}$$

But as $\Delta t \to 0$,

$$\frac{\Delta \theta}{\Delta t} \to \frac{d\theta}{dt},$$

$\therefore$ average angular velocity $\to \dfrac{d\theta}{dt}$.

$\dfrac{d\theta}{dt}$ is called **angular velocity** and is denoted by ω (the Greek letter omega). Therefore

$$\boldsymbol{\omega = \frac{d\theta}{dt}.}$$

[In motion in a straight line

$$\text{average velocity} = \frac{\text{distance}}{\text{time}},$$

and if a distance Δs is travelled in a time Δt,

$$\text{average velocity} = \frac{\Delta s}{\Delta t}.$$

But $\dfrac{\Delta s}{\Delta t} \to \dfrac{\mathrm{d}s}{\mathrm{d}t}$ as $\Delta t \to 0$ and so the velocity at an instant is given by

$$v = \frac{\mathrm{d}s}{\mathrm{d}t}.$$

In this way there is a parallel between linear motion and angular motion.]

If a particle moves in a circle of radius r with speed v and angular velocity ω about the centre, the relation between r, v, ω can be obtained from one of the results obtained in §16.1. If s is the distance of the particle measured along the circumference of the circle from a fixed point,

$$s = r\theta.$$

Differentiating with respect to time (remember r is constant),

$$\frac{\mathrm{d}s}{\mathrm{d}t} = r\frac{\mathrm{d}\theta}{\mathrm{d}t}.$$

$$\therefore\ v = r\omega.$$

Remember that ω must be measured in radians/unit time. Three sets of possible units for v, r, ω are shown in the table below:

	v	r	ω
(i)	m/s	m	rad/s
(ii)	km/h	km	rad/h
(iii)	cm/min	cm	rad/min

Example 2. *A belt runs round a pulley attached to the shaft of a motor. If the belt runs at 0·75 m/s and the radius of the pulley is 6 cm, find the angular velocity of the pulley* (i) *in rad/s*, (ii) *in rev/min*.

(i) Using the result $v = r\omega$,

$$\omega = \frac{75}{6} = 12\!\cdot\!5 \text{ rad/s}.$$

(ii) $12 \cdot 5$ rad/s $= \dfrac{12 \cdot 5}{2\pi}$ rev/s,

$= \dfrac{12 \cdot 5}{2\pi} \times 60$ rev/min,

$\simeq 120$ rev/min.

(The sign $\simeq$ means "is approximately equal to".)   Therefore the angular velocity is $12 \cdot 5$ rad/s or approximately 120 rev/min.

Exercise 16d

(Use the result $v = r\omega$ where you can.)

1. Express the angular velocity of the minute hand of a clock in:
   (i) rev/min,   (ii) deg/s,   (iii) rad/s.

2. A wheel is turning at 200 rev/min.   Express this angular velocity in:
   (i) deg/s,   (ii) rad/s.

3. A cook can rotate the handle of her egg whisk 32 times in 5 seconds.   Each time the handle rotates, the paddles rotate four times.   At what speed are the paddles rotating in:
   (i) rev/min,   (ii) rad/s?

4. The Earth rotates on its axis approximately $366\frac{1}{4}$ times in a year.   Calculate its angular velocity in rad/h, correct to three significant figures.

5. The cutters of a well-known electric shaver rotate about 3000 times a minute, and the distance from the axis to the tip of the cutter is $0 \cdot 65$ cm.   Find:
   (i) the angular velocity of the cutter in rad/s,
   (ii) the speed of the tip of a cutter in cm/s.

6. When I dial 0 on the telephone, the dial rotates through $334°$ in $1\frac{1}{2}$ s approximately.   What is the average angular speed of the dial in rad/s, and what is the speed of a point on the circumference of the dial if its diameter is 8 cm?

7. A motor runs at 1200 rev/min.   What is its angular velocity in rad/s?   If the shaft of the motor is $2 \cdot 5$ cm in

diameter, at what speed is a point on the circumference of the shaft moving?

8. A point on the rim of a wheel of diameter 2·5 m is moving at a speed of 44 m/s relative to the axis. At what rate in: (i) rad/s, (ii) rev/min, is the wheel turning?

9. If a cotton reel drops 1·76 m in 0·7 s, the end of the cotton being held still, at what average angular velocity, in rev/min, is the reel turning, if its diameter is 3 cm?

10. A belt runs round two pulleys of diameters 26·25 cm and 15 cm. If the larger rotates 700 times in a minute, find the angular velocity of the smaller in rad/s.

11. The diameter of a gramophone record is 30 cm. At what speed in m/s is a point on the circumference moving when the turn-table is rotating at 78 rev/min? What would be the speed if the turn-table was set to turn at $33\frac{1}{3}$ rev/min by mistake?

12. The Earth moves round the sun approximately in a circle of radius 150 000 000 km. Find its angular speed in rad/s, and obtain its speed along its orbit in km/s.

13. Taking the Earth to be a sphere of radius 6300 km which rotates about its axis once in 23·93 hours what error will be made in calculating the velocity of a point on the equator, if it is assumed that the Earth rotates once in 24 hours? Express your answer in km/h, correct to two significant figures.

Small angles

16.4 A glance at Fig. 16.5 will show the reader that, for small acute angles, $\tan \theta$, θ and $\sin \theta$ are practically equal. That this is so is borne out by seven-figure tables:

Angle in degrees	10°	5°	1°
θ (radians)	0·174 532 9	0·087 266 5	0·017 453 3
$\tan \theta$	0·176 327 0	0·087 488 7	0·017 455 1
$\sin \theta$	0·173 648 2	0·087 155 7	0·017 452 4

We shall now consider this geometrically.

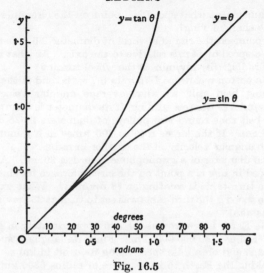

Fig. 16.5

In Fig. 16.6, the chord AB subtends an angle θ at the centre of a circle of radius r, and the tangent at B meets

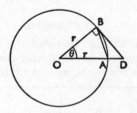

Fig. 16.6

OA at D. Consider the three areas $\triangle$AOB, sector AOB, $\triangle$DOB.

    (i) In $\triangle$AOB, two sides of length r include an angle θ, therefore its area is $\frac{1}{2}r^2 \sin \theta$.

(ii) From §16.1, the area of sector AOB is $\frac{1}{2}r^2\theta$.

(iii) In $\triangle$DOB, B is a right angle, therefore $BD = r\tan\theta$ and so its area is $\frac{1}{2}r^2\tan\theta$.

From the figure it can be seen that

$$\triangle\text{AOB} < \text{sector AOB} < \triangle\text{DOB}.$$
$$\therefore \ \tfrac{1}{2}r^2\sin\theta < \tfrac{1}{2}r^2\theta < \tfrac{1}{2}r^2\tan\theta.$$

But if we divide each by $\frac{1}{2}r^2$ the order of magnitude is unchanged, therefore

$$\sin\theta < \theta < \tan\theta,$$

providing θ is acute, as the figure requires. Again, if we divide each term by $\sin\theta$, the order of magnitude is unchanged, therefore

$$\frac{\sin\theta}{\sin\theta} < \frac{\theta}{\sin\theta} < \frac{\tan\theta}{\sin\theta}.$$

But $\tan\theta = \dfrac{\sin\theta}{\cos\theta}$, therefore

$$1 < \frac{\theta}{\sin\theta} < \frac{1}{\cos\theta}.$$

Now as $\theta \to 0$, $\cos\theta \to 1$,

$$\therefore \ \frac{1}{\cos\theta} \to 1.$$

Thus $\dfrac{\theta}{\sin\theta}$ lies between 1 and a function which approaches 1 as $\theta \to 0$.

$$\therefore \ \frac{\theta}{\sin\theta} \to 1 \quad \text{as } \theta \to 0.$$

This limit (or, more strictly, $(\sin\theta)/\theta \to 1$ as $\theta \to 0$) is required in the next section for the differentiation of $\sin x$.

Another way of expressing the statement that $\theta/\sin\theta \to 1$ as $\theta \to 0$, is to say that, for small values of θ,

$$\sin\theta \simeq \theta.$$

An approximation for $\cos \theta$ is obtained from the identity
$$\cos \theta = 1 - 2 \sin^2 \tfrac{1}{2}\theta.$$

If θ is small, $\sin \tfrac{1}{2}\theta \simeq \tfrac{1}{2}\theta$, therefore
$$\cos \theta \simeq 1 - 2(\tfrac{1}{2}\theta)^2.$$

Therefore, for small values of θ,
$$\boldsymbol{\cos \theta \simeq 1 - \tfrac{1}{2}\theta^2.}$$

Example 3. *Find the approximate value of* $\dfrac{1 - \cos 2\theta}{\theta \tan \theta}$ *when θ is small.*

We cannot put $\theta = 0$, as the numerator and denominator would both be zero.

Since $\cos \theta \simeq 1 - \tfrac{1}{2}\theta^2$,
$$\cos 2\theta \simeq 1 - \tfrac{1}{2}(2\theta)^2 = 1 - 2\theta^2.$$

Therefore the numerator $\simeq 2\theta^2$. But the denominator $\simeq \theta^2$, since $\tan \theta \simeq \theta$. Therefore, when θ is small,
$$\frac{1 - \cos 2\theta}{\theta \tan \theta} \simeq \frac{2\theta^2}{\theta^2}.$$

$$\therefore \quad \frac{1 - \cos 2\theta}{\theta \tan \theta} \simeq 2 \quad \text{when } \theta \text{ is small.}$$

Qu. 4. Find approximations for the following functions when θ is small.

(i) $\dfrac{\sin 3\theta}{2\theta}$,

(ii) $\dfrac{\sin 4\theta}{\sin 2\theta}$,

(iii) $\dfrac{1 - \cos \theta}{\theta^2}$,

(iv) $\dfrac{\theta \sin \theta}{1 - \cos 2\theta}$,

(v) $\dfrac{\sin (\alpha + \theta) \sin \theta}{\theta}$,

(vi) $\dfrac{\sin (\alpha + \theta) - \sin \alpha}{\theta}$,

(vii) $\dfrac{\sin \theta \tan \theta}{1 - \cos 3\theta}$,

(viii) $\sin \theta \operatorname{cosec} \tfrac{1}{2}\theta$,

(ix) $\dfrac{\tan (\alpha + \theta) - \tan \alpha}{\theta}$.

Derivatives of sin x and cos x

16.5. The graph of sin x may be sketched:

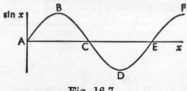

Fig. 16.7

and from it may be obtained a rough graph of its gradient. The gradient is zero at B, D, F, positive from A to B and from D to F, and negative from B to D, giving a graph of this sort:

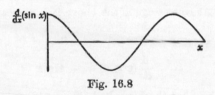

Fig. 16.8

Qu. 5. Does this resemble any graph you have met so far?

Qu. 6. Express sin A − sin B in factors.

The derivative of sin x will now be found from first principles.

Let $y = \sin x$, where x is in radians.

$$\therefore \ y + \Delta y = \sin (x + \Delta x).$$

Subtracting,

$$\Delta y = \sin (x + \Delta x) - \sin x,$$

$$= 2 \cos \left(\frac{2x + \Delta x}{2} \right) \sin \frac{\Delta x}{2}.$$

$$\therefore \frac{\Delta y}{\Delta x} = \frac{2 \cos\left(\dfrac{2x + \Delta x}{2}\right) \sin \dfrac{\Delta x}{2}}{\Delta x},$$

$$= \frac{\cos\left(\dfrac{2x + \Delta x}{2}\right) \sin \dfrac{\Delta x}{2}}{\dfrac{\Delta x}{2}}.$$

But as $\Delta x \to 0$, $\cos\left(\dfrac{2x + \Delta x}{2}\right) \to \cos x$ and $\dfrac{\sin \dfrac{\Delta x}{2}}{\dfrac{\Delta x}{2}} \to 1.$

$$\therefore \frac{dy}{dx} = \cos x.$$

$$\therefore \frac{d}{dx}(\sin x) = \cos x.$$

Qu. 7. At what stage in the above is it necessary to have x in radians?

Qu. 8. Prove from first principles that

$$\frac{d}{dx}(\cos x) = -\sin x$$

Remember that these results hold only if x is in radians.

Example 4. *Differentiate* (i) $\sin(2x + 3)$, (ii) $\cos^2 x$, (iii) $\sin x°$.

(i) Let $y = \sin(2x + 3)$, $t = 2x + 3$, then $y = \sin t$.

$$\therefore \frac{dy}{dt} = \cos t, \qquad \frac{dt}{dx} = 2.$$

But $\dfrac{dy}{dx} = \dfrac{dy}{dt} \cdot \dfrac{dt}{dx} = (\cos t)2,$

$$\therefore \frac{d}{dx}\{\sin(2x + 3)\} = 2\cos(2x + 3).$$

(ii) Let $y = \cos^2 x$, $t = \cos x$, then $y = t^2$.

$$\frac{dy}{dx} = \frac{dy}{dt} \cdot \frac{dt}{dx} = 2t(-\sin x) = -2 \cos x \sin x.$$

$$\therefore \frac{d}{dx}(\cos^2 x) = -\sin 2x.$$

(iii) Let $y = \sin x°$. Now $x° = \frac{\pi}{180}x$ radians,

$$\therefore y = \sin \frac{\pi}{180}x°.$$

Put $t = \frac{\pi}{180}x$, then $\qquad y = \sin t$.

$$\frac{dy}{dx} = \frac{dy}{dt} \cdot \frac{dt}{dx} = (\cos t)\frac{\pi}{180} = \frac{\pi}{180} \cos \frac{\pi}{180}x° = \frac{\pi}{180} \cos x°.$$

$$\therefore \frac{d}{dx}(\sin x°) = \frac{\pi}{180} \cos x°.$$

Qu. 9. Differentiate (i) $\cos 3x$, (ii) $\sin^2 x$, (iii) $2 \sin 2x$, (iv) $\cos^3 x$.

Example 5. *Integrate* (i) $\cos 2x$, (ii) $3 \sin \frac{1}{2}x$.

The method used here is to change cos to sin, or sin to cos, and to determine the coefficient by differentiation:

(i) $\dfrac{d}{dx}(\sin 2x) = 2 \cos 2x.$

$$\therefore \frac{d}{dx}\left(\tfrac{1}{2} \sin 2x\right) = \cos 2x.$$

$$\therefore \int \cos 2x\, dx = \tfrac{1}{2} \sin 2x + c.$$

(ii) $\dfrac{d}{dx}(3 \cos \tfrac{1}{2}x) = -\tfrac{3}{2} \sin \tfrac{1}{2}x.$

$$\therefore \frac{d}{dx}(-2.3 \cos \tfrac{1}{2}x) = 3 \sin \tfrac{1}{2}x.$$

$$\therefore \int 3 \sin \tfrac{1}{2}x\, dx = -6 \cos \tfrac{1}{2}x + c.$$

Exercise 16e

1. Differentiate:

(i) $\cos 2x$,   (ii) $\sin 6x$,   (iii) $\cos (3x - 1)$,

(iv) $\sin (2x - 3)$,   (v) $- 3 \cos 5x$,   (vi) $2 \sin 4x$,

(vii) $- 4 \sin \frac{3}{2}x$,   (viii) $2 \sin \frac{1}{2}(x + 1)$,   (ix) $\sin x^2$.

2. Integrate:

(i) $\sin 3x$,   (ii) $\cos 3x$,   (iii) $2 \sin 4x$,

(iv) $2 \cos 2x$,   (v) $- \frac{1}{2} \sin 6x$,   (vi) $6 \cos 4x$,

(vii) $\sin (2x + 1)$,   (viii) $3 \cos (2x - 1)$,

(ix) $\frac{2}{3} \sin \frac{1}{2}x$.

3. Differentiate:

(i) $\sin^2 x$,   (ii) $4 \cos^2 x$,   (iii) $\cos^3 x$,

(iv) $2 \sin^3 x$,   (v) $3 \cos^4 x$,   (vi) $\sqrt{(\sin x)}$,

(vii) $\sqrt{(\cos x)}$,   (viii) $\cos^2 3x$,   (ix) $\sin^2 2x$,

(x) $- 2 \sin^3 3x$,   (xi) $3 \sin^4 2x$,   (xii) $\sqrt{(\sin 2x)}$.

4. Differentiate:

(i) $x \cos x$,   (ii) $x \sin 2x$,   (iii) $x^2 \sin x$,

(iv) $\sin x \cos x$,   (v) $\dfrac{\sin x}{x}$,   (vi) $\dfrac{\cos 2x}{x}$,

(vii) $\dfrac{x}{\sin x}$,   (viii) $\dfrac{x^2}{\cos x}$,   (ix) $\dfrac{\sin x}{\cos x}$,

(x) $\cot x$,   (xi) $\dfrac{1}{\cos x}$,   (xii) $\operatorname{cosec} x$.

5. A particle moves in a straight line such that its velocity in m/s, t s after passing through a fixed point O, is $3 \cos t - 2 \sin t$. Find:

(i) its distance from O after $\frac{1}{2}\pi$ s,

(ii) its acceleration after π s,

(iii) the time when its velocity is first zero.

6. A particle is moving in a straight line in such a way that its distance from a fixed point O, t s after the motion begins, is $\cos t + \cos 2t$ cm. Find:

(i) the time when the particle first passes through O,

(ii) the velocity of the particle at this instant,

(iii) the acceleration when the velocity is zero.

7. The distance of a particle from a fixed point O is given by $s = 3 \cos 2t + 4 \sin 2t$. Show that the velocity v and the acceleration a are given by $v^2 + 4s^2 = 100$, $a + 4s = 0$. Hence find:

  (i) the greatest distance of the particle from O,

  (ii) the acceleration at this instant.

8. The velocity at time t of a particle moving in a straight line is $6 \cos 2t + \cos t$, and when $t = 0$, the particle is at O. Find:

  (i) the time when v is first zero,

  (ii) the distance from O at this instant,

  (iii) the acceleration at the same instant.

9. Find the area between the curve $y = \sin 3x$ and the x-axis between $x = 0$ and $x = \frac{1}{3}\pi$.

10. Sketch the curve $y = 1 + \cos x$ from $x = -\pi$ to $x = \pi$, and find the area enclosed by the curve and the x-axis between these limits.

11. Find the maximum value of $y = x + \sin 2x$ which is given by a value of x between 0 and $\frac{1}{2}\pi$. Sketch the graph of y for acute values of x and find the area bounded by the curve, the x-axis and the line $x = \frac{1}{2}\pi$.

12. Find the maximum value of $y = 2 \sin x - x$ which is given by a value of x between 0 and $\frac{1}{2}\pi$. Sketch the graph of y for values of x from 0 to π, and find the area between the curve, the x-axis and the line $x = \frac{1}{2}\pi$.

13. Show that $\dfrac{d}{dx}(\frac{1}{2}x - \frac{1}{4}\sin 2x) = \sin^2 x$ and deduce that

$$\int_0^\pi \sin^2 x \, dx = \tfrac{1}{2}\pi.$$

14. Express $\cos^2 x$ in terms of $\cos 2x$, and hence show that

$$\int \cos^2 x \, dx = \tfrac{1}{2}x + \tfrac{1}{4}\sin 2x + c.$$

15. Show that $\cos^3 x = \frac{1}{4}(\cos 3x + 3 \cos x)$, and deduce that

$$\int \cos^3 x \, dx = \tfrac{1}{12}\sin 3x + \tfrac{3}{4}\sin x + c = \sin x - \tfrac{1}{3}\sin^3 x + c.$$

16. By expressing $\sin^3 x$ in terms of $\sin x$ and $\sin 3x$, show that

$$\int \sin^3 x \, dx = \tfrac{1}{12}\cos 3x - \tfrac{3}{4}\cos x + c = \tfrac{1}{3}\cos^3 x - \cos x + c.$$

17. Express $2 \cos 5x \cos 3x$ as a sum of two cosines and hence evaluate

$$\int_0^{\frac{1}{4}\pi} 2 \cos 5x \cos 3x \, \mathrm{d}x$$

Derivatives of tan *x*, cot *x*, sec *x*, cosec *x*

16.6. Using the derivatives of $\sin x$ and $\cos x$, those of the four other trigonometrical ratios can be obtained by writing

$$\tan x = \frac{\sin x}{\cos x}, \quad \cot x = \frac{\cos x}{\sin x},$$

$$\sec x = \frac{1}{\cos x}, \quad \mathrm{cosec} \, x = \frac{1}{\sin x}.$$

This is left as an exercise for the reader, if he has not already done No. 4 (ix)–(xii) of Exercise 16e. The results are:

$$\frac{\mathrm{d}}{\mathrm{d}x}(\tan x) = \sec^2 x, \quad \frac{\mathrm{d}}{\mathrm{d}x}(\cot x) = -\operatorname{cosec}^2 x,$$

$$\frac{\mathrm{d}}{\mathrm{d}x}(\sec x) = \sec x \tan x, \quad \frac{\mathrm{d}}{\mathrm{d}x}(\operatorname{cosec} x) = -\operatorname{cosec} x \cot x.$$

NOTE: (i) the similarity of the pairs of formulae on the two lines.

(ii) the associations between $\tan x$ and $\sec x$, and between $\cot x$ and $\operatorname{cosec} x$. The same associations occur in the identities $1 + \tan^2 x = \sec^2 x$, $\cot^2 x + 1 = \operatorname{cosec}^2 x$.

(iii) that the derivatives of ratios beginning with "co"—$\cos x$, $\cot x$, $\operatorname{cosec} x$—all have a negative sign.

Example 6. *Find, without using trigonometrical tables, an approximation for* $\tan 45° \, 10'$.

We know that $\tan 45° = 1$, and so the value of $\tan 45° \, 10'$ can be found approximately by the method of small increases (§6.6).

Let $y = \tan x$, $\therefore \dfrac{dy}{dx} = \sec^2 x$.

But $\Delta y \simeq \dfrac{dy}{dx} \Delta x$, $\therefore \Delta y \simeq \sec^2 x \, \Delta x$.

Put $x = \frac{1}{4}\pi$ radians and $\Delta x = \dfrac{1}{6} \times \dfrac{\pi}{180}$ radians, then

$$\Delta y \simeq \sec^2 \tfrac{1}{4}\pi \times \dfrac{\pi}{1080} = 2 \times \dfrac{\pi}{1080} = 0.005\ 817\ldots$$

$$\therefore \tan 45° 10' \simeq 1.0058.$$

The reader must not expect to get answers correct to four places of decimals every time he uses this method. That such accuracy has been obtained in this example, is due to the small size of Δx. It is beyond the scope of this book to give a method of determining the accuracy of the answer, but it may be mentioned that, *in this case*, a better approximation is given by $1 + 2\Delta x + 2(\Delta x)^2$.

Exercise 16f

1. Differentiate:

(i) $\tan 2x$,   (ii) $\cot 3x$,   (iii) $3 \sec 2x$,

(iv) $2 \operatorname{cosec} \frac{1}{2}x$,   (v) $-\tan(2x + 1)$,

(vi) $\frac{1}{3} \sec(3x - 2)$,   (vii) $-2 \cot(3x + 2)$,

(viii) $\cot x^2$,   (ix) $\tan \sqrt{x}$.

2. Differentiate:

(i) $\tan^2 x$,   (ii) $\sec^2 x$,   (iii) $2 \cot^3 x$,

(iv) $3 \operatorname{cosec}^2 x$,   (v) $-\tan^2 2x$, *   (vi) $\frac{1}{2} \cot^2 3x$,

(vii) $\frac{1}{6} \sec^3 2x$,   (viii) $-2 \operatorname{cosec}^4 x$,   (ix) $\sqrt{(\tan x)}$.

3. Differentiate:

(i) $x \tan x$,   (ii) $\sec x \tan x$,   (iii) $x^2 \cot x$,

(iv) $3x \operatorname{cosec} x$,   (v) $\operatorname{cosec} x \cot x$,   (vi) $\dfrac{\tan x}{x}$,

(vii) $\dfrac{\sec x}{x^2}$,   (viii) $\sin x - x \cos x$,

(ix) $x \sec^2 x - \tan x$.

---

* The following method of working often overcomes the initial difficulty some students find with this type of "function of a function":

$$\dfrac{d}{dx}(3 \sin^4 5x) = \dfrac{d}{dx}\left\{3(\sin 5x)^4\right\} = 3.4(\sin 5x)^3 . \cos 5x . 5 = 60 \sin^3 5x \cos 5x.$$

4. Integrate:

(i) $\sec^2 2x$,   (ii) $3 \sec x \tan x$,   (iii) $- \operatorname{cosec}^2 \frac{1}{2}x$,

(iv) $\frac{1}{3} \operatorname{cosec} 3x \cot 3x$,   (v) $2 \sec^2 x \tan x$,

(vi) $\dfrac{1}{\cos^2 x}$,   (vii) $\dfrac{\sin x}{\cos^2 x}$,

(viii) $\dfrac{1}{\sin^2 2x}$,   (ix) $\dfrac{\cos 2x}{\sin^2 2x}$.

5. Sketch the graph of the curve $y = \sec^2 x - 1$ between $x = -\frac{1}{2}\pi$ and $x = \frac{1}{2}\pi$. Calculate the area enclosed by the curve, the x-axis and the line $x = \frac{1}{4}\pi$.

6. Find the volume generated by revolving the area bounded by the x-axis, the lines $x = \pm \frac{1}{4}\pi$ and the curve $y = \sec x$ about the x-axis.

7. Find the minimum values of the following functions which are given by values of x between 0 and $\frac{1}{2}\pi$.

(i) $\tan x + 3 \cot x$,   (ii) $\sec x + 8 \operatorname{cosec} x$,

(iii) $6 \sec x + \cot x$.

8. Find, without using trigonometrical tables, approximations for:

(i) $\tan 60° 30'$, given $\sqrt{3} = 1\cdot7321$,

(ii) $\sec^2 45° 06'$,   (iii) $\cot 45° 20'$,

(iv) $\tan 29° 50'$, given $\sqrt{3} = 1\cdot7321$.

9. The height of a tree is calculated by measuring the angle of elevation of the top from a point on the ground 45 m from the foot. If the angle of elevation is measured as 22°, but is only correct to the nearest degree, what error can there be in the calculated height of the tree?

10. The horizontal distance between the ends of a ramp inclined at 8° 20′ to the horizontal is calculated from the distance along the ramp, 20·3 m, and the inclination, which is taken to be 8°. What will be the error?

11. By expressing $\tan^2 x$ in terms of $\sec^2 x$, show that

$$\int \tan^2 x \, dx = \tan x - x + c.$$

12. Express $\cot^2 x$ in terms of $\operatorname{cosec}^2 x$ and hence integrate $\cot^2 x$.

Exercise 16g (Miscellaneous)

1. Convert to degrees: (i) $\dfrac{2\pi}{5}$, (ii) $\dfrac{5\pi}{6}$, (iii) $\dfrac{3\pi}{8}$, (iv) $\dfrac{7\pi}{12}$.

2. Convert to radians, leaving π in your answers: (i) $330°$, (ii) $50°$, (iii) $75°$, (iv) $24°$.

3. Use tables to find the values of: (i) sin 2 rad, (ii) sec 0·5 rad, (iii) tan 1·32 rad, (iv) cos 2·98 rad.

4. The area of a sector of a circle, diameter 7 cm, is 18·375 cm². What is the length of the arc of the sector?

5. A sector with an area of $\frac{2}{3}$ cm² is bounded by an arc of length $\frac{5}{6}$ cm. What is the radius of the circle? Also find the angle contained by the sector, giving your answer in degrees and minutes.

6. A chord AB subtends a right angle at the centre of a circle of radius r. BC is a chord in the minor segment, inclined at 15° to BA. Show that the area bounded by the two chords and the arc AC is $\frac{1}{2}r^2(\frac{1}{3}\pi + \frac{1}{2}\sqrt{3} - 1)$.

7. The common chord of two circles of radii 13 cm and 37 cm is 24 cm long. Calculate the area common to both circles.

8. Show by means of a graph, that the equation $\theta = 1 + \sin\theta$ has only one root. Solve the equation graphically.

9. Draw the graph of cos 2θ for values of θ from $-\frac{1}{2}\pi$ to $\frac{1}{2}\pi$. Use your graph to solve the equation $\cos^2\theta = \frac{1}{4}(1 + \theta)$.

10. Draw the graph of cos 3θ + cos θ for values of θ from 0 to π, and find the roots of the equation

$$2\cos 3\theta + 2\cos\theta + 1 = 0$$

in this range.

11. A radar scanner rotates at a speed of 30 rev/min. Express this angular velocity in rad/s.

12. What is the angular velocity of the hour hand of a clock: (i) in rev/min, (ii) in rad/s?

13. A wheel of diameter 3 m is rotating with an angular velocity of 420 rev/min. Find:
    (i) the angular velocity of the wheel in rad/s,
    (ii) the velocity of a point on the circumference in km/h.
    [Take $\pi = 22/7$.]

14. A lift goes down a distance of 6 m in $3\frac{1}{2}$ s, and a cable to the counter-weight passes over a pulley of diameter 0·5 m. What is the average angular velocity of the pulley while the lift is in motion?

15. In order to investigate the effect of acceleration on the human body, a man is placed in a cabin which is made to travel in a circle of radius 10 m. If the speed of the cabin reaches 160 km/h, what is its angular velocity in rev/min at that instant?

16. Find approximations for the following when θ is small:

  (i) $\dfrac{\sin \theta \tan \theta}{\theta^2}$,   (ii) $\dfrac{1 - \cos 2\theta}{\theta \sin 3\theta}$,

  (iii) $\dfrac{\cos (\theta + a) - \cos a}{\theta}$.

17. Show that, if θ is small,

  (i) $\sin (\frac{1}{6}\pi + \theta) \simeq \frac{1}{2} + \frac{1}{2}\sqrt{3}\theta - \frac{1}{4}\theta^2$,

  (ii) $\cos (\frac{1}{4}\pi + \theta) \simeq \frac{1}{2}\sqrt{2}(1 - \theta - \frac{1}{2}\theta^2)$.

18. Differentiate:

  (i) $\sin 3x$,   (ii) $\tan \frac{1}{2}x$,   (iii) $\cos x^2$,

  (iv) $\sqrt{(\cos x)}$,   (v) $2 \operatorname{cosec}^3 x$,   (vi) $4 \sin^2 \frac{1}{2}x$,

  (vii) $- 3 \sec^3 2x$,   (viii) $\sqrt{(\sin 2x)}$,   (ix) $3 \tan^2 2x$.

19. Integrate:

  (i) $\cos 2x$,   (ii) $\sin (2x - 1)$,   (iii) $3 \cos \frac{1}{2}x$,

  (iv) $\sec^2 \frac{1}{2}x$,   (v) $\operatorname{cosec} x \cot x$,

  (vi) $\sec 2x \tan 2x$,   (vii) $\dfrac{\cos x}{\sin^2 x}$,

  (vi.i) $\dfrac{1}{\cos^2 2x}$ (ix) $x \sin x^2$.

20. Differentiate:

  (i) $x \sin x$,   (ii) $\sin x \cos 2x$,   (iii) $x^2 \tan^2 x$,

  (iv) $\dfrac{\sec x}{x}$,   (v) $\dfrac{\cos 2x}{\sin 3x}$,   (vi) $\sin x \tan 2x$,

  (vii) $\dfrac{\sin x}{x^2}$,   (viii) $2 \cos x + 2x \sin x - x^2 \cos x$.

21. If $x = a \sec \theta$, $y = b \tan \theta$, show that

$$\frac{dy}{dx} = \frac{b}{a} \operatorname{cosec} \theta \quad \text{and} \quad \frac{d^2y}{dx^2} = -\frac{b}{a^2} \cot^3 \theta.$$

22. If $x = a \cos \theta$, $y = b \sin \theta$, show that

$$\frac{d^2y}{dx^2} = -\frac{b}{a^2} \operatorname{cosec}^3 \theta.$$

23. A particle travels in a straight line in such a way that its distance from a fixed point O after time t is

$$3 \cos 2t + 4 \sin 2t.$$

Find

(i) the distance of the particle from O when it first comes to rest instantaneously,

(ii) its acceleration at this instant,

(iii) its maximum velocity.

24. A particle is moving in a straight line with velocity $\sin 2t + 7 \sin t$ cm/s, t s after passing through a fixed point O. Find:

(i) the maximum velocity of the particle,

(ii) the greatest distance of the particle from O.

25. Evaluate:

(i) $\displaystyle\int_0^{\frac{1}{2}\pi} \sin 2x \, dx$,

(ii) $\displaystyle\int_{-\frac{1}{3}\pi}^{\frac{1}{4}\pi} \sec^2 x \, dx$,

(iii) $\displaystyle\int_0^{\pi} \sin^2 x \, dx$,

(iv) $\displaystyle\int_0^{\frac{1}{4}\pi} \cos 3x \sin 5x \, dx$.

26. Find, without using trigonometrical tables, approximations for:

(i) $\cos 60° \, 15'$;   (ii) $\sin 45° \, 10'$, given $\sqrt{2} = 1·4142$.

27. Show that the maximum value of $2 \sin 2t + 31 \sin t$ is $\frac{63}{16}\sqrt{63}$.

28. Find the minimum value of the function $\frac{1}{2} \tan x - 4 \sin x$.

29. Find the turning values of the function $\tan x - 6 \operatorname{cosec} x$ and determine the natures of the turning points.

CHAPTER 17

LOCI

17.1. In coordinate geometry, the properties of curves (including straight lines) are investigated by algebra, and the first step in such an investigation is to represent the curve by an equation. So far the reader has only been shown how to find equations of straight lines, and the purpose of the first sections of this chapter is to show some ways in which other equations may be found.

In order to find the equation of a curve, we need to know some condition which must be satisfied by points which lie on it. For instance, we may be told that a point P is equidistant from two fixed points A and B. Then we know from elementary geometry that the locus of P is the perpendicular bisector of AB.* The idea of a locus is twofold, and is illustrated by this example: (i) every point on the perpendicular bisector is equidistant from A and B, (ii) every point which is equidistant from A and B lies on the perpendicular bisector.

In terms of algebra, a locus is represented by an equation such that (i) every point whose coordinates satisfy the equation lies on the locus, (ii) the coordinates of every point which lies on the locus satisfy the equation.

If, for example, A and B are the points $(3, 2)$ and $(5, - 1)$, we may readily find the equation connecting the coordinates (x, y) of a point P which is equidstant from A and B. Expressed geometrically, the condition to be satisfied by P is

$$PA = PB,$$

---

* It is to be understood in this book that we are only concerned with the geometry of a plane.

however, since we shall use Pythagoras' theorem to express PA and PB in terms of x and y, it is neater to square this equation, obtaining

$$PA^2 = PB^2.$$

Now $$PA^2 = (x - 3)^2 + (y - 2)^2$$

and $$PB^2 = (x - 5)^2 + (y + 1)^2,$$

therefore the equation which must be satisfied by the coordinates of P is

$$(x - 3)^2 + (y - 2)^2 = (x - 5)^2 + (y + 1)^2,$$

i.e.

$$x^2 - 6x + 9 + y^2 - 4y + 4$$
$$= x^2 - 10x + 25 + y^2 + 2y + 1.$$

Therefore the equation of the locus of points equidistant from $(3, 2)$ and $(5, -1)$ is $4x - 6y - 13 = 0$.

Because of the close connection between the locus and the equation connecting the coordinates of points lying on the locus, the equation itself is often referred to as the locus.

NOTE. When drawing graphs it is often useful to take different scales on the two axes, but in coordinate geometry the scales must be the same or the figures will be distorted. Thus the circle in Fig. 17.1 would appear elliptical (squashed) if the scales were different.

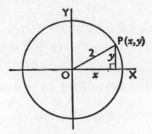

Fig. 17.1

Example 1. *Find the equation of the circle of radius two units with its centre at the origin.*

Let $P(x, y)$ be any point on the locus. Then by Pythagoras' theorem (see Fig. 17.1)

$$x^2 + y^2 = 4.$$

Therefore the equation of the circle is $x^2 + y^2 = 4$.

Example 2. *Find the locus of a point* P, *whose distance from the point* A$(-1, 2)$ *is twice its distance from the origin.*

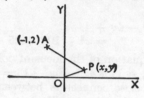

Fig. 17.2

Let $P(x, y)$ be a point on the locus, then

$$PA = 2PO.$$

PA and PO may be written down by the method of §1.2, but as both expressions involve a square root, it is neater to square first, giving

$$PA^2 = 4PO^2.$$
$$\therefore (x + 1)^2 + (y - 2)^2 = 4(x^2 + y^2).$$
$$\therefore x^2 + 2x + 1 + y^2 - 4y + 4 = 4x^2 + 4y^2.$$

Therefore the locus of P is

$$3x^2 + 3y^2 - 2x + 4y - 5 = 0.$$

Exercise 17a

1. Find the equation of a circle with centre at the origin and radius 5 units.

2. What is the locus of a point which moves so that its distance from the point (3, 1) is 2 units?

3. What is the locus of a point which is equidistant from the origin and the point ($-2, 5$)?

4. What is the locus of a point which moves so that its distance from the point ($-2, 1$) is equal to its distance from the point (3, -2)?

5. What is the distance of the point (x, y) from the line $x = -1$? Find the locus of a point which is equidistant from the origin and the line $x = -1$.

6. Find the locus of a point which is equidistant from the point (0, 1) and the line $y = -1$.

7. Find the locus of a point which moves so that its distance from the point A($-2, 0$) is three times its distance from the origin.

8. A point P moves so that its distance from A(2, 1) is twice its distance from B($-4, 5$). What is the locus of P?

9. Find the locus of a point which moves so that its distance from the point (8, 0) is twice its distance from the line $x = 2$.

10. Find the locus of a point which moves so that its distance from the point (2, 0) is half its distance from the line $x = 8$.

11. Find the locus of a point which moves so that the sum of the squares of its distances from the points ($-2, 0$) and (2, 0) is 26 units.

12. Find the locus of a point which moves so that it is equidistant from the point ($a, 0$) and the line $x = -a$.

13. A is the point (1, 0), and B is the point ($-1, 0$). Find the locus of a point P which moves so that PA $+$ PB $= 4$.

14. A is the point (1, 0), and B is the point ($-1, 0$). Find the locus of a point P which moves so that PA $-$ PB $= 2$.

15. A rectangle is formed by the axes and the lines $x = 4$ and $y = 6$. Find the locus of a point which moves so that the sum of the squares of its distances from the axes is equal to the sum of the squares of its distances from the other two sides.

16. A is the point (1, 0), and B is the point ($-1, 0$). Find the locus of a point which moves so that the sum of the squares of its distances from A and B is equal to four times the square of its distance from the origin.

17.2. Example 3. *Show that the equation of the circle on the line joining* A(3, − 5) *and* B(2, 6) *as diameter is* $(y + 5)(y − 6) + (x − 3)(x − 2) = 0$.

If P(x, y) is any point on the locus, then PA is perpendicular to PB, therefore the product of their gradients is − 1.

$$\therefore \frac{y + 5}{x − 3}.\frac{y − 6}{x − 2} = − 1.$$

Hence, on simplification of this equation, we obtain the equation of the circle in the form

$$(y + 5)(y − 6) + (x − 3)(x − 2) = 0.$$

Example 4. *A variable point* P *moves on the curve* $y^2 = 4x$ *and* A *is the point* (1, 0). *Find the locus of the mid-point of* AP.

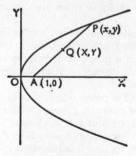

Fig. 17.3

Let P be the point (x, y), and let Q(X, Y) be the mid-point of AP. Then the coordinates of Q are given by

$$X = \frac{x + 1}{2}, \qquad Y = \frac{y}{2}.$$

Since P lies on the given curve, we have

$$y^2 = 4x,$$

but $x = 2X - 1$ and $y = 2Y$,

therefore $4Y^2 = 4(2X - 1)$.

Therefore the locus of the mid-point of AP is

$$y^2 = 2x - 1.$$

Example 5. *A straight line* AB *of length* 10 *units is free to move with its ends on the axes. Find the locus of a point* P *on the line at a distance of* 3 *units from the end on the x-axis.*

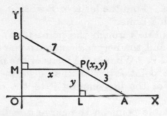

Fig. 17.4

Let the coordinates of P be (x, y). [No relationship between x and y is obvious, so try a construction.] Let the feet of the perpendiculars from P to the x- and y-axes be L and M. Then PL $= y$, PM $= x$ and triangles BMP and PLA are similar. (See Fig. 17.4.) Therefore

$$\frac{BM}{PL} = \frac{BP}{PA}.$$

$$\therefore BM = \tfrac{7}{3}y.$$

Using Pythagoras' theorem in triangle BMP,

$$(\tfrac{7}{3}y)^2 + x^2 = 7^2.$$

Therefore the locus of P is

$$49y^2 + 9x^2 = 441.$$

Exercise 17b

1. Find the equations of the circles on the diameters whose ends are:

    (i) $(-3, 2)$ and $(4, -5)$;   (ii) $(\frac{1}{2}, 1)$ and $(-\frac{3}{4}, 4)$;

    (iii) $(0, a)$ and $(a, 0)$;   (iv) (x_1, y_1) and (x_2, y_2).

2. P is a point on a line of length 12 units, which moves so that its ends lie on the axes. Find the locus of P when it is:

    (i) the mid-point of the line.

    (ii) the point of trisection of the line nearer to the y-axis.

3. L and M are the feet of perpendiculars from a point P on to the axes. Find the locus of P when it moves so that LM is of length 4 units.

4. A variable line through the point $(3, 4)$ cuts the axes at Q and R, and the perpendiculars to the axes at Q and R intersect at P. What is the locus of the point P?

5. A variable point P lies on the curve $xy = 12$. Q is the mid-point of the line joining P to the origin. Find the locus of Q.

6. P is a variable point on the curve $y = 2x^2 + 3$, and O is the origin. Q is the point of trisection of OP nearer the origin. Find the locus of Q.

7. A line parallel to the x-axis cuts the curve $y^2 = 4x$ at P and the line $x = -1$ at Q. Find the locus of the mid-point of PQ.

8. Variable lines through the points $O(0, 0)$ and $A(2, 0)$ intersect at right angles at the point P. Show that the locus of the mid-point of OP is $y^2 + x(x - 1) = 0$.

9. Find the locus of a point which moves so that the sum of the squares of its distances from the lines $x + y = 0$ and $x - y = 0$ is 4.

10. A is the point $(1, 0)$, B is the point $(2, 0)$ and O is the origin. A point P moves so that angle BPO is a right angle, and Q is the mid-point of AP. What is the locus of Q?

11. A line parallel to the y-axis meets the curve $y = x^2$ at P and the line $y = x + 2$ at Q. Find the locus of the mid-point of PQ.

12. M is a variable point on the x-axis, and A is the point $(2, 3)$. A line through A, perpendicular to AM, meets the y-axis at N. Perpendiculars to the axes at M and N meet at P. Find the locus of the point P.

13. M and N are points on the axes, and the line MN passes through the point (3, 2). P is a variable point which moves so that the mid-point of the line joining P to the origin is the mid-point of MN. Find the locus of the point P.

14. A straight line LM, of length 4 units, moves with L on the line $y = x$ and M on the x-axis. Find the locus of the mid-point of LM.

15. A straight line LM meets the x-axis in M and the line $y = x$ in L, and passes through the point (6, 4). What is the locus of the mid-point of LM?

16. M is a point on the x-axis, and A is the point (4, 6). A perpendicular to AM through A meets the y-axis at N. Find the locus of the mid-point of MN.

17. Find the locus of a point which moves so that the distance between the feet of the perpendiculars from it to the lines $y = x$ and $y = 0$ is 1 unit.

18. A and B are the points $(-a, 0)$ and $(2a, 0)$. A point P moves such that the line joining P to the origin bisects the angle BPA. Find the locus of P.

19. A variable circle touches the line $x = 5$ and the circle with centre (0, 0) and radius 4. Find the locus of its centre,

   (i) if the circles touch externally,

   (ii) if they touch internally.

Tangents and normals *

17.3. If a tangent touches a curve at the point P, the line through P perpendicular to the tangent is called a **normal**.

Example 6. *Find the equations of the tangent and normal to the curve $y = 3x^2 - 8x + 5$, at the point whose abscissa is 2.*

[The equation of a line can be found from its gradient and the coordinates of a point through which it passes. Therefore we begin by finding these.]

$$y = 3x^2 - 8x + 5.$$

* This section makes use of calculus.

Therefore the gradient of the tangent, $\dfrac{dy}{dx}$, is given by

$$\frac{dy}{dx} = 6x - 8.$$

At the point of contact $x = 2$, (abscissa means x-coordinate) and so

$$\frac{dy}{dx} = 6 \times 2 - 8 = 4.$$

The ordinate (or y-coordinate) of the point of contact may be found by substituting $x = 2$ in the equation of the curve:

$$y = 3 \times 2^2 - 8 \times 2 + 5 = 1.$$

Therefore the coordinates of the point of contact are $(2, 1)$.

Using the equation of a line in the form

$$y - y_1 = m(x - x_1),$$

the equation of the tangent is

$$y - 1 = 4(x - 2),$$

i.e. $4x - y - 7 = 0.$

The normal is perpendicular to the tangent, and so its gradient is $-\frac{1}{4}$. Therefore its equation may be written

$$y - 1 = -\tfrac{1}{4}(x - 2),$$

i.e. $x + 4y - 6 = 0.$

Thus the equations of the tangent and normal to the curve $y = 3x^2 - 8x + 5$ at the point $(2, 1)$ are respectively $4x - y - 7 = 0$ and $x + 4y - 6 = 0$.

NOTE. It should be emphasized that, when the equation of the tangent was found, the gradient of the curve *at* $(2, 1)$ was used. If we had taken the gradient to be $6x - 8$, the equation $y - 1 = (6x - 8)(x - 2)$ would not have represented a straight line.

Example 7. *Find the equations of the tangents to the curve $xy = 6$ which are parallel to the line $2y + 3x = 0$.*

The gradient of the line $2y + 3x = 0$ is $-\frac{3}{2}$. Therefore we must find at what points on the curve $xy = 6$ the gradient is $-\frac{3}{2}$.

$$y = \frac{6}{x}.$$

$$\therefore \frac{dy}{dx} = -\frac{6}{x^2}.$$

If $\frac{dy}{dx} = -\frac{3}{2}$, $\qquad -\frac{6}{x^2} = -\frac{3}{2}.$

$$\therefore 3x^2 = 12, \quad \text{and so} \quad x^2 = 4.$$

$$\therefore x = \pm 2.$$

When $x = 2$, $y = \frac{6}{2} = 3$; and when $x = -2$, $y = -\frac{6}{2} = -3$. Thus the gradient of the curve is $-\frac{3}{2}$ at the points $(2, 3)$ and $(-2, -3)$.

The equations of the tangents may be found from the form $y - y_1 = m(x - x_1)$:

$$y - 3 = -\tfrac{3}{2}(x - 2) \quad \text{and} \quad y + 3 = -\tfrac{3}{2}(x + 2).$$

Therefore the equations of the tangents to the curve $xy = 6$ which are parallel to the line $2y + 3x = 0$ are $3x + 2y - 12 = 0$ and $3x + 2y + 12 = 0$.

17.4. Sometimes questions about tangents may be solved without using the calculus. Fig. 17.5 shows a curve with a

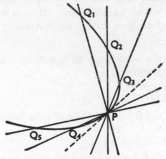

Fig. 17.5

chord PQ passing through a fixed point P and a variable point Q. When P and Q are distinct, we must obtain distinct roots when the equations of the curve and PQ are solved simultaneously; and when P and Q coincide, producing a tangent, there will be a repeated root.

Example 8. *Show that if the line $y = mx + c$ is a tangent to the curve $4x^2 + 3y^2 = 12$, then $c^2 = 3m^2 + 4$.*

[If the line $y = mx + c$ is a tangent, then the point of contact must be given by an equation with a repeated root.]

Substituting $y = mx + c$ in the equation $4x^2 + 3y^2 = 12$, we obtain

$$4x^2 + 3(mx + c)^2 = 12.$$

$$\therefore \ 4x^2 + 3m^2x^2 + 6mxc + 3c^2 = 12.$$

$$\therefore \ (4 + 3m^2)x^2 + 6mcx + 3c^2 - 12 = 0.$$

Now if the equation $ax^2 + bx + c = 0$ has equal roots then $b^2 = 4ac$. Therefore if $y = mx + c$ is a tangent,

$$36m^2c^2 = 4(4 + 3m^2)(3c^2 - 12).$$

$$\therefore \ 9m^2c^2 = 12c^2 - 48 + 9m^2c^2 - 36m^2.$$

$$\therefore \ 12c^2 = 36m^2 + 48.$$

Therefore if $y = mx + c$ is a tangent to the curve $4x^2 + 3y^2 = 12$, then $c^2 = 3m^2 + 4$.

This means that the line $y = mx \pm \sqrt{(3m^2 + 4)}$ will touch the curve for all values of m. Hence we may find the tangents parallel to $y = 2x$ by substituting $m = 2$, which gives $y = 2x \pm 4$.

Qu. 1. Find the equations of the tangents to the curve $4x^2 + 3y^2 = 12$ which are:

(i) parallel to $y = x$,

(ii) inclined at $60°$ to the x-axis.

Qu. 2. Solve the following pairs of simultaneous equations:

(i) $y = x$, $y^2 = x^3 + x^2$;

(ii) $y = 2x$, $y^2 = x^3 + x^2$.

What is the significance of the repeated root in each case?

Exercise 17c

1. Find the equations of the tangents and normals to the following curves at the points indicated:

   (i) $y = x^2$, $(2, 4)$;

   (ii) $y = 3x^2 - 2x + 1$, where $x = 1$;

   (iii) $y = x + \dfrac{1}{x}$, $(-1, -2)$;

   (iv) $y^2 = 4x$, $(1, -2)$;

   (v) $y = x^2 - 2x$, where $x = -2$;

   (vi) $xy = 4$, where $y = 2$;

   (vii) $y^3 = x^2$, $(1, 1)$.

2. Show that the following lines touch the given curves and find the coordinates of the points of contact:

   (i) $y^2 = 8x$, $y - 2x - 1 = 0$;

   (ii) $x^2 + y^2 = 8$, $x - y - 4 = 0$;

   (iii) $xy = 4$, $x + 9y - 12 = 0$;

   (iv) $9x^2 - 4y^2 = 36$, $5x - 2y + 8 = 0$.

3. At what points does the parabola $y = x^2 - 4x + 3$ cut the x-axis? Find the equations of the tangents and normals at these points.

4. Find the equations of the tangents at the points of intersection of the line $y = x + 1$ and the parabola

$$y = x^2 - x - 2.$$

5. Find the equations of the normals to the curve $y = x^2 - 1$ at the points where it cuts the x-axis. What are the coordinates of the point of intersection of these normals?

6. Find the coordinates of the points of intersection of the parabolas $y^2 = x$ and $x^2 = y$. What are the equations of the tangents to the curves at these points?

7. What is the equation of the normal to the curve $y = x^2 - 4x - 12$ at the point where it cuts the y-axis? Where does this normal meet the x-axis?

8. Find the equations of the tangents to the curve $y = x^3 - 3x^2$ which are parallel to the line $y = 9x$.

9. Find the equations of the tangents to the hyperbola $xy = 4$, which are inclined at $135°$ to the x-axis.

10. Show that the equation of the tangent to the parabola $y = \frac{1}{4}x^2$ at the point (h, k) may be written $2(y + k) = xh$. [HINT: (h, k) lies on the curve, therefore $k = \frac{1}{4}h^2$.]

11. Find the point of intersection of the tangents to the parabola $y = 2x^2 + 3x$ at the ends of the chord $y = 2x + 1$.

12. Show that the equation of the tangent to the parabola $y = x^2$ at the point (h, k) may be written

$$y - 2hx + h^2 = 0.$$

Find the values of h for which the tangent passes through the point $(1, 0)$, and obtain the equations of these tangents.

13. Show that the equation of the tangent to the rectangular hyperbola $xy = c^2$ at the point (h, k) may be written $xk + yh - 2c^2 = 0$. Find the equation of the tangent which passes through the point $(0, c)$.

14. Show that the following pairs of loci touch each other, and find the coordinates of the points of contact:

  (i) $y = x^2$, $x^2 + y^2 - y = 0$;
  (ii) $x^2 + y^2 = 2$, $xy = 1$;
  (iii) $x^2 + y^2 = 4$, $x^2 + y^2 - 8x - 6y + 16 = 0$.

15. Show that, if the line $y = x + c$ is a tangent to the circle $x^2 + y^2 = 4$, then $c^2 = 8$.

16. Prove that the condition that the line $y = mx + c$ should touch the ellipse $x^2 + 4y^2 = 4$ is $c^2 = 4m^2 + 1$.
Hence find the equations of the tangents to the ellipse which are parallel to the line $3x - 8y = 0$.

17. Show that the line $y = mx + c$ touches the hyperbola $b^2x^2 - a^2y^2 = a^2b^2$ if $c^2 = a^2m^2 - b^2$.
Hence find the equations of the tangents to the hyperbola $9x^2 - 25y^2 = 225$ which are parallel to the line $x - y = 0$.

18. Find the condition that the line $lx + my + n = 0$ should touch the ellipse $b^2x^2 + a^2y^2 = a^2b^2$.

Exercise 17d (Miscellaneous)

1. Find the locus of a point which is equidistant from the points $(4, -1)$ and $(3, 7)$.

2. Find the locus of a point which is equidistant from the y-axis and the point $(4, 0)$.

3. A point P moves so that its distance from the point $(5, 0)$ is half its distance from the line $x - 8 = 0$. Find the locus of P.

4. Find the locus of a point which moves so that its distance from the origin is three times its distance from the line $x = a$.

5. Find the locus of a point which moves so that its distance from $(2, 0)$ is twice its distance from $(-1, 0)$. Show that a point P, which moves so that the sum of the squares of the distances from P to the origin and the point $(-4, 0)$ is 16, describes the same locus.

6. If A is the point $(2, 0)$ and B is $(-3, 0)$, find the locus of a point P which moves so that $AP^2 + 2BP^2 = 22$.

7. Find the equation of the circle on the line joining (a, b) to (c, d) as diameter.

8. A straight line of length 24 units moves with its ends on the axes. Find the locus of a point on the line which is:

   (i) 12 units from the end on the x-axis,
   (ii) 6 units from the end on the x-axis.

9. A straight line of length 6 units moves with its ends A and B on the axes. Perpendiculars to the axes, erected at the points A and B, meet at P. Find the locus of P.

10. A and B are points on the axes OX and OY, and P is the mid-point of AB. Find the locus of P if the area of triangle AOB is 8 units.

11. A variable line through the point (a, b) cuts the axes at L and M, and the perpendiculars to the axes at L and M meet at P. What is the locus of P?

12. P is a variable point on the curve $4x^2 + y^2 = 36$ and A is the point $(1, 0)$. Find the locus of the mid-point of AP.

13. Find the gradient of the curve $y = 9x - x^2$ at the point where $x = 1$. Find the equation of the tangent to the curve at this point. Where does this tangent meet the line $x = y$?

14. Find the equation of the normal to the parabola $y = \frac{1}{4}x^2$ at the point $(4, 4)$. Find also the coordinates of the point at which this normal meets the parabola again, and show that the length of the chord so formed is $5\sqrt{5}$.

15. Find the equations of the tangents to the rectangular hyperbola $xy = 4$ at the points $(2, 2)$, $(6, \frac{2}{3})$. Show that they intersect on the line $3y = x$.

16. Find the gradient of the curve $y = 4x^2 - 7x + 5$ at each of the points where it is cut by the line $y = 2$. Find the equations of the tangents at these points and show that they meet on the line $15x = 7y$.

17. Find the equation of the normal to the parabola $y = \frac{1}{4}x^2$ which is parallel to $y = 3x$, and find the coordinates of the point on the parabola at which it is the normal.

18. Prove that the line $y = mx + a/m$ touches the parabola $y^2 = 4ax$. Find the equation of the tangent to the parabola $y^2 = 2x$ which is perpendicular to the straight line $2y + 7x = 4$.

19. The gradient of a curve at the point (x, y) is $1 - 2/x^2$. Find the equation of the curve if it passes through the point $(2, 4)$.

   Find the point of contact of the tangent which is parallel to the tangent at $(2, 4)$; also find the equations of both these tangents.

20. A point P whose x-coordinate is a is taken on the line $y = 3x - 7$. If Q is the point $(4, 1)$, show that

$$PQ^2 = 10a^2 - 56a + 80.$$

   Find the value of a which will make this expression a minimum. Hence show that the coordinates of N, the foot of the perpendicular from Q on to the line, are $(2\frac{4}{5}, 1\frac{2}{5})$. Find the equation of QN.

21. A curve $y = f(x)$ has a gradient $3x^2 - 2x - 1$. If the curve passes through the point $(1, 1)$, find its equation.

   Find the equations of the tangent and normal to the curve at the point where $x = 2$.

22. Show that the line $y = mx + c$ touches the ellipse

$$\frac{x^2}{a^2} + \frac{y^2}{b^2} = 1$$

if $c^2 = a^2m^2 + b^2$.

   Find the equations of the tangents to the ellipse $4x^2 + 9y^2 = 1$ which are perpendicular to $y = 2x + 3$.

CHAPTER 18

THE CIRCLE

18.1. The work of previous chapters will now be applied to the circle, and we begin by obtaining the equation of a circle, radius r, with its centre at the origin.

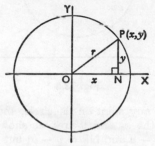

Fig. 18.1

We require an equation connecting the coordinates (x, y) of any point P on the circle. (See Fig. 18.1.) Let N be the foot of the perpendicular from P to the x-axis, so that $ON = x$ and $NP = y$.

Then by Pythagoras' theorem,

$$ON^2 + NP^2 = r^2.$$
$$\therefore\ x^2 + y^2 = r^2.$$

Therefore the equation of the circle, radius r, with its centre at the origin is

$$x^2 + y^2 = r^2.$$

18.2. The last section gives the simplest form in which the equation of a circle can be written, but now, to be quite

367

general, consider the circle, radius r, whose centre is at the point $C(a, b)$.

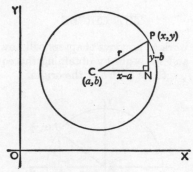

Fig. 18.2

Let $P(x, y)$ be any point on the circle, and draw CN and NP parallel to the x- and y-axes, as shown in Fig. 18.2.

Now $CN = x - a$ and $NP = y - b$; but by Pythagoras' theorem in triangle CNP,

$$CN^2 + NP^2 = CP^2.$$
$$\therefore (x - a)^2 + (y - b)^2 = r^2.$$

Therefore the equation of the circle, radius r, whose centre is at (a, b) is

$$(x - a)^2 + (y - b)^2 = r^2.$$

18.3.   Using this result, the equation of the circle with centre at $(4, -1)$ and radius 2 may be written down as

$$(x - 4)^2 + (y + 1)^2 = 2^2.$$

Expanding the squares:

$$x^2 - 8x + 16 + y^2 + 2y + 1 = 4.$$

Collecting the terms:

$$x^2 + y^2 - 8x + 2y + 13 = 0.$$

The equation of a circle is usually given in this form. Note that

(i) the coefficients of x^2 and y^2 are equal,

(ii) the only other terms are linear (such as may occur in the equation of a straight line).

Qu. 1. Express the equation $(x - a)^2 + (y - b)^2 = r^2$ in the form $x^2 + y^2 + 2gx + 2fy + c = 0$. Write down g, f, c, in terms of a, b, r.

Example 1. *Find the radius and the coordinates of the centre of the circle* $2x^2 + 2y^2 - 8x + 5y + 10 = 0$.

[We may find the centre and radius if the equation is expressed in the form $(x - a)^2 + (y - b)^2 = r^2$.]

Divide both sides of the equation of the circle

$$2x^2 + 2y^2 - 8x + 5y + 10 = 0$$

by 2, in order to make the coefficients of x^2 and y^2 equal to 1:

$$x^2 + y^2 - 4x + \tfrac{5}{2}y + 5 = 0.$$

Rearrange the terms, grouping those in x and y:

$$x^2 - 4x \quad + y^2 + \tfrac{5}{2}y \quad = -5.$$

Complete the squares (add the squares of half the coefficients of x and y to both sides):

$$x^2 - 4x + 4 + y^2 + \tfrac{5}{2}y + (\tfrac{5}{4})^2 = -5 + 4 + \tfrac{25}{16}.$$
$$\therefore (x - 2)^2 + (y + \tfrac{5}{4})^2 = \tfrac{9}{16}.$$
$$\therefore (x - 2)^2 + (y + \tfrac{5}{4})^2 = (\tfrac{3}{4})^2.$$

Comparing this with the equation of the circle, radius r, centre (a, b):

$$(x - a)^2 + (y - b)^2 = r^2,$$

we obtain $\quad a = 2, \quad b = -\tfrac{5}{4}, \quad r = \tfrac{3}{4}.$

Therefore the radius is $\tfrac{3}{4}$ and the centre is at the point $(2, -\tfrac{5}{4})$.

Example 2. *Find the equations of the circles which pass through the points* A(0, 2) *and* B(0, 8), *and which touch the x-axis.*

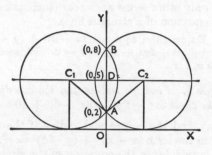

Fig. 18.3

Fig. 18.3 suggests a method. The centre of the circle must lie on the perpendicular bisector of the chord AB, i.e. on the line $y = 5$.

Now the circle touches the x-axis, therefore the radius is 5.

If D is the point (0, 5) and C is the centre of either circle, then triangle ADC is right-angled and DC = 4 by Pythagoras' theorem. Therefore the centres of the circles are $(-4, 5)$ and $(4, 5)$ and so their equations are

$$(x \pm 4)^2 + (y - 5)^2 = 5^2.$$

Therefore the equations of the circles are

$$x^2 + y^2 \pm 8x - 10y + 16 = 0.$$

Exercise 18a

1. Find the equations of the circles with the following centres and radii:

(i) centre (2, 3), radius 1; (ii) centre $(-3, 4)$, radius 5;
(iii) centre $(\frac{2}{3}, -\frac{1}{3})$, radius $\frac{2}{3}$;

  (iv) centre $(0, -5)$, radius 5;
  (v) centre $(3, 0)$, radius $\sqrt{2}$;
  (vi) centre $(-\frac{1}{4}, \frac{1}{3})$, radius $\frac{1}{2}\sqrt{2}$.

2. Find the radii and the coordinates of the centres of the following circles:

  (i) $x^2 + y^2 + 4x - 6y + 12 = 0$,
  (ii) $x^2 + y^2 - 2x - 4y + 1 = 0$,
  (iii) $x^2 + y^2 - 3x = 0$,
  (iv) $x^2 + y^2 + 3x - 4y - 6 = 0$,
  (v) $2x^2 + 2y^2 + x + y = 0$,
  (vi) $36x^2 + 36y^2 - 24x - 36y - 23 = 0$,
  (vii) $x^2 + y^2 - 2ax - 2by = 0$,
  (viii) $x^2 + y^2 + 2gx + 2fy + c = 0$.

3. Which of the following equations represent circles?

  (i) $x^2 + y^2 - 5 = 0$,   (ii) $x^2 + y^2 + 10 = 0$,
  (iii) $3x^2 + 2y^2 + 6x - 8y + 100 = 0$,
  (iv) $ax^2 + ay^2 = 1$,
  (v) $x^2 + y^2 + 8x + xy + 4 = 0$,
  (vi) $x^2 + y^2 + bxy = 1$,   (vii) $x^2 + y^2 + c = 0$,
  (viii) $x^2 + dy^2 - 8x + 10y + 50 = 0$.

Which of them can represent circles if suitable values are given to the constants a, b, c, d?

4. Find the equation of the circle whose centre is at the point $(2, 1)$ and which passes through the point $(4, -3)$.

5. A circle, whose centre is at the point $(-1, 2)$, passes through the mid-point of the line joining the points $(-1, 7)$ and $(5, 5)$. What is the equation of the circle?

6. The points $(8, 4)$ and $(2, 2)$ are the ends of a diameter of a circle. Find the coordinates of the centre, and the radius. Deduce the equation of the circle.

7. Find the equation of the circle, centre $(4, 5)$, which passes through the point where the line $5x - 2y + 6 = 0$ cuts the y-axis.

8. Find the equation of the circle whose centre lies on the line $y = 4$, and which passes through the points $(2, 0)$ and $(6, 0)$.

9. What is the equation of the circle, centre $(2, -3)$, which touches the x-axis?

10. Find the equation of the circle which passes through the points A(2, 0), B(0, 2), C($-$ 4, 0). [HINT: what is the equation of the perpendicular bisector of AB?]

11. Find the radii of the two circles, with centres at the origin, which touch the circle $x^2 + y^2 - 8x - 6y + 24 = 0$.

12. Show that the distance of the centre of the circle $x^2 + y^2 - 6x - 4y + 4 = 0$ from the y-axis is equal to the radius. What does this prove about the y-axis and the circle?

13. Find the equations of the circles which touch the x-axis, have radius 5, and pass through the point (0, 8).

14. What is the equation of the circle whose centre lies on the line $x - 2y + 2 = 0$, and which touches the positive axes?

15. A circle passes through the points A($-$ 5, 2), B($-$ 3, $-$4), C(1, 8). Find the point of intersection of the perpendicular bisectors of AB and BC. What is the equation of the circle?

16. The circle $x^2 + y^2 + 2gx + 2fy + c = 0$ passes through the points A($-$ 1, $-$ 2), B(1, 2), C(2, 3). Write down three equations which must be satisfied by g, f, c. Solve these equations and write down the equation of the circle ABC.

17. Find the equations of the circles which pass through:

(i) the origin, ($-$ 1, 3), ($-$ 4, 2);

(ii) (3, 1), (8, 2), (2, 6);

(iii) (6, $-$ 5), (2, $-$7), ($-$ 6, $-$ 1).

18. A point moves so that its distance from the origin is twice its distance from the point (3, 0). Show that the locus is a circle, and find its centre and radius.

19. A is the point (3, $-$ 1), and B is the point (5, 3). Show that the locus of a point P, which moves so that $PA^2 + PB^2 = 28$, is a circle. Find its centre and radius.

20. A point P moves so that the sum of the squares of the distances of P from the lines $2y - x - 4 = 0$ and $y + 2x - 7 = 0$ is equal to 4 units. Find the locus of P.

Tangents

18.4. Elementary geometry will frequently help to simplify working in coordinate geometry, as the reader may have found in the last exercise. It provides a simple way of obtaining the equation of a tangent at a given point on a given circle: using the theorem that a tangent is perpendicular to the radius through the point of contact. This method will be employed in the next example.

Example 3. *Verify that the point* $(3, 2)$ *lies on the circle* $x^2 + y^2 - 8x + 2y + 7 = 0$, *and find the equation of the tangent at this point.*

Substituting the coordinates $(3, 2)$ into the equation $x^2 + y^2 - 8x + 2y + 7 = 0$,

$$\text{L.H.S.} = 9 + 4 - 24 + 4 + 7 = 0 = \text{R.H.S.}$$

Therefore $(3, 2)$ lies on the circle.

[The gradient of the tangent can be found from the gradient of the radius through $(3, 2)$; and, in order to find this, we obtain the coordinates of the centre of the circle.]

The equation of the circle may be written

$$x^2 - 8x \qquad + y^2 + 2y \qquad = -7.$$
$$\therefore \; x^2 - 8x + 16 + y^2 + 2y + 1 = -7 + 16 + 1.$$
$$\therefore \; (x - 4)^2 + (y + 1)^2 = 10.$$

Therefore the centre of the circle is $(4, -1)$. Hence the gradient of the radius through $(3, 2)$ is $\dfrac{-1 - 2}{4 - 3} = -3$.

Therefore the gradient of the tangent is $\frac{1}{3}$. Using the formula $y - y_1 = m(x - x_1)$, the equation of the tangent at $(3, 2)$ is

$$y - 2 = \tfrac{1}{3}(x - 3).$$
$$\therefore \; 3y - 6 = x - 3.$$

Therefore the equation of the tangent to the circle at $(3, 2)$ is $x - 3y + 3 = 0$.

18.5   Example 4. *Find the length of the tangents from the point* (7, 5) *to the circle* $x^2 + y^2 - 4x - 6y + 9 = 0$.

[Fig. 18.4 suggests a method. The tangent is perpendicular to the radius through the point of contact, so t can be found by Pythagoras' theorem if d and r are known.]

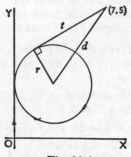

Fig. 18.4

In Fig. 18.4, the radius, length r, is perpendicular to the tangent, length t, from the point (7, 5). If the distance of (7, 5) from the centre of the circle is d, then by Pythagoras' theorem $d^2 = t^2 + r^2$, or

$$t^2 = d^2 - r^2.$$

To find the coordinates of the centre of the circle $x^2 + y^2 - 4x - 6y + 9 = 0$:

$$x^2 - 4x + 4 + y^2 - 6y + 9 = 4.$$
$$\therefore (x - 2)^2 + (y - 3)^2 = 2^2.$$

Therefore the centre is (2, 3) and the radius is 2.

Now, by Pythagoras' theorem,

$$d^2 = (7 - 2)^2 + (5 - 3)^2 = 25 + 4 = 29.$$

But $r^2 = 4$,

$$\therefore t^2 = 29 - 4 = 25.$$

Therefore the length of the tangents from (7, 5) to the circle is 5.

Qu. 2. Calculate the lengths of the tangents to the circle in Example 4 from (i) (4, 3), (ii) (2, 2). What do you conclude from these results? If in doubt, mark these points in a figure containing the circle.

Exercise 18b

1. Verify that the given points lie on the following circles and find the equations of the tangents to the circles at these points.

   (i) $x^2 + y^2 + 6x - 2y = 0$,   (0, 0);

   (ii) $x^2 + y^2 - 8x - 2y = 0$,   (3, 5);

   (iii) $x^2 + y^2 + 2x + 4y - 12 = 0$,   (3, -1);

   (iv) $x^2 + y^2 + 2x - 2y - 8 = 0$,   (2, 2);

   (v) $2x^2 + 2y^2 - 8x - 5y - 1 = 0$,   (1, -1).

2. Find the lengths of the tangents from the given points to the following circles.

   (i) $x^2 + y^2 + 4x - 6y + 10 = 0$,   (0, 0);

   (ii) $x^2 + y^2 - 4x - 8y - 5 = 0$,   (8, 2);

   (iii) $x^2 + y^2 + 6x + 10y - 2 = 0$,   (-2, 3);

   (iv) $x^2 + y^2 - 10x + 8y + 5 = 0$,   (5, 4);

   (v) $x^2 + y^2 = a^2$, (x_1, y_1);

   (vi) $x^2 + y^2 + 2gx + 2fy + c = 0$,   (0, 0).

3. The tangent to the circle $x^2 + y^2 - 2x - 6y + 5 = 0$ at the point (3, 4) meets the x-axis at M. Find the distance of M from the centre of the circle.

4. Find the equations of the tangents to the circle

$$x^2 + y^2 - 6x + 4y + 5 = 0$$

at the points where it meets the x-axis.

5. The tangent to the circle $x^2 + y^2 - 4x + 6y - 77 = 0$ at the point (5, 6) meets the axes at A and B. Find the coordinates of A and B. Deduce the area of triangle AOB.

6. Find the length of the tangents from the origin to the circle $x^2 + y^2 - 10x + 2y + 13 = 0$. Use this answer to show that these two tangents and the radii through the points of contact form a square.

7. Find the length of the tangents to the circle

$$x^2 + y^2 + 10x - 6y - 8 = 0$$

from the centre of the circle $x^2 + y^2 - 4x = 0$.

8. Find the length of the tangents to the circle

$$x^2 + y^2 - 4 = 0$$

from the point $P(X, Y)$; and deduce the equation of the locus of P, when it moves so that the length of the tangents to the circle is equal to the distance of P from the point $(1, 0)$.

9. Show that the length of the tangents to the circle

$$x^2 + y^2 - 4x - 6y + 12 = 0$$

from the point $P(X, Y)$ is $\sqrt{(X^2 + Y^2 - 4X - 6Y + 12)}$. Find the locus of P when it moves so that the length of the tangents to the circle is equal to its distance from the origin.

10. Show that the point (x_1, y_1) is outside, on or inside the circle $x^2 + y^2 + 2gx + 2fy + c = 0$, according as to whether $x_1^2 + y_1^2 + 2gx_1 + 2fy_1 + c$ is positive, zero or negative.

11. Find the coordinates of the centres of the circles of radius 3 which touch the y-axis and whose centres lie on the line $3x = 4y$. What are the equations of these circles?

12. Prove that the line $x - y - 3 = 0$ is a common tangent to the circles $x^2 + y^2 - 2x - 4y - 3 = 0$ and

$$x^2 + y^2 + 4x - 2y - 13 = 0.$$

What are the coordinates of the point in which it meets the other common tangent?

The intersection of two circles

18.6. Example 5. *Find the equation of the common chord of the circles* $x^2 + y^2 - 4x - 2y + 1 = 0$ *and* $x^2 + y^2 + 4x - 6y - 10 = 0.$

The coordinates of the points of intersection A and B of the circles satisfy the two equations

$$x^2 + y^2 - 4x - 2y + 1 = 0,$$
$$x^2 + y^2 + 4x - 6y - 10 = 0.$$

Therefore, by subtraction, the coordinates of A and B satisfy the equation

$$-8x + 4y + 11 = 0.$$

But this equation represents a straight line, and it is satisfied by the coordinates of A and B, therefore it is the equation of the common chord.

Two circles may not intersect but, by subtracting one equation from the other, the equation of a line may still be obtained. What then does the line represent? Qu. 3 suggests an answer.

Qu. 3. What are the squares of the lengths of the tangents from the point $P(X, Y)$ to the circles $x^2 + y^2 - 1 = 0$, $x^2 + y^2 - 6x - 8y + 21 = 0$? What is the locus of P such that the lengths of the tangents from P to the circles are equal?

Qu. 4. Write down the equation of the line joining the origin to the point of intersection of the lines

$$17x - 15y + 7 = 0, \quad 19x - 13y + 7 = 0.$$

18.7. If the tangents to two circles at their points of intersection are perpendicular, the circles are said to be **orthogonal.** Since the radius through a point of contact is perpendicular to the tangent, it follows that the tangent to one circle is a radius of the other. Thus if the centres of two orthogonal circles of radii R and r are a distance d apart, it follows by Pythagoras' theorem that

$$d^2 = R^2 + r^2.$$

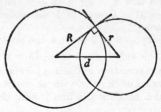

Fig. 18.5

Example 6. *Show that the circles*

$$x^2 + y^2 - 6x + 4y + 2 = 0$$

and

$$x^2 + y^2 + 8x + 2y - 22 = 0$$

are orthogonal.

The centres of the circles are $(3, -2)$ and $(-4, -1)$, and their radii are $\sqrt{11}$ and $\sqrt{39}$.

The sum of the squares of the radii is 50, and the square of the distance between the centres is $7^2 + 1^2 = 50$, therefore the circles are orthogonal.

Exercise 18c   (Miscellaneous)

1. Show that the common chord of the circles $x^2 + y^2 = 4$ and $x^2 + y^2 - 4x - 2y - 4 = 0$ passes through the origin.

2. Find the coordinates of the point where the common chord of the circles $x^2 + y^2 - 4x - 8y - 5 = 0$ and

$$x^2 + y^2 - 2x - 4y - 5 = 0$$

meets the line joining their centres.

3. Show that the following pairs of circles are orthogonal:
   (i) $x^2 + y^2 - 6x - 8y + 9 = 0, x^2 + y^2 = 9$;
   (ii) $x^2 + y^2 - 4x + 2 = 0, x^2 + y^2 + 6y - 2 = 0$;
   (iii) $x^2 + y^2 - 6y + 8 = 0, x^2 + y^2 - 4x + 2y - 14 = 0$;
   (iv) $x^2 + y^2 + 10x - 4y - 3 = 0$,
   $\qquad x^2 + y^2 - 2x - 6y + 5 = 0$.

4. Prove that the line $y = 2x$ is a tangent to the circle $x^2 + y^2 - 8x - y + 5 = 0$ and find the coordinates of the point of contact.

5. Show that the line $x - 2y + 12 = 0$ touches the circle $x^2 + y^2 - x - 31 = 0$ and find the coordinates of the point of contact.

6. The line $2x + 2y - 3 = 0$ touches the circle

$$4x^2 + 4y^2 + 8x + 4y - 13 = 0$$

at A.   Find the equation of the line joining A to the origin.

7. Find the equation of the circle whose centre is at the point $(5, 4)$ and which touches the line joining the points $(0, 5)$ and $(4, 1)$.

8. Find the equation of the tangent to the circle

$$x^2 + y^2 - 2x + y - 5 = 0$$

at the point $(3, -2)$. If this tangent cuts the axes at A and B, find the area of triangle OAB.

9. Find the length of the tangents to the circle

$$x^2 + y^2 - 2x + 4y - 3 = 0$$

from the centre of the circle $x^2 + y^2 + 6x + 8y - 1 = 0$.

10. A tangent is drawn from the point $(-a, 0)$ to a variable circle, centre $(a, 0)$. What is the locus of the point of contact?

11. Prove that the circles $x^2 + y^2 + 3x + y = 0$ and $x^2 + y^2 - 6x - 2y = 0$ touch each other. Find the co-ordinates of the point of contact and the equation of their common tangent at that point.

12. Show that the line $y = x + 1$ touches the circle $x^2 + y^2 - 8x - 2y + 9 = 0$. What is the equation of the other tangent to the circle from the point $(0, 1)$?

13. A circle passing through the point $(4, 0)$ is orthogonal to the circle $x^2 + y^2 = 4$. Find the locus of the centre of the variable circle.

14. The circle $x^2 + y^2 - 2x - 4y - 5 = 0$ has centre C, and is cut by the line $y = 2x + 5$ at A and B. Show that BC is perpendicular to AC and find the area of the triangle ABC.

15. Find the equation of the circle which passes through the points $(0, 2)$, $(8, -2)$, $(9, 5)$. Verify that it also passes through the point $(2, 6)$.

16. Find the coordinates of the points A and B in which the line $x - 3y = 0$ meets the circle

$$x^2 + y^2 - 10x - 5y + 25 = 0.$$

Find also the coordinates of the point T where the circle touches the axis OX and verify that $OA . OB = OT^2$.

17. A triangle has vertices $(0, 6)$, $(4, 0)$, $(6, 0)$. Find the equation of the circle through the mid-points of the sides and show that it passes through the origin.

18. Two circles have their centres on the line $y + 3 = 0$ and touch the line $3y - 2x = 0$. If the radii of the circles are

$\sqrt{13}$, find the coordinates of their centres and also their equations. [HINT: use similar triangles.]

19. A and B have coordinates $(-3, 0)$ and $(3, 0)$. Show that the locus of a point P which moves such that $PB = 2PA$ is a circle with centre $(-5, 0)$ and radius 4.

20. A variable circle through the origin touches the circle $x^2 + y^2 - 8x + 12 = 0$. Find the locus of its centre.

21. Show that the line $y = mx + c$ touches the circle $x^2 + y^2 = a^2$ if $c^2 = a^2(1 + m^2)$.

22. The point (h, k) lies on the circle $x^2 + y^2 = a^2$. What equation connects h and k? Show that the equation of the tangent to the circle at (h, k) may be written $hx + ky = a^2$.

23. A circle touching the line $x = 4$ is orthogonal to the circle $x^2 + y^2 = 4$. Find the locus of the centre of the variable circle.

24. Find the equation, centre and radius of the circle described by a point P which moves so that its distances from the points $A(-2, 0)$ and $B(3, 0)$ are connected by the equation $3PA = 2PB$.

25. A and B are the points $(a, 0)$ and $(2a, 0)$. P is a point which moves so that $PB^2 - 2PA^2 = a^2$. Show that the locus of P is a circle, and find its radius.

26. A point P moves so that the sum of the squares of its distances from the lines

$$3x - 2y + 6 = 0 \quad \text{and} \quad 2x + 3y + 4 = 0$$

is 13 units. Find the locus of P.

CHAPTER 19

FURTHER TOPICS IN COORDINATE GEOMETRY

19.1. Straight lines occur so often in coordinate geometry that it is worth while learning to write down their equations by a quick method. Example 9 in Chapter 1 was done by two methods, and what follows is an extension of the second.

Example 1. *Find the equation of the line with gradient* $-\frac{2}{3}$, *which passes through the point* $(1, -4)$.

[Think: the line has equation $y = -\frac{2}{3}x + c$, therefore it may be written

$$3y + 2x = \text{constant}.$$

Now since the line passes through $(1, -4)$, the constant may be found by substituting these coordinates in the left-hand side.]

The equation of the line is

$$3y + 2x = -12 + 2,$$

i.e. $\qquad\qquad 2x + 3y + 10 = 0.$

NOTE. Check that the line (i) has gradient $-\frac{2}{3}$, (ii) passes through $(1, -4)$.

Qu. 1. Write down the equations of the lines with the given gradients which pass through the given points:

(i) gradient 1, through $(3, 2)$;
(ii) gradient -2, through $(1, -3)$;
(iii) gradient $\frac{1}{2}$, through $(0, -6)$;
(iv) gradient $-\frac{1}{3}$, through $(-2, 5)$;
(v) gradient $-\frac{4}{7}$, through $(3, -6)$;
(vi) gradient $\frac{3}{4}$, through $(-1, 1)$;

   (vii) gradient $-\frac{5}{6}$, through $(-3, -4)$;

 (viii) gradient $\frac{4}{5}$, through $(-2, 5)$;

  (ix) gradient $1/t$, through $(at^2, 2at)$;

   (x) gradient $-t$, through $(at^2, 2at)$;

  (xi) gradient $-\cot\theta$, through $(a\cos\theta, a\sin\theta)$;

 (xii) gradient $-1/t^2$, through $(ct, c/t)$.

Given the equation of a line, it is easy to write down the equation of a perpendicular line through a given point. For example, if we require the equation of the line perpendicular to $4x + 5y + 7 = 0$ which passes through $(6, -5)$, we interchange the coefficients of x and y, changing one of the signs, and balance the equation as before. Thus the perpendicular is $5x - 4y = 50$.

Qu. 2. Write down the equations of the perpendiculars to:

   (i) $3x + 2y - 1 = 0$, through $(2, 2)$;

  (ii) $4x - 3y + 7 = 0$, through the origin;

 (iii) $5x + 6y + 11 = 0$, through $(-3, 5)$;

 (iv) $3x - 2y - 7 = 0$, through the mid-point of the line joining $(3, 4)$ and $(-5, 2)$;

  (v) $ty - x = at^2$, through (h, k);

 (vi) $ax + by + c = 0$, through (x_1, y_1);

 (vii) $t^2y + x = 2ct$, through $(ct, c/t)$.

19.2.   Example 2. *Find the equation of the line joining the points $(a, 0)$, $(0, b)$.*

The gradient of the line is $-b/a$. Therefore, using the method of the last section, its equation is $ay + bx = ab$.

Dividing through by ab, the equation becomes

$$\frac{x}{a} + \frac{y}{b} = 1,$$

which is known as the **intercept** form of the equation of a line.

Qu. 3. Write down the equations of the lines which make the following intercepts on the x- and y-axes respectively:
(i) $3, 2$; (ii) $-1, 2$; (iii) $\frac{1}{2}, \frac{1}{5}$; (iv) $-\frac{1}{3}, \frac{1}{4}$.

Qu. 4. Write down the equations of the lines joining the following pairs of points: (i) $(0, 2)$, $(3, 0)$; (ii) $(-1, 0)$, $(0, 5)$; (iii) $(-\frac{1}{2}, 0)$, $(0, \frac{2}{3})$.

Qu. 5. The perpendicular from the origin to a straight line is of length p and makes an angle α with the x-axis. (See Fig. 19.1.) What intercepts does the line make on the axes? Write down the equation of the line.

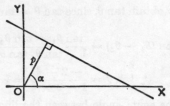

Fig. 19.1

Angles between two lines

19.3. It is sometimes necessary to find angles between two straight lines, and the method used here involves one of the results of §13.2.

Example 3. *Find the acute angle between the lines* $2x + y - 6 = 0$ *and* $3x - 2y + 2 = 0$.

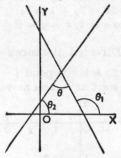

Fig. 19.2

If the lines $2x + y - 6 = 0$ and $3x - 2y + 2 = 0$ make angles θ_1 and θ_2 with the x-axis, then it follows that one of the angles between the two lines $\theta = \theta_1 - \theta_2$, since (see Fig. 19.2) the exterior angle of a triangle is equal to the sum of the two interior opposite angles.

Now the gradients of the two lines are -2 and $\frac{3}{2}$, therefore

$$\tan \theta_1 = -2, \qquad \tan \theta_2 = \tfrac{3}{2}.$$

Hence we may obtain $\tan \theta$, since $\tan \theta = \tan(\theta_1 - \theta_2)$ and from §13.2,

$$\tan(\theta_1 - \theta_2) = \frac{\tan \theta_1 - \tan \theta_2}{1 + \tan \theta_1 \tan \theta_2}.$$

$$\therefore \tan \theta = \frac{-2 - \tfrac{3}{2}}{1 - 2 \times \tfrac{3}{2}} = \tfrac{7}{4}.$$

Therefore the acute angle between the lines is $\tan^{-1} \tfrac{7}{4}$.

[$\tan^{-1} x$ is an abbreviation for "the angle whose tangent is x".]

Qu. 6. Find the angles between the following pairs of straight lines. Use $\tan^{-1}$ in your answers.

(i) $y = x - 6$, $y = 3x + 1$;

(ii) $2x + y + 3 = 0$, $x + y = 0$;

(iii) $3x + 4y - 2 = 0$, $2x - 3y - 4 = 0$;

(iv) $4x - 2y + 3 = 0$, $6x - 2y + 1 = 0$;

(v) $3x + 5y = 0$, $2x - 3y + 1 = 0$;

(vi) $x \sin \alpha - y \cos \alpha = 1$, $x - y = 0$.

Directed distances

19.4. In Fig. 19.3, A is the point $(-2, 0)$ and B is the point $(3, 0)$. If we start at A, and move to B, we travel a

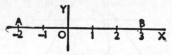

Fig. 19.3

distance of 5 units along the x-axis in the *positive* direction; and if we move from B to A, we travel a distance of 5 units in the *negative* direction. This may be expressed more briefly by saying "the distance from A to B is $+5$ units, and the distance from B to A is -5 units". The phrase "the distance from A to B" is abbreviated to $\overrightarrow{AB}$ and "the distance from B to A" is written $\overrightarrow{BA}$. Thus in Fig. 19.3,

$$\overrightarrow{AB} = 5 \qquad \text{and} \qquad \overrightarrow{BA} = -5.$$

Qu. 7. C, D, E, F, are the points $(-5, 0)$, $(-3, 0)$, $(2, 0)$, $(6, 0)$. Write down the distances (i) $\overrightarrow{EF}$, (ii) $\overrightarrow{FE}$, (iii) $\overrightarrow{DE}$, (iv) $\overrightarrow{DC}$, (v) $\overrightarrow{FD}$, (vi) $\overrightarrow{DF}$, (vii) $\overrightarrow{EC}$, (viii) $\overrightarrow{CD}$, (ix) $\overrightarrow{FC}$.

It has been assumed in calculating gradients and distances that the distance $\overrightarrow{M_1M_2}$ between the points $(x_1, 0)$ and $(x_2, 0)$ is $x_2 - x_1$, but the proof has been deferred until this stage.

From the definition of $\overrightarrow{AB}$ it follows that

$$\overrightarrow{M_1M_2} = \overrightarrow{M_1O} + \overrightarrow{OM_2}.$$

But $\overrightarrow{M_1O} = -x_1$ and $\overrightarrow{OM_2} = x_2$,

$$\therefore \ \overrightarrow{M_1M_2} = -x_1 + x_2.$$

This may be verified in the six cases shown in Fig. 19.4.

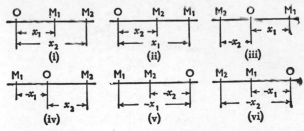

Fig. 19.4

Example 4. A *and* B *are the points* $(-3, 5)$ *and* $(6, 2)$. *Find* (i) *the coordinates of the point* P *which divides* AB *internally in the ratio* $7:3$, (ii) *the coordinates of the point* Q *which divides* AB *externally in the ratio* $7:3$.

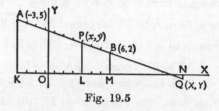

Fig. 19.5

(i) Divide AB into 10 equal parts, then P lies on the seventh point of division from A, and the third from B. In Fig. 19.5, AK, PL, BM are ordinates. Draw lines parallel to the y-axis through the points of division of AB, then, by the intercept theorem, KL is divided into 7 equal parts and LM into 3 equal parts. Therefore

$$\frac{AP}{PB} = \frac{KL}{LM} = \frac{7}{3}.$$

Let P have coordinates (x, y), then

$$\overrightarrow{KL} = x + 3 \quad \text{and} \quad \overrightarrow{LM} = 6 - x.$$

$$\therefore \frac{x + 3}{6 - x} = \frac{7}{3}.$$

$$\therefore 3x + 9 = 42 - 7x.$$

$$\therefore 10x = 33.$$

$$\therefore x = 3\tfrac{3}{10}.$$

Similarly, $$\frac{y - 5}{2 - y} = \frac{7}{3}.$$

$$\therefore y = 2\tfrac{9}{10}.$$

Therefore P is the point $(3\tfrac{3}{10}, 2\tfrac{9}{10})$.

(ii) Divide AB into 4 equal parts, and mark off 3 equal parts on AB produced, then Q lies on the third point of division. Let N be the foot of the perpendicular from Q to the x-axis. As before, using the intercept theorem, KN may be divided into 7 equal parts, 4 of which lie on KM and 3 on MN. Now this time $\overrightarrow{KN}$ is positive and $\overrightarrow{NM}$ is negative, so we write

$$\frac{AQ}{QB} = \frac{KN}{NM} = -\frac{7}{3}.$$

Let Q have coordinates (X, Y), then

$$\frac{X + 3}{6 - X} = -\frac{7}{3}.$$

$$\therefore \; 3X + 9 = -42 + 7X.$$

$$\therefore \; X = 12\tfrac{3}{4}.$$

Similarly

$$\frac{Y - 5}{2 - Y} = -\frac{7}{3}.$$

$$\therefore \; Y = -\tfrac{1}{4}.$$

Therefore Q is the point $(12\tfrac{3}{4}, -\tfrac{1}{4})$.

NOTE. In the example above, APBQ is a straight line which is not parallel to either axis, and the signs of $\overrightarrow{AP}$, $\overrightarrow{PB}$, $\overrightarrow{AQ}$, $\overrightarrow{QB}$ are uncertain. However, $\overrightarrow{AP}$ and $\overrightarrow{PB}$ are both in the same direction and so their ratio $\overrightarrow{AP} : \overrightarrow{PB}$ is positive; on the other hand $\overrightarrow{AQ}$ and $\overrightarrow{QB}$ are in opposite directions and so their ratio $\overrightarrow{AQ} : \overrightarrow{QB}$ is negative.

For any straight line AB (see Fig. 19.6) $\overrightarrow{AP} : \overrightarrow{PB}$ is positive

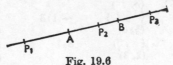

Fig. 19.6

when P divides AB internally and it is negative when P divides AB externally. Thus by giving the ratio AP:PB a sign it becomes unnecessary to specify whether P divides AB internally or externally.

Exercise 19a

1. Find the coordinates of the points which divide the lines joining the following points in the given ratios:

   (i) $(3, 2)$, $(7, 9)$, $5:3$;

   (ii) $(-2, 5)$, $(6, -1)$, $2:3$;

   (iii) $(-1, -6)$, $(7, 3)$, $1:3$;

   (iv) $(3, 2)$, $(5, 7)$, $-1:2$;

   (v) $(-1, -3)$, $(-4, -7)$, $-3:2$;

   (vi) $(5, 1)$, $(-3, 6)$, $-3:4$;

   (vii) $(a, 2a)$, $(2b, b)$, $3:5$;

   (viii) $(0, 3)$, $(1, 5)$, $p:q$;

   (ix) (x_2, y_2), (x_1, y_1), $m_1:m_2$.

2. Find the coordinates of the centroids, G, of the following triangles by writing down the mid-points of one side of each triangle and using the result that G trisects the medians.

   (i) $(2, 3)$, $(5, 7)$, $(6, -1)$;

   (ii) $(-3, 2)$, $(4, -6)$, $(8, -2)$;

   (iii) $(5, 1)$, $(-3, -7)$, $(-7, 2)$;

   (iv) $(0, 0)$, (a, b), (c, d);

   (v) (x_1, y_1), (x_2, y_2), (x_3, y_3).

3. The line AB is divided in the ratio $\lambda:1$ by the point P. Complete the following table:

	A	B	λ	P
(i)	$(5, 2)$	$(3, -6)$		$(\ \ , -3)$
(ii)	$(\ \ , \ \ )$	$(4, -5)$	$-2:1$	$(2, 9)$
(iii)	$(-4, 7)$	$(3, \ \ )$		$(-1\frac{1}{4}, 5\frac{1}{4})$
(iv)	$(3, 4)$	$(\ \ , \ \ )$	$-1:3$	$(8, 5\frac{1}{2})$
(v)	$(\ \ , \frac{1}{2})$	$(\frac{2}{3}, 0)$		$(\frac{4}{5}, -\frac{1}{10})$
(vi)	$(3, \ \ )$	$(\ \ , 5)$	$8:7$	$(4\frac{3}{5}, 3\frac{4}{5})$
(vii)	$(5, 1)$	$(2, 3)$		$(17, \ \ )$

4. Use the method of §19.1 to find the equations of the follow-straight lines:

   (i) with gradient 3, through $(4, 3)$;

   (ii) with gradient $-\frac{1}{4}$, through $(2, -1)$;

   (iii) with gradient $\frac{4}{5}$, through $(1, 1)$;

   (iv) with gradient $-\frac{3}{5}$, through $(0, -3)$;

   (v) joining $(3, 2)$ and $(2, -4)$;

   (vi) joining $(1, 3)$ and $(-3, -6)$;

   (vii) joining $(-1, 2)$ to the mid-point of $(3, 5)$ and $(5, -1)$;

 (viii) through $(2, 1)$, perpendicular to $2x - y = 0$;

   (ix) through $(-1, 3)$ perpendicular to $3x + 4y - 2 = 0$;

   (x) the altitude through A of the triangle A$(1, 3)$, B$(2, -1)$, C$(3, 5)$;

   (xi) the altitude through B of the triangle in (x);

   (xii) through (h, k), perpendicular to $t^2 y + x = 2ct$.

5. Find the angles between the following pairs of straight lines, using $\tan^{-1}$ in your answers.

   (i) $y = x + 7$, $y = 2x - 4$;

   (ii) $x + y - 3 = 0$, $y - 3x + 2 = 0$;

   (iii) $4x + 3y - 7 = 0$, $3x + 4y - 1 = 0$;

   (iv) $2x + 3y + 5 = 0$, $3x - 5y = 0$;

   (v) $y = m_1 x$, $y = m_2 x$;

   (vi) $a_1 x + b_1 y + c_1 = 0$, $a_2 x + b_2 y + c_2 = 0$.

6. Deduce the condition that lines with gradients m_1 and m_2 should be perpendicular, from the expression for $\tan(\theta_1 - \theta_2)$ in Example 3.

7. A point P lies on the curve $y^2 = 4x$, and A is the point $(5, 0)$. Find the locus of the point which divides AP in the ratio $2 : 3$.

8. A point P lies on the circle of radius 2 whose centre is at the origin, and A is the point $(4, 0)$. Find the locus of a point which divides AP in the ratio $-1 : 2$.

9. Find the locus of the centroid of the triangle A$(3, 0)$, B$(0, 6)$, P, where P moves on the parabola $y = x^2$.

10. The centroid of the triangle A$(0, 0)$, B$(3, 0)$, C, lies on the circle $x^2 + y^2 = 1$. What is the locus of C?

11. Find the gradients of the lines which meet the line $3y = x$ at $45°$.

12. Show that the bisector of the acute angle between $y = x + 1$ and the x-axis has gradient $\sqrt{2} - 1$. What is the gradient of the bisector of the obtuse angle?

Definition. If two loci meet, the angle between the loci at a point of intersection is the angle between their tangents at that point.

13. Find the acute angles between the following pairs of loci:

    (i) $y = 2x - 5$, $x^2 + y^2 = 25$ at $(4, 3)$;

    (ii) $xy = 12$, $x^2 + y^2 = 25$ at $(-4, -3)$;

    (iii) $y = x^2 - 4x$, $y = 6 - x^2$;

    (iv) $y = x^3 - 3x^2$, $y = 2x - 4x^2$.

14. Show that the rectangular hyperbolas $xy = 1$ and $x^2 - y^2 = 1$ are orthogonal.

15. Show that the ellipse $16x^2 + 25y^2 = 400$ and the hyperbola $4x^2 - 5y^2 = 20$ are orthogonal.

16. The line AB, where A and B are the points $(4, 0)$ and $(2, 0)$, subtends an angle of $45°$ at a point P. Find the locus of P.

Polar coordinates

19.5. If someone asks me at Harrow to tell him where Enfield is, I may reply that it is about 19 km East and 9 km North, or I might tell him that it is roughly 21 km away on a bearing N $60°$ E. These two descriptions of the position of Enfield correspond to the two systems of coordinates used in this book. The first has been used previously, but the description of a point by means of a bearing and the distance from a fixed point is one which is useful for certain purposes.

Let O be a fixed point and OX a fixed line (see Fig. 19.7), then if P is any point, let OP $= r$ and angle POX $= \theta$, then

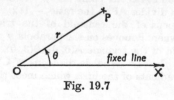

Fig. 19.7

r and θ are called the **polar coordinates** of the point P, and the coordinates may be written (r, θ). O is called the origin and OX the initial line.

It should be noticed that, while a bearing is usually measured in a clockwise sense from North, in mathematics the polar coordinate θ is normally represented in an anti-clockwise sense.

Thus in Fig. 19.8, the coordinates of A are $(2, 30°)$ and

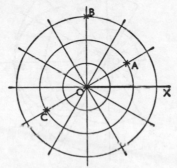

Fig. 19.8

B is $(3, 90°)$. A point may be described in different ways, for instance C may be written as $(2, 210°)$, $(2, -150°)$, $(-2, 30°)$ and so on. If, for any reason, a unique way of referring to each point is required, r may be taken to be positive and θ to lie in the range $-180° < \theta \leqslant 180°$.

Example 5. *Sketch the curve whose polar equation is* $r = a(1 + 2 \cos \theta)$.

Take values of θ, and calculate $1 + 2 \cos \theta$, as below.

θ	0°	30°	45°	60°	90°	120°	135°	150°	180°
$2 \cos \theta$	2	1·732	1·414	1	0	-1	$-1·414$	$-1·732$	-2
$1 + 2 \cos \theta$	3	2·732	2·414	2	1	0	$-0·414$	$-0·732$	-1

Plot these values.   (See Fig. 19.9.)   Now if α is any angle, $\cos(-\alpha) = \cos\alpha$, therefore the same values of r will be obtained for negative values of θ.   Thus the curve may be completed.

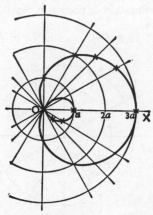

Fig. 19.9

Example 6.   *Find the polar equation of a line such that the perpendicular to it from the origin is of length p and makes an angle α with the initial line.*

In Fig. 19.10, N is the foot of the perpendicular from the origin to the line, and let P be any point (r, θ) on the line.

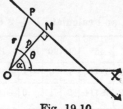

Fig. 19.10

In the triangle ONP, N is a right angle and angle PON $= \theta - \alpha$ (or $\alpha - \theta$).

$$\therefore\ r \cos (\theta - \alpha) = p \quad (\text{or } r \cos (\alpha - \theta) = p).$$

Therefore, in either case, the polar equation of the line is $r \cos (\theta - \alpha) = p$.

Relations between polar and Cartesian coordinates

19.6. In Fig. 19.11, P is the point (x, y) in Cartesian coordinates and (r, θ) in polar coordinates, and PM is an ordinate.

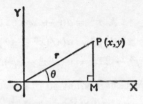

Fig. 19.11

Now, by the definitions of cosine and sine given in §12.1,

$$\cos \theta = \frac{x}{r}, \qquad \sin \theta = \frac{y}{r}.$$

Therefore x and y are given in terms of r and θ by the equations

$$x = r \cos \theta, \qquad y = r \sin \theta.$$

Squaring and adding these equations,

$$r^2 \cos^2 \theta + r^2 \sin^2 \theta = x^2 + y^2.$$

$$\therefore\ r^2 = x^2 + y^2.$$

Dividing the two equations,

$$\frac{r \sin \theta}{r \cos \theta} = \frac{y}{x}.$$

$$\therefore\ \tan \theta = \frac{y}{x}.$$

Therefore r and θ are given in terms of x and y by the equations

$$r = \sqrt{(x^2 + y^2)}, \qquad \theta = \tan^{-1} \frac{y}{x}.$$

Example 7. *Find the Cartesian equations of:*

(i) $r = a(1 + 2 \cos \theta)$, (ii) $r \cos (\theta - \alpha) = p$.

(i) $\qquad\qquad\qquad r = a(1 + 2 \cos \theta).$

[The $\cos \theta$ suggests the relation $x = r \cos \theta$, so multiply through by r.]

$$\therefore r^2 = a(r + 2r \cos \theta).$$
$$\therefore x^2 + y^2 = a(\sqrt{(x^2 + y^2)} + 2x).$$
$$\therefore x^2 + y^2 - 2ax = a\sqrt{(x^2 + y^2)}.$$

Therefore the Cartesian equation of $r = a(1 + 2 \cos \theta)$ is

$$(x^2 + y^2 - 2ax)^2 = a^2(x^2 + y^2).$$

(ii) $\qquad\qquad\qquad r \cos (\theta - \alpha) = p.$

$\cos (\theta - \alpha)$ may be expanded (see §13.2),

$$\therefore r \cos \theta \cos \alpha + r \sin \theta \sin \alpha = p.$$

Therefore the Cartesian equation of $r \cos (\theta - \alpha) = p$ is

$$x \cos \alpha + y \sin \alpha = p.$$

NOTE. The perpendicular from the origin to this line is of length p and makes an angle α with the x-axis. This form of the equation of a straight line is known as the **normal** or **perpendicular** form.

Example 8. *Find the polar equation of the circle whose Cartesian equation is $x^2 + y^2 = 4x$.*

$$x^2 + y^2 = 4x.$$

Put $x = r \cos \theta$, $y = r \sin \theta$, then

$$r^2 \cos^2 \theta + r^2 \sin^2 \theta = 4r \cos \theta.$$
$$\therefore \quad r^2 = 4r \cos \theta.$$

Therefore the polar equation of the circle is

$$r = 4 \cos \theta.$$

Exercise 19b

1. Sketch the curves given by the following polar equations:

 (i) $r = a(1 + \cos \theta)$,   (ii) $r = a \cos 2\theta$,

 (iii) $r = a(1 - \sin \theta)$,   (iv) $r = a \sin 3\theta$,

 (v) $r = a \sec \theta$,   (vi) $r = a \tan \theta$,

 (vii) $r = a \cos \dfrac{\theta}{2}$,   (viii) $r = a(1 + \sin 2\theta)$.

2. Find the polar equations of the following loci:

 (i) a circle, centre at the origin, radius a;

 (ii) a straight line through the origin, inclined at an angle α to the initial line;

 (iii) a straight line perpendicular to the initial line, at a distance a from the origin;

 (iv) a straight line parallel to the initial line at a distance a;

 (v) a circle on the line joining the origin to $(a, 0)$ as a diameter;

 (vi) a circle, radius a, touching the initial line at the origin and lying above it;

 (vii) a circle, radius a, centre on the initial line at a distance c from the origin;

 (viii) a point which moves so that its distance from the origin is equal to its distance from the straight line $x = 2a$.

3. P_1 is the point (r_1, θ_1), P_2 is (r_2, θ_2) and $\theta_2 > \theta_1$. Show that the area of the triangle OP_1P_2 is $\frac{1}{2}r_1r_2 \sin(\theta_2 - \theta_1)$. Deduce that if the Cartesian coordinates of P_1 and P_2 are (x_1, y_1) and (x_2, y_2), then the area of OP_1P_2 is

$$\tfrac{1}{2}(x_1y_2 - x_2y_1).$$

4. Deduce from the result of No. 3, that the area of the triangle $P_1(x_1, y_1)$, $P_2(x_2, y_2)$, $P_3(x_3, y_3)$ is

$$\tfrac{1}{2}\{(x_2y_3 - x_3y_2) + (x_3y_1 - x_1y_3) + (x_1y_2 - x_2y_1)\}.$$

[If new axes are drawn at (x_3, y_3), the coordinates of P_1 and P_2 referred to them are $(x_1 - x_3, y_1 - y_3)$ and $(x_2 - x_3, y_2 - y_3)$.]

5. Obtain the polar equations of the following loci:

(i) $x^2 + y^2 = a^2$,   (ii) $x^2 - y^2 = a^2$,

(iii) $y = 0$,   (iv) $y^2 = 4a(a - x)$,

(v) $x^2 + y^2 - 2y = 0$,   (vi) $xy = c^2$.

6. Obtain the Cartesian equations of the following loci:

(i) $r = 2$,   (ii) $r = a(1 + \cos \theta)$,

(iii) $r = a \cos \theta$,   (iv) $r = a \tan \theta$,

(v) $r = 2a(1 + \sin 2\theta)$,   (vi) $2r^2 \sin 2\theta = c^2$,

(vii) $\dfrac{l}{r} = 1 + e \cos \theta$,   (viii) $r = 4a \cot \theta \operatorname{cosec} \theta$.

7. Express the following straight lines in the form

$$x \cos \alpha + y \sin \alpha = p.$$

State the distance of each line from the origin and give the angle which the perpendicular from the origin makes with the x-axis.

(i) $x + \sqrt{3}y = 2$,   (ii) $x - y = 4$,

(iii) $3x + 4y - 10 = 0$,   (iv) $5x - 12y + 26 = 0$,

(v) $x + 3y - 2 = 0$,   (vi) $ax + by + c = 0$.

The distance of a point from a line

19.7.   Given a point $P_1(x_1, y_1)$ and the line

$$ax + by + c = 0,$$

we shall first find the distance, r, of P_1 from a point P_2 on the line, such that $\overrightarrow{P_1P_2}$ makes an angle α with the x-axis. (See Fig. 19.12.)

P_2 has coordinates $(x_1 + r \cos \alpha, y_1 + r \sin \alpha)$, but, as P_2 lies on the line $ax + by + c = 0$, its coordinates satisfy the equation, therefore

$$a(x_1 + r \cos \alpha) + b(y_1 + r \sin \alpha) + c = 0.$$

$$\therefore r(a \cos \alpha + b \sin \alpha) = -(ax_1 + by_1 + c).$$

$$\therefore r = -\frac{ax_1 + by_1 + c}{a \cos \alpha + b \sin \alpha}.$$

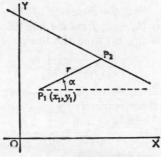

Fig. 19.12

Now take the case when P_1P_2 is perpendicular to the line $ax + by + c = 0$. The gradient of $ax + by + c = 0$ is $-a/b$, therefore the gradient of P_1P_2 is b/a.

$$\therefore \tan \alpha = \frac{b}{a}.$$

$$\therefore \sec^2 \alpha = 1 + \frac{b^2}{a^2} = \frac{a^2 + b^2}{a^2}.$$

$$\therefore \cos \alpha = \pm \frac{a}{\sqrt{(a^2 + b^2)}}$$

and, since $\tan \alpha = \dfrac{b}{a}$,

$$\sin \alpha = \pm \frac{b}{\sqrt{(a^2 + b^2)}}.$$

$$\therefore\ a \cos \alpha + b \sin \alpha = \pm \left(\frac{a^2}{\sqrt{(a^2 + b^2)}} + \frac{b^2}{\sqrt{(a^2 + b^2)}} \right)$$

$$= \pm \sqrt{(a^2 + b^2)}.$$

Therefore the perpendicular distance of (x_1, y_1) from the line $ax + by + c = 0$ is

$$\pm \frac{ax_1 + by_1 + c}{\sqrt{(a^2 + b^2)}}.$$

Example 9. *Find the distances of the points* (i) $(1, 3)$, (ii) $(-3, 4)$, (iii) $(4, -2)$ *from the line* $2x + 3y - 6 = 0$.

The distance of (x_1, y_1) from the line $ax + by + c = 0$ is

$$\pm \frac{ax_1 + by_1 + c}{\sqrt{(a^2 + b^2)}}.$$

Therefore the distances of $(1, 3)$, $(-3, 4)$, $(4, -2)$ from $2x + 3y - 6 = 0$ are respectively

$$\text{(i)}\ \pm \frac{2 \times 1 + 3 \times 3 - 6}{\sqrt{(2^2 + 3^2)}},$$

$$\text{(ii)}\ \pm \frac{2 \times (-3) + 3 \times 4 - 6}{\sqrt{(2^2 + 3^2)}},$$

$$\text{(iii)}\ \pm \frac{2 \times 4 + 3 \times (-2) - 6}{\sqrt{(2^2 + 3^2)}};$$

which reduce to

$$\text{(i)}\ \pm \frac{5}{\sqrt{13}}, \quad \text{(ii)}\ \pm \frac{0}{\sqrt{13}}, \quad \text{(iii)}\ \pm \frac{-4}{\sqrt{13}}.$$

Therefore the distances are respectively $\frac{5}{13}\sqrt{13}, 0, \frac{4}{13}\sqrt{13}$.

NOTE. The necessity for the $\pm$ sign in the formula is illustrated by the example above. The line

$$2x + 3y - 6 = 0$$

divides the plane into two parts: above the line

$$y > -\tfrac{2}{3}x + 2,$$

and the expression $2x + 3y - 6$ is positive; below the line $y < -\frac{2}{3}x + 2$, and $2x + 3y - 6$ is negative.

The formula is more easily remembered if two points are noticed: (i) the numerator is obtained by substituting the coordinates of the point into the equation of the line (remember that the perpendicular distance is zero if the point lies on the line), (ii) the square root in the denominator is of the sum of the squares of the coefficients.

Qu. 8. Find the distances of the given points from the following lines:

    (i) $(3, 2)$, $3x - 4y + 4 = 0$;

   (ii) $(2, -1)$, $5x + 12y = 0$;

  (iii) $(0, -3)$, $x + 5y + 2 = 0$;

  (iv) $(2, 5)$, $x + y - 1 = 0$;

   (v) $(-4, 2)$, $3y = 5x - 6$;

  (vi) $(2, 1)$, $y = \frac{2}{3}x + \frac{1}{3}$;

 (vii) $(0, a)$, $3y = 4x$;

(viii) (p, q), $3x + 4y - 3p = 0$;

  (ix) (X, Y), $12x - 5y + 7 = 0$;

   (x) (x_1, y_1), $8x = 15y$.

Example 10. *Find the equations of the bisectors of the angles between the lines $4x + 3y - 12 = 0$ and $y = 3x$.*

[The angle bisectors are the locus of a point which is equidistant from the two lines, and this provides a method of finding their equations.]

Let $P(X, Y)$ be a point on the locus, then the distances of P from the lines $4x + 3y - 12 = 0$ and $y - 3x = 0$ are

$$\pm \frac{4X + 3Y - 12}{\sqrt{(4^2 + 3^2)}} \quad \text{and} \quad \pm \frac{Y - 3X}{\sqrt{(3^2 + 1^2)}}.$$

But P is equidistant from the two lines, therefore

$$\frac{4X + 3Y - 12}{5} = \pm \frac{Y - 3X}{\sqrt{10}}.$$

[One $\pm$ sign has been dropped, since there are only two

distinct equations: one given by the same sign each side, the other by different signs.]

Simplifying these equations we obtain

$$4\sqrt{10}X + 3\sqrt{10}Y - 12\sqrt{10} = 5Y - 15X,$$
$$4\sqrt{10}X + 3\sqrt{10}Y - 12\sqrt{10} = -5Y + 15X.$$

Therefore the equations of the angle bisectors of the lines are

$$(4\sqrt{10} + 15)x + (3\sqrt{10} - 5)y - 12\sqrt{10} = 0,$$
$$(4\sqrt{10} - 15)x + (3\sqrt{10} + 5)y - 12\sqrt{10} = 0.$$

Example 11. *Find the equations of the tangents to the circle $x^2 + y^2 - 4x - 2y - 8 = 0$ which are parallel to the line $3x + 2y = 0$.*

[This will be done by using the result that the perpendicular distance of a tangent from the centre of a circle is equal to the radius.]

The required tangents are parallel to the line $3x + 2y = 0$, therefore their equations may be written in the form

$$3x + 2y + c = 0,$$

where c is a constant to be determined for each tangent.

To find the centre and radius of the circle

$$x^2 + y^2 - 4x - 2y - 8 = 0,$$
$$\therefore \ x^2 - 4x + 4 + y^2 - 2y + 1 = 8 + 4 + 1.$$
$$\therefore \ (x - 2)^2 + (y - 1)^2 = 13.$$

Therefore the centre is $(2, 1)$ and the radius is $\sqrt{13}$.

Now the distance of the point (x_1, y_1) from the line $ax + by + c = 0$ is $\pm (ax_1 + by_1 + c)/\sqrt{(a^2 + b^2)}$, therefore the distance of the centre of the circle $(2, 1)$ from the line $3x + 2y + c = 0$ is

$$\pm \frac{3 \times 2 + 2 \times 1 + c}{\sqrt{(3^2 + 2^2)}}.$$

But if the line is a tangent, this distance is equal to the radius, therefore

$$\pm \frac{8+c}{\sqrt{13}} = \sqrt{13}.$$

$$\therefore \pm (8 + c) = 13.$$

Taking the positive sign, $8 + c = 13$, and so $c = 5$. With the negative sign, $-8 - c = 13$, and so $c = -21$.

Therefore the equations of the tangents parallel to $3x + 2y = 0$ are $3x + 2y + 5 = 0$ and $3x + 2y - 21 = 0$.

Exercise 19c

1. Write down the distances of the given points from the following lines:

    (i) $(2, 5)$, $4x + 3y - 2 = 0$;

    (ii) $(-1, 3)$, $12x - 5y = 0$;

    (iii) $(-2, 0)$, $4x + y - 2 = 0$;

    (iv) $(3, 5)$, $x - y + 2 = 0$;

    (v) $(-1, 7)$, $2x = 5y + 1$;

    (vi) $(0, 0)$, $3x = 4y + 6$;

    (vii) $(2, 3)$, $y = \frac{4}{3}x + \frac{1}{3}$;

   (viii) $(1, 4)$, $\dfrac{x}{2} + \dfrac{y}{3} = 1$;

    (ix) $(0, 0)$, $x \cos \alpha + y \sin \alpha = p$;

    (x) (X, Y), $5x - 12y + 1 = 0$;

    (xi) $(c, 2c)$, $8x = 15y$;

    (xii) (x_1, y_1), $y = \frac{3}{4}x - \frac{1}{2}$.

2. Find the equations of the bisectors of the angles between:

    (i) $3x + 4y - 7 = 0$, $y - 1 = 0$;

    (ii) $4x - 3y + 1 = 0$, $3x - 4y + 3 = 0$;

    (iii) $5x + 12y = 0$, $12x + 5y - 4 = 0$;

    (iv) $x + y - 1 = 0$, the x-axis.

3. Find the equations of the bisectors of the acute angles between:

    (i) $3x - 4y + 2 = 0$, $x + 3 = 0$;

    (ii) $5x + 12y + 9 = 0$, $5x - 12y + 6 = 0$;

    (iii) $x + y + 1 = 0$, $x = 7y$.

[Draw figures to determine which equations give the required lines.]

4. What is the locus of a point which moves so that it is equidistant from the point $(2, -3)$ and the line $x + 2y = 0$?

5. Find the locus of a point which is equidistant from the line $3x - 4y + 7 = 0$ and the point $(3, 4)$ on the line.

6. Find the locus of a point which moves so that the sum of the squares of the perpendiculars from it to the x-axis and the line $y = x$ is 1.

7. What is the locus of a point which moves so that its distance from $(2, 2)$ is half its distance from $x + y + 4 = 0$?

8. Find the equations of the tangents to the circle

$$x^2 + y^2 - 4x - 6y + 8 = 0$$

which are parallel to the line $y = 2x$.

9. Find the equations of the tangents to the circle

$$x^2 + y^2 + 4x + 8y - 5 = 0$$

which are parallel to the line $4y - 3x = 0$.

10. Show that the line $8x + 6y + 7 = 0$ touches the circle $x^2 + y^2 - 3x - 2y - 3 = 0$, and find the equation of the parallel tangent.

11. Show that the line $3x + 2y = 0$ touches the circle $x^2 + y^2 + 6x + 4y = 0$, and find the equations of the perpendicular tangents.

12. Find the equation of the circle in the first quadrant with radius 2 which touches the y-axis and the line

$$3y - 4x - 3 = 0.$$

13. Find the equation of the circle in the first quadrant with radius $2\sqrt{65}$ which touches the lines $8y = x$ and $4y = 7x$.

14. Show that the circles $x^2 + y^2 - 8x - 4 = 0$ and $x^2 + y^2 + 4x - 6y + 8 = 0$ touch externally. Find the equation of the tangent to the larger circle which is parallel to the common tangent to the two circles at their point of contact.

15. Prove that the line $y = mx + c$ touches the circle $x^2 + y^2 = a^2$ if $c^2 = a^2(1 + m^2)$. Also find the condition that the line $lx + my + n = 0$ should touch the circle.

16. Find the equation of the incircle of the triangle bounded by the x-axis and the lines $4x - 3y = 0$, $4x + 3y - 48 = 0$.

Parameters

19.8. Consider a circle, radius a, centre at the origin. (See Fig. 19.13.) Let $P(x, y)$ be any point on the circle, and let angle POX be θ, then

$$x = a \cos \theta, \qquad y = a \sin \theta.$$

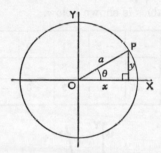

Fig. 19.13

These equations, which give the coordinates of any point on the curve in terms of θ, are called **parametric equations**, and θ is called a **parameter**.

If we wish to refer to a particular point on the curve, a single number, the corresponding value of θ, will determine it. Thus $\theta = 60°$ gives the point $\left(\frac{1}{2}a, \frac{\sqrt{3}}{2}a\right)$. On the other hand, if we were given a value of x, say $\frac{1}{2}a$, there are two corresponding points: $\left(\frac{1}{2}a, \frac{\sqrt{3}}{2}a\right)$ and $\left(\frac{1}{2}a, -\frac{\sqrt{3}}{2}a\right)$. Another advantage of parameters is that we may write down the coordinates of a general point on the curve $(a \cos \theta, a \sin \theta)$. If we wrote (x_1, y_1), we should also have to bear in mind the equation $x_1^2 + y_1^2 = a^2$.

Another example of parameters was used in §19.7. The point $(x_1 + r \cos \alpha, y_1 + r \sin \alpha)$ lies on the straight line through (x_1, y_1) with gradient α, and in this case the parameter, r, is a distance. However, it is not always possible to give an easy interpretation of a parameter in terms of angles or distances.

Example 12. *Plot the graph of the curve given parametrically by the equations $x = t^2 - 4$, $y = t^3 - 4t$, for values of t from -3 to $+3$.*

A table of values is shown below.

t	-3	-2	-1	0	1	2	3
$x = t^2 - 4$	5	0	-3	-4	-3	0	5
$y = t^3 - 4t$	-15	0	3	0	-3	0	15

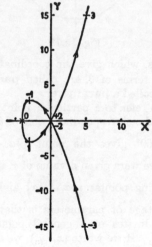

Fig. 19.14

The graph has been plotted in Fig. 19.14, and the values of the parameter, t, have been written against the corresponding points. The arrows indicate the direction of motion of a point on the curve as t increases from -3 to $+3$.

Example 13. *Sketch the curve given parametrically by the equations* $x = \sin \theta$, $y = \sin 2\theta$.

A few values of θ will give all the points we need.

θ	0	45°	90°	135°	180°
$x = \sin \theta$	0	0·7071	1	0·7071	0
$y = \sin 2\theta$	0	1	0	-1	0

Plotting these points and joining them by a curve we obtain the part of the curve in Fig. 19.15 which lies to the right of the y-axis.

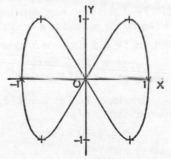

Fig. 19.15

Now $\sin (-\alpha) = -\sin \alpha$, so that negative values of θ change the signs of x and y. Therefore the rest of the curve may be drawn in symmetrically.

Qu. 9. Sketch the locus given by $x = t^2$, $y = 1 - t^2$, for real values of t. Is it the line $x + y = 1$?

19.9. The graph of the curve given parametrically by the equations $x = t^2 - 4$, $y = t^3 - 4t$ was plotted for values of t from -3 to $+3$ in Example 12. The question may well have risen in the reader's mind, "What is the equation connecting x and y?" This can be found by eliminating t from the equations

$$x = t^2 - 4, \qquad y = t^3 - 4t.$$

Notice that $y = tx$. Therefore we may substitute $t = y/x$ in either of the equations above. Choosing the simpler,

$$x = \frac{y^2}{x^2} - 4.$$

$$\therefore x^3 = y^2 - 4x^2.$$

Therefore the Cartesian equation of the locus is

$$y^2 = x^2(x + 4).$$

Example 14. *Find the Cartesian equation of the locus given parametrically by the equations $x = \sin \theta$, $y = \sin 2\theta$.*

$y = \sin 2\theta$, but $\sin 2\theta = 2 \sin \theta \cos \theta$, therefore

$$y = 2 \sin \theta \cos \theta.$$

$$\therefore y^2 = 4 \sin^2 \theta \cos^2 \theta.$$

Now $x = \sin \theta$, therefore $1 - x^2 = \cos^2 \theta$, and so the Cartesian equation of the locus is

$$y^2 = 4x^2(1 - x^2).$$

The process of obtaining parametric equations from a given Cartesian equation is not so easy as the reverse, but one method is illustrated in the next example.

Example 15. *Obtain parametric equations for the locus $y^2 = x^3 - x^2$.*

Put $y = tx$ in the equation $y^2 = x^3 - x^2$, then

$$t^2x^2 = x^3 - x^2.$$
$$\therefore \ t^2 = x - 1.$$
$$\therefore \ x = t^2 + 1.$$

Therefore the locus may be represented by the parametric equations $x = t^2 + 1, y = t^3 + t$.

NOTE. This method is not suitable for all equations, but it works well when the terms are of degree n and $n - 1$.

Exercise 19d

1. Plot the curves given parametrically by the equations:

(i) $x = t^2 + 1, y = t + 2$; from $t = -3$ to $t = +3$.

(ii) $x = t^2, y = t^3$; from $t = -3$ to $t = +3$.

(iii) $x = t, y = \dfrac{1}{t}$; taking $t = \pm 4, \pm 3, \pm 2, \pm 1, \pm \frac{1}{2}, \pm \frac{1}{3}, \pm \frac{1}{4}$.

(iv) $x = 1 + t, y = 3 - 2t$;

(v) $x = at^2, y = 2at$;

(vi) $x = t^2, y = \dfrac{1}{t}$;

(vii) $x = \dfrac{2 + 3t}{1 + t}, y = \dfrac{3 - 2t}{1 + t}$;

(viii) $x = 3\left(t + \dfrac{1}{t}\right), y = 2\left(t - \dfrac{1}{t}\right)$;

(ix) $x = 3 \cos \theta, y = 2 \sin \theta$;

(x) $x = \sin \theta, y = \cos 2\theta$;

(xi) $x = 4 \sec \theta, y = 3 \tan \theta$;

(xii) $x = \cos 3\theta, y = \cos \theta$;

(xiii) $x = \sin 3\theta, y = \cos \theta$;

(xiv) $x = \cos 2\theta, y = \sin 3\theta$.

2. Find the values of the parameters and the other coordinates of the given points on the following curves:

(i) $x = t, y = \dfrac{2}{t}$; where $y = 1\frac{1}{2}$.

(ii) $x = at^2, y = 2at$; where $x = \frac{9}{4}a$.

(iii) $x = \dfrac{1 + t}{1 - t}, y = \dfrac{2 + 3t}{1 - t}$; where $y = -\frac{4}{3}$.

(iv) $x = a \cos \theta, y = b \sin \theta$; where $x = \frac{1}{2}a$.

3. Find the Cartesian equations of the loci in Nos. 1 (i)–(xii).

4. By substituting $y = tx$, find parametric equations for the loci whose Cartesian equations are:

    (i) $y^4 = x^5$,   (ii) $y = x^2 + 2x$,

    (iii) $y^2 = x^2 + 2x$,   (iv) $x^2 = x^3 - y^3$,

    (v) $x^3 + y^3 = 3xy$,   (vi) $y = 2x + 1$,

    (vii) $x^2 + y^2 = x$,   (viii) $xy = x - y$.

5. Show that the parametric equations:

    (i) $x = 1 + 2t, y = 2 + 3t$,

    (ii) $x = \dfrac{1}{2t - 3}, y = \dfrac{t}{2t - 3}$,

both represent the same straight line, and find its Cartesian equation.

6. Show that the line given parametrically by the equations $x = \dfrac{2 - t}{1 + 2t}, y = \dfrac{3 + t}{1 + 2t}$ passes through the points $(6, 7)$ and $(- 2, - 1)$. Find the values of t corresponding to these points.

7. P is the variable point $(t^2, 3t)$ and O is the origin. Find the coordinates of Q, the mid-point of OP, and hence obtain the locus of Q as P varies.

8. P is the variable point $(at^2, 2at)$ on the parabola $y^2 = 4ax$, and Q is the foot of the perpendicular from P to the y-axis. Find the locus of the mid-point of PQ.

9. The line joining the origin to the variable point $P(t, 1/t)$ meets the line $x = 1$ at Q. Find the locus of the mid-point of PQ.

10. Find the coordinates of the points nearest to the origin on the curve $x = t, y = 1/t$. What is their distance from the origin?

11. Find the coordinates of the points on the curve $x = at^2$, $y = 2at$ where the distance from the point $(5a, - 2a)$ is stationary. Distinguish between maxima, minima and points of inflexion.

12. Find the equations of the chords joining the points with parameters p and q on the following curves:

   (i) $x = t^2, y = 2t$;   (ii) $x = t, y = -\dfrac{1}{t}$;

   (iii) $x = t^3, y = t$;   (iv) $x = t + \dfrac{1}{t}, y = 2t$.

13. Determine the point on the parabola $x = at^2,\ y = 2at$ where the distance to the line $x - y + 4a = 0$ is least and find the least distance.

14. Find the values of t at the points of intersection of the line $2x - y - 4 = 0$ with the parabola $x = t^2, y = 2t$ and give the coordinates of these points.

15. Find the points of intersection of the parabola $x = t^2$, $y = 2t$ with the circle $x^2 + y^2 - 6x + 4 = 0$.

16. Obtain an equation for the parameters of the points where the circle $x^2 + y^2 = a^2$ cuts the rectangular hyperbola $x = ct, y = c/t$. Deduce the radius of the circle, centre at the origin, which touches the hyperbola.

17. S is the point $(3, 0)$ and P is any point on the ellipse $x = 5 \cos \theta,\ y = 4 \sin \theta$. Show that the distance of P from S is three-fifths of the distance of P from the line $x = \frac{25}{3}$.

18. Show that the sum of the distances of any point on the curve $x = 5 \cos \theta,\ y = 4 \sin \theta$ from the points $(-3, 0)$ and $(3, 0)$ is constant.

19. Show that the line $x - y \sin \theta = \cos \theta$ touches the rectangular hyperbola $x^2 - y^2 = 1$. Show also that the area of the triangle formed by this tangent and the lines $y = \pm x$ is 1 unit.

Example 16. *Find the equation of the tangent to the rectangular hyperbola $xy = c^2$ at the point $P(ct, c/t)$, and show that, if this tangent meets the axes at Q and R, then P is the mid-point of QR.*

The gradient of the curve is given by

$$\frac{dy}{dx} = \frac{dy}{dt} \Big/ \frac{dx}{dt}.$$

But $y = \dfrac{c}{t}$, $\therefore \dfrac{dy}{dt} = -\dfrac{c}{t^2}$,

and $x = ct$, $\therefore \dfrac{dx}{dt} = c$.

$$\therefore \frac{dy}{dx} = \frac{-c/t^2}{c} = -\frac{1}{t^2}.$$

Therefore the equation of the tangent at P is

$$yt^2 + x = 2ct.$$

This tangent meets the axes at $Q(2ct, 0)$ and $R(0, 2c/t)$ therefore $P(ct, c/t)$ is the mid-point of QR.

Example 17.   *Find the coordinates of the points where the line $4x - 5y + 6a = 0$ cuts the curve given parametrically by $(at^2, 2at)$.*

If the line $4x - 5y + 6a = 0$ meets the curve at the point $(at^2, 2at)$, then its coordinates must satisfy the equation of the line.   Therefore

$$4at^2 - 10at + 6a = 0.$$
$$\therefore 2t^2 - 5t + 3 = 0.$$
$$\therefore (2t - 3)(t - 1) = 0.$$
$$\therefore t = \tfrac{3}{2} \quad \text{or} \quad 1.$$

Therefore the coordinates of the points of intersection are $(\tfrac{9}{4}a, 3a)$ and $(a, 2a)$.

Exercise 19e

1. Find the equations of the tangents and normals to the following curves at the given points:

   (i) $x = t^2, y = t^3, (1, -1)$;
   (ii) $x = t^2, y = 1/t, (\tfrac{1}{4}, 2)$;
   (iii) $x = at^2, y = 2at, (a, -2a)$;
   (iv) $x = ct, y = c/t, (-c, -c)$;
   (v) $x = t^2 - 4, y = t^3 - 4t, (-3, -3)$;
   (vi) $x = 3\cos\theta, y = 2\sin\theta, (\tfrac{3}{2}, \sqrt{3})$.

2. Find the equations of the tangents and normals to the following curves at the point whose parameter is t:

   (i) $x = t^3, y = 3t^2$; (ii) $x = at^2, y = 2at$;
   (iii) $x = 4t^3, y = 3t^4$; (iv) $x = ct, y = c/t$;
   (v) $x = a \cos t, y = b \sin t$;
   (vi) $x = a \sec t, y = b \tan t$.

3. Find the equations of the chords joining the points whose parameters are p and q on the following curves. Deduce the equations of the tangents at the points p by finding the limiting equations of the chords as q approaches p.

   (i) $x = t^2, y = 2t$; (ii) $x = 1/t, y = t^2$;
   (iii) $x = ct, y = c/t$; (iv) $x = a \cos t, y = b \sin t$.
   [HINT: cancel a factor of $p - q$ in the gradients.]

4. Find the equation of the normal to the parabola $x = at^2$, $y = 2at$ at the point $(4a, 4a)$. Find also the coordinates of the point where the normal meets the curve again.

5. Find the coordinates of the point where the normal to the rectangular hyperbola $x = ct, y = c/t$ at $(2c, \frac{1}{2}c)$ meets the curve again.

6. Find the coordinates of the point where the tangent to the curve $x = 1/t, y = t^2$ at $(1, 1)$ meets the curve again.

7. Find the equation of the tangent to the parabola $y^2 = 4ax$ at the point $(at^2, 2at)$. For what values of t does the tangent pass through the point $(8a, 6a)$? Write down the equations of the tangents to the parabola from $(8a, 6a)$.

8. Find the equations of the tangents to the hyperbola $x = ct, y = c/t$ from the point $(\frac{3}{8}c, \frac{1}{6}c)$.

9. Find the equations of the normals to the parabola $x = at^2, y = 2at$ from the point $(14a, -16a)$.

10. The normal to the hyperbola $x = ct, y = c/t$ at the point P with parameter p meets the curve again at Q. Find the coordinates of Q.

11. The points P, Q, R with parameters p, q, r lie on the hyperbola $x = ct, y = c/t$ and the altitude through P of triangle PQR meets the curve at S. Show that the parameter of S is $-1/pqr$ and deduce that the orthocentre of triangle PQR lies on the curve.

12. Show that the cycloid $x = a(\theta - \sin \theta), y = a(1 - \cos \theta)$ has gradient $\cot \frac{1}{2}\theta$. Find the gradients and the co-

ordinates of the points given by $\theta = 0, \pi, 2\pi$. Sketch the curve.

13. Show that, if a tangent to the curve $x = 1/t$, $y = t^2$ meets the axes OX and OY in A and B, then $PB = 2AP$.

14. Show that the tangent at the point t on the astroid $x = a \cos^3 t$, $y = a \sin^3 t$ is the line

$$y \cos t + x \sin t = a \sin t \cos t.$$

Show that the tangent meets the axes in points whose distance apart is a.

15. The points P and Q on the cissoid $x = a/(1 + t^2)$, $y = at^3/(1 + t^2)$ have parameters p and q. Show that, if angle POQ is a right angle, $pq = -1$, and prove that the locus of the mid-point of PQ is $x = \frac{1}{2}a$.

The parabola

19.10. As no new method is required, work on the parabola is given in the form of exercises. It is intended that any result proved may be used in later questions.

DEFINITION

The locus of a point equidistant from a given point and a given line is called a **parabola**. The given point is the focus and the given line the **directrix**.

Exercise 19f

1. Use compasses and graph paper to plot a parabola from the definition.

2. Given a parabola, take axes with the x-axis through the focus, perpendicular to the directrix, and the origin where the x-axis meets the curve. Let the focus be $(a, 0)$ and show that the equation of the parabola is $y^2 = 4ax$. [It follows from the definition that the equation of the directrix is $x = -a$.]

3. Verify that the point $(at^2, 2at)$ lies on the parabola $y^2 = 4ax$ for all values of t, and that every point on the parabola is given thus.

4. Find the equations of the tangent and normal to the parabola $y^2 = 4ax$ at the point $(at^2, 2at)$.

In Fig. 19.6, the tangent and normal at the point P on the parabola $y^2 = 4ax$ meet the x-axis at T and G, and the y-axis at T' and G'. PN is an ordinate. S is the focus. LD is the directrix and L is the foot of the perpendicular from P to the directrix.

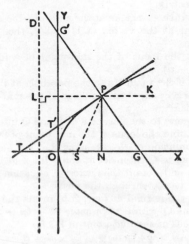

Fig. 19.16

5. Show that ST' = T'L and deduce that:

(i) ∠LPT' = ∠SPT' [use the definition of the curve],
(ii) ∠SPG = ∠KPG.

[This proves the optical property of the parabola, that light from a point source at the focus is reflected in rays parallel to the axis.]

6. Show that L, T', S are collinear (i.e. lie on a straight line), and that LS is perpendicular to PT.

7. Show that TS = SP = SG.

8. Show that LPST is a rhombus and that LPGS is a parallelogram.

9. Show that NG = 2a.

10. If the parameters of the points P and Q are p and q, show that the tangents to the parabola meet at the point $(apq, a(p + q))$.

11. If PQ passes through the focus prove that, with the notation of No. 10, $pq = -1$.

12. Show that the tangents at the ends of a focal chord meet on the directrix.

13. Show that if the tangents at the ends of a focal chord meet the tangent at the vertex at U and V, then $\angle$ USV is a right angle.

14. Show that the locus of the mid-point of a focal chord is $y^2 = 2a(x - a)$.

15. Show that if the tangents to the parabola at P and Q meet on the line $x = ah$, then the locus of the mid-point of the chord PQ is $y^2 = 2a(x + ah)$.

16. If the tangents to the parabola at P and Q intersect on the line $y = k$, find the locus of the mid-point of PQ.

17. With the notation of No. 10, show that the normals at the ends of a chord PQ intersect at the point Z $(a(p^2 + q^2 + 1), a(p + q))$, and obtain the Cartesian equation of the locus of Z if PQ passes through the focus.

18. Show that the normal at $(ap^2, 2ap)$ meets the curve again at the point Q whose parameter is $-(p + 2/p)$. Show that the locus of the mid-point of PQ is

$$y^4 + 2a(2a - x)y^2 + 8a^4 = 0.$$

19. Find the values of t for which the normal at $(at^2, 2at)$ passes through the point $(5a, 2a)$. Hence find the equations of the normals to the parabola from $(5a, 2a)$.

20. Find the equations of the tangents to the parabola from the point $(4a, 5a)$.

21. Show that the tangent to the curve

$$y = -2at^3, x = 3at^2 + 2a$$

at the point with parameter t is also the normal to the parabola $y^2 = 4ax$ at the point $(at^2, 2at)$.

Exercise 19g (Miscellaneous)

1. Find the coordinates of the centroid of the triangle A(2, 5), B(3, -4), C(5, 2).

2. Masses of 3 kg, 4 kg, 5 kg are placed at the points $(0, 4)$, $(4, 0)$, $(8, 5)$. Find the coordinates of their centre of gravity.

3. A point P moves on the hyperbola $xy = 1$, and A is the point $(1, 0)$. Find the locus of a point Q which divides AP internally in the ratio $3:2$.

4. Find the acute angle between the lines $2x + 3y - 5 = 0$ and $7x - 4y = 0$.

5. Find the angles of intersection of the curves $y = x^3 - 3x^2$ and $y = 5x - 4x^2 - 3$.

6. Show that the ellipse $4x^2 + 9y^2 = 36$ and the hyperbola $2x^2 - 3y^2 = 6$ are orthogonal.

7. Sketch the curve whose polar equation is $r = a \cos 3\theta$.

8. Sketch the curve whose polar equation is $r = a(1 + \sin \theta)$ and from this obtain a sketch of the curve $r(1 + \sin \theta) = a$.

9. Find the polar equation of a parabola, taking the focus as the origin and the axis as the initial line.

10. Calculate the area of the triangle $A(2, 5)$, $B(3, -1)$, $C(4, 6)$.

11. Find the polar equation of $(x^2 + y^2 + ax)^2 = a^2(x^2 + y^2)$ and the Cartesian equation of $r(1 + \sin \theta) = a$.

12. Express the equation $7x - 24y - 10 = 0$ in perpendicular form and state the distance of the line from the origin.

13. Find the equations of the bisectors of the angles between:
    (i) $6x - 7y + 11 = 0$, $2x + 9y - 3 = 0$;
    (ii) $7x - y = 3$, $x + y = 2$.

14. Find the locus of a point which moves so that its distance from the line $y + x - 2 = 0$ is equal to its distance from the point $(-1, -1)$.

15. Find the equation of the incircle of the triangle formed by the lines $y - x - 4 = 0$, $7y - x + 14 = 0$, $y + x - 6 = 0$.

16. Find the equations of the tangents to the circle
$$x^2 + y^2 - 12x - 14y + 75 = 0$$
which are parallel to the line $3y - x = 0$.

17. Sketch the curves given parametrically by:
    (i) $x = at^3$, $y = at^4$;
    (ii) $x = \cos \theta$, $y = \sin 2\theta$.

18. Find the Cartesian equations of the loci in No. 17.

19. A straight line through the point $(1, 1)$ and the variable point $P(t, 1/t)$ meets the y-axis at Q.   Find the locus of the mid-point of PQ.

20. The chord PQ of the hyperbola $x = ct, y = c/t$ meets the axes at A and B.   Show that the mid-point of PQ is also the mid-point of AB.

21. Find the equations of the tangent and normal to the curve $x = t^3 - t^2, y = t^2 - 1$ at the point $(4, 3)$.

22. Find the equation of the tangent at $P(t^2, 1/t)$ to the curve $xy^2 = 1$.   If the tangent meets the x-axis at Q, find the locus of the mid-point of PQ.

23. The tangent at $P(t^2, 1/t)$ to the curve $xy^2 = 1$ meets the y-axis at A, the x-axis at B and the curve again at Q. Show that $AP:PB:BQ = 1:2:1$.

24. Find the equations of the tangents to the parabola $x = at^2, y = 2at$ from the point $(5a, 6a)$.

25. Find the coordinates of the point where the normal to the parabola $x = at^2, y = 2at$ at $(9a, 6a)$ meets the curve again.

26. Show that if the tangent at $P(t, t^3)$ to the curve $y = x^3$ meets the curve again at Q, then the y-axis divides PQ in the ratio $1:2$.

27. A tangent to the rectangular hyperbola $x = ct, y = c/t$ meets the axes OX and OY at A and B.   Show that the area of triangle AOB is constant.

28. Show that if the tangent to a parabola at P meets the axis at T, and N is the foot of the perpendicular from P to the axis, then TN is bisected by the vertex.

29. If the points P, Q, R on the parabola $x = at^2, y = 2at$ have parameters p, q, r, and if the chord QR is parallel to the tangent at P, show that $q + r = 2p$. Show also that the line joining P to the mid-point of QR is parallel to the axis.

30. P and Q are the points $(cp, c/p)$, $(cq, c/q)$ on the rectangular hyperbola $xy = c^2$.   Show that the mid-point of PQ, and the point of intersection of the tangents at P and Q lie on the same straight line through the origin, $x = pqy$.

CHAPTER 20

VARIATION AND EXPERIMENTAL LAWS

Variation

20.1. "Variation" in its mathematical sense is concerned with certain ways in which one variable depends on one or more others. The idea is bound up with ratio and proportion which the reader will have met in elementary arithmetic. Some readers may need to revise these ideas and to appreciate their power for the first time.

Proportion arises in arithmetic in a number of ways. For instance the circumference C of a circle is proportional to its radius r: this is usually expressed in the form of an equation

$$C = 2\pi r.$$

Sometimes a graph shows us that two variables are in proportion; for example Fig. 20.1 shows the "travel graph" of a car moving at a steady speed of 50 km/h along a road. Note that: (a) the gradient of the graph is uniform, (b) the straight line passes through the origin.

Yet another aspect of proportion, and indeed the most basic, is used in arithmetic when we use ratios.

To summarize, if y is proportional to x, then:

(i) $y = kx$, where k is some constant,

(ii) the graph of y against x is a straight line through the origin,

417

(iii) if x_1, y_1 and x_2, y_2 are corresponding values of x and y, then

$$\frac{y_1}{y_2} = \frac{x_1}{x_2}.$$

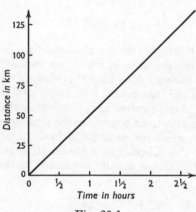

Fig. 20.1

Note that any one of these statements follows from either of the others. The equivalence of (i) and (ii) is familiar from the work of Chapter 1. The equivalence of (ii) and (iii) can be seen by writing (iii) in the form

$$\frac{y_1}{x_1} = \frac{y_2}{x_2}$$

which shows that (x_1, y_1) and (x_2, y_2) lie on the same straight line through the origin.

In the last paragraph we used the phrase,

"y is proportional to x".

Sometimes another phrase with exactly the same meaning is used instead, namely,

"y varies as x".

Other examples of variation will already be familiar to the reader. For instance, the area A of a circle is given in terms of its radius r by the equation

$$A = \pi r^2.$$

Here A is *not* proportional to r, but it *is* proportional to r^2 and we express this by saying that "A *varies* as the square of r".

Another example from mensuration is the volume V of a sphere in terms of its radius r. The equation connecting V, r is

$$V = \tfrac{4}{3}\pi r^3.$$

Again, V is *not* proportional to r, but it *is* proportional to r^3 and we express this by saying that "V varies as the cube of r".

Kinematics provides another example. If a distance of 60 km is travelled at a constant speed u km/h, the time t h is given by the equation

$$t = \frac{60}{u}.$$

We may say that t and u are inversely proportional, or we may express this by saying "t varies inversely as u".

The "inverse square law" may be familiar to the reader: one example of this is the force F exerted by the Earth on a given satellite at distance r from the centre of the Earth. The equation connecting F and r is

$$F = \frac{k}{r^2}, \qquad \text{where } k \text{ is a constant.}$$

This may also be expressed by saying that "F varies inversely as the square of r".

Qu. 1. Express the following equations as statements involving the word "varies":

(i) $s = 16t^2$ (ii) $V = \pi r^3$,   (iii) $y = \dfrac{10}{x^2}$,

(iv) $T = \dfrac{\pi}{2\sqrt{2}}\sqrt{l}$,   (v) $p = \dfrac{200}{V}$,   (vi) $T^2 = d^3$.

Suppose that a number of spheres are made out of wood of uniform density. Then, unless we know the density of the wood, we cannot calculate the weight W of one of these spheres from its diameter d. We can, however, say that

$$W \text{ varies as the cube of } d$$

or write $W = kd^3$,   where k is a constant.

Further, if W_1, d_1 and W_2, d_2 are the weights and diameters of two of the spheres,

$$W_1 = kd_1{}^3$$
$$W_2 = kd_2{}^3$$

so that, by division, $\dfrac{W_1}{W_2} = \dfrac{d_1{}^3}{d_2{}^3}.$

Now, if we know the weight and radius of one of the spheres, this last equation provides us with a very convenient way of calculating the weight of any other when its diameter is known.

Example 1. *A number of spheres are made out of wood of uniform density. A sphere with diameter 7 cm weighs 0·11 kg. How much will a sphere of diameter 9 cm weigh?*

As we have seen above, W varies as d^3. Hence if

W_1, d_1 and W_2, d_2 are the weights and diameters of the two spheres,

$$\frac{W_1}{W_2} = \frac{d_1{}^3}{d_2{}^3}.$$

It is often helpful to tabulate the data and it is worth noting that the algebra of the question is simplified if we place the *quantity to be found* in the line labelled (1):

	weight (kg)	diameter (cm)
(1)	W_1	9
(2)	0·11	7

Then substituting into the equation $\dfrac{W_1}{W_2} = \dfrac{d_1{}^3}{d_2{}^3}$,

$$\frac{W_1}{0\cdot11} = \frac{3^3}{7^3}.$$

$$\therefore\ W_1 = 0\cdot11 \times \frac{3^3}{7^3},$$

$$= 0\cdot23 \text{ to 2 sig. fig.}$$

Therefore the sphere of diameter 9 cm weighs 0·23 kg, correct to 2 significant figures.

[Example 1 illustrates the power of the method: an alternative way of tackling this question would have been to find the density of the wood from the data numbered (2) in the table.]

Qu. 2. In Example 1, what is the effect on W of (i) doubling d, (ii) trebling d?

We saw on page 420 (just above Example 1), that from the statement,

" W varies as the cube of d ",

could be deduced the equation

$$\frac{W_1}{W_2} = \frac{d_1{}^3}{d_2{}^3}$$

which connects corresponding values W_1, d_1 and W_2, d_2. It is important for the following work to be able to convert a statement to an equation quickly and easily, so some more examples of this process follow.

If we are given that

y varies as the square of x,

this is simply another way of saying

y is proportional to x^2.

From this it follows immediately (see pp. 417–18) that:

(i) $y = kx^2$, where k is some constant,
(ii) the graph of y against x^2 is a straight line through the origin,
(iii) if x_1, y_1 and x_2, y_2 are corresponding values of x and y, then

$$\frac{y_1}{y_2} = \frac{x_1{}^2}{x_2{}^2}.$$

On the other hand, if we are given that

"y varies *inversely* as the square of x",

this is simply another way of saying

"y is proportional to $\dfrac{1}{x^2}$",

from which it follows immediately that:

(i) $y = \dfrac{k}{x^2}$, where k is some constant,

(ii) the graph of y against $\dfrac{1}{x^2}$ is a straight line through the origin,

(iii) if x_1, y_1 and x_2, y_2 are corresponding values of x and y, then

$$\frac{y_1}{y_2} = \frac{\dfrac{1}{x_1{}^2}}{\dfrac{1}{x_2{}^2}},$$

or, multiplying numerator and denominator of the right-hand side by $x_1{}^2 x_2{}^2$,

$$\frac{y_1}{y_2} = \frac{x_2{}^2}{x_1{}^2}.$$

Note that, in this case of *inverse* variation, the x's are upside down compared with the y's. "Inverse" comes from the same root as "invert", one meaning of which is to "turn upside down".

Qu. 3. Write down equations (a) with k's, (b) with suffixes, similar to those in the last three paragraphs for the statements:

(i) p varies as q,

(ii) p varies inversely as q,

(iii) v varies as the cube of x,

(iv) u varies as the square root of l,

(v) F varies as the square of c,

(vi) H varies inversely as the square of d,

(vii) T varies inversely as the square root of g,

(viii) A varies as the nth power of s,

(ix) the cube of A varies as the square of v.

Example 2. *The length l of a simple pendulum varies as the square of the period T (time to swing to and fro). A pendulum 0.994 m long has a period of approximately 2 s, find*

(i) *the length of a pendulum whose period is* 3 *s*, (ii) *an equation connecting l and T.*

(i) Tabulating the data:

	length (m)	period (s)
(1)	*l*	3
(2)	0·994	2

l varies as T^2.   Therefore

$$\frac{l_1}{l_2} = \frac{T_1{}^2}{T_2{}^2}.$$

$$\therefore \frac{l}{0·994} = \frac{3^2}{2^2}.$$

$$\therefore \quad l = 0·994 \times \frac{9}{4},$$

$$= 2·236.$$

Therefore the length of a pendulum whose period is 3 s is 2·24 m.

(ii) Tabulating the data again, we enter *l* and T in the row numbered (1):

	length (m)	period (s)
(1)	*l*	T
(2)	0·994	2

Substituting in the same equation as before,

$$\frac{l}{0·994} = \frac{T^2}{2^2}.$$

$$\therefore l = 0·248(5)T^2.$$

Therefore the equation connecting l, T is

$$l \simeq 0.25T^2.$$

Qu. 4. In Example 2, find the effect on l of (i) doubling T, (ii) trebling T. What is the effect on T of doubling l?

Qu. 5. Find the period of a pendulum whose length is 0·3 m from the data of Example 2. Time ten swings to and fro of such a pendulum and compare this with your answer.

Example 3. *The weight w N* of an astronaut varies inversely as the square of his distance d from the centre of the Earth. If an astronaut's weight on Earth is 792 N, what will his weight be at a height of 230 km above the Earth? Take the radius of the Earth to be 6370 km.*

We tabulate the data:

	weight (N)	distance from the centre of the Earth (km)
(1)	w	$6370 + 230 = 6600$
(2)	792	6370

Now w varies inversely as d^2, so if w_1, d_1 and w_2, d_2 are corresponding values,

$$\frac{w_1}{w_2} = \frac{d_2{}^2}{d_1{}^2}.$$

$$\therefore \frac{w}{792} = \frac{6370^2}{6600^2}.$$

$$\therefore w = 792 \times \frac{6370^2}{6600^2},$$

$$= 737\cdot7.$$

Therefore the astronaut's weight would be 738 N.

* The newton (N) is the absolute unit of force in S.I. units. The magnitude of 1 kg.wt varies with the value of g, since 1 kg.wt gives to a mass of 1 kg an acceleration of g m/s². In contrast, 1 N gives to the same mass a fixed acceleration of 1 m/s², by definition. Hence in a context of varying gravitational pull we use this constant, or absolute, unit of force the newton.

To find the height above the Earth at which his weight would be halved, we again tabulate the data:

	weight (N)	distance from the centre of the Earth (km)
(1)	396	d
(2)	792	6370

Again using

$$\frac{w_1}{w_2} = \frac{d_2{}^2}{d_1{}^2}$$

for the new $w_1, d_1,$

$$\frac{396}{792} = \frac{6370^2}{d^2}.$$

$$\therefore d^2 = 2 \times 6370^2.$$

$$\therefore d = \sqrt{2} \times 6370,$$

$$= 9010.$$

Therefore the height above the Earth at which his weight would be halved is $9010 - 6370$ km $= 2640$ km.

Qu. 6. Find an equation in the form $w = \dfrac{k}{d^2}$ connecting the weight of the astronaut in Example 3 and his distance from the centre of the Earth.

Qu. 7. With the equation of Qu. 6, find the effect on w of (i) doubling d, (ii) trebling d.

Qu. 8. Discuss whether the first of the following pairs of variables varies as some power of the second and, if so, state what power:

(i) the cost c of 100 copies of a book and the price p of one,

(ii) the cost C of a square of plywood and its side a,

  (iii) the weight w of a spherical lead shot and its radius r,

  (iv) the length l of a rectangle of given area and its breadth b,

  (v) the surface area S of a scale model and its length l,

  (vi) the area A of an equilateral triangle and its side a,

  (vii) the side a of an equilateral triangle and its area A,

(viii) the volume V of a regular tetrahedron and its side a,

  (ix) the side a of a regular tetrahedron and its volume V.

Exercise 20a

1. The area of a circular sector containing a given angle varies as the square of the radius of the circle. If the area of the sector is 2 cm² when the radius is 1·6 cm, find the area of the sector containing the same angle when the radius of the circle is 2·7 cm.

2. The distance of the horizon d km varies as the square root of the height h m of the observer above sea level. An observer at a height of 100 m above sea level sees the horizon at a distance of 35·7 km. Find the distance of the horizon from an observer 70 m above sea level.

    Also find an equation connecting d and h.

3. The length l cm of a simple pendulum varies as the square of its period T s. A pendulum with period 2 s is 99·4 cm long; find the length of a pendulum whose period is 2·5 s.

    What equation connects l and T?

4. Assuming that the length of paper in a roll of given dimensions varies inversely as the thickness of the paper, find the increase in length when the thickness of paper in a 100 m roll is decreased from 0·25 mm to 0·20 mm.

5. A certain type of hollow plastic sphere is designed in such a way that the mass varies as the square of the diameter. Three spheres of this type are made: one has mass 0·10 kg and diameter 9 cm; a second has diameter 14 cm; and a third has mass 0·15 kg. Find the mass of the second, the diameter of the third, and an equation connecting the mass m kg and the diameter d cm of spheres of this type.

6. The circumference C inches of a circle of radius r inches is given by the formula $C = 2\pi r$; if C_1, r_1 and C_2, r_2 are corresponding values of C, r,

$$\frac{C_1}{C_2} = \frac{r_1}{r_2}. \tag{1}$$

   (i) What formula gives the circumference C cm of a circle of radius r m? Does equation (1) still hold?

   (ii) Given that 1 inch = 2·54 cm, what equation gives the circumference C cm of a circle of radius r inches? Does equation (1) still hold?

7. Boyle's law states that, under certain conditions, the pressure exerted by a given mass of gas is inversely proportional to the volume occupied by it. The gas inside a cylinder is compressed by a piston in such a way that Boye's law may legitimately be applied. When this happens, the volume is decreased from 200 cm³ to 70 cm³. If the original pressure of the gas is $9·8 \times 10^4$ N/m², find the final pressure of the gas.

8. The number of square lino tiles needed to surface the floor of a hall varies inversely as the square of the length of a side of the tile used. If 2016 tiles of side 0·4 m would be needed to surface the floor of a certain hall, how many tiles of side 0·3 m would be required?

9. If the volume of a model 10 cm long is 72 cm³, what is the volume of a similar model 6 cm long? What is the length of a similar model with volume 100 cm³?

10. The maximum speed of yachts of normal dimensions varies as the square root of their length. If a yacht of 20 m can maintain a maximum speed of 12 k, find the maximum speed of a yacht 15 m long. Obtain an equation connecting a yacht's maximum speed v k and its length l m.

11. For similar printing type, the number of characters on a given size of page varies inversely as the square of the height of the type. On a certain page 2200 characters of height 6 mm could be printed. How many characters of similar type of height 5 mm could be printed on the page?

When 7000 characters have to be printed on the page with similar type, what height would the type be if the height is a multiple of 0·1 mm?

12. (i) If y varies as x^3 and x varies as t^2, does y vary as any power of t? [HINT: Write the statements y varies as x^3, x varies as t^2 as equations with constants k, K.]

   (ii) p varies inversely as q; q varies as the square of r. Does p vary as any power of r?

13. When I drive round a certain corner at 18 km/h, the sideways frictional force between the tyres of my car and the road is 1050 N. The sideways frictional force F N varies as the square of the speed v km/h. Find an equation connecting F, v and use it to find:

   (i) the total sideways frictional force at 27 km/h,
   (ii) the speed at which the sideways frictional force is equal to 6170 N which is half the weight of the loaded car.

14. Assuming that the power H kW developed by a certain car travelling on a level road varies as the cube of the speed v km/h, find an equation connecting H, v for this car, given that it develops 50 kW at 65 km/h. Find the power developed by it at 30 km/h along a level road.

15. The speed of a certain point on a high-speed centrifuge varies as the angular velocity of the centrifuge, and the acceleration of this point varies as the square of the angular velocity. Find the percentage changes in the speed and acceleration of the point when the angular velocity is increased from 56 000 rev/min to 60 000 rev/min.

16. The cube of the surface area of a regular icosahedron varies as the square of its volume. By what factor will the surface area of a regular icosahedron be increased if its volume is doubled?

17. The period T s of a given pendulum varies inversely as the square root of the acceleration due to gravity g m/s^2 at the location of the pendulum. Find the percentage change in the period of a pendulum moved from Greenwich, where $g = 9·812$ m/s^2, to New York where $g = 9·802$ m/s^2.

[HINT: Use the first two terms of the expansion of $(1 + x)^{\frac{1}{4}}$.]

18. The volume and areas of similar solids vary respectively as the cubes and squares of their linear dimensions. Some similar solids are placed in an upward current of air. Assuming that the upthrust of the air current varies as the surface area of the solid and that the weight of the solid varies as its volume, show that some of the solids will rise if their linear dimensions are small enough.

19. The square of the period (time to go round its orbit) of an Earth satellite varies as the cube of its mean distance from the centre of the Earth. The period of the Moon is 28 days and its mean distance from the centre of the Earth is 380 000 km. Find the period, to the nearest minute, of an Earth satellite whose mean distance from the surface of the Earth is 470 km, given that the radius of the Earth is 6370 km.

   Also find an equation giving the period of an Earth satellite T hours in terms of its mean distance d km from the centre of the Earth.

20. Like and unlike poles of two bar magnets repel and attract each other respectively with a force which varies inversely as the square of the distance between the poles. The poles of each of two bar magnets are at a distance $2d$ apart. The magnets are placed in line with two unlike poles of the magnets at a distance d apart. They are then placed in line with two unlike poles at a distance $2d$ apart. By what factor is the attractive force between the magnets decreased?

Joint variation

20.2. So far we have only considered examples of variation where one variable, say y, varies as some power of another variable, say x. But there are many examples in science, engineering and everyday life when one variable depends on two or more others. For example, the volume

V of a right circular cylinder is given in terms of its radius r and height h by the formula

$$V = \pi r^2 h.$$

If we consider a metal rod of uniform circular cross-section which can be cut into lengths, we have a case of this law in which the radius is constant and so

the volume varies as the length

or, using the symbol "$\propto$" as an abbreviation for "varies as",

$$V \propto h.$$

On the other hand, if circular discs are cut out of sheet metal or plywood, h will be constant and so

the volume varies as the square of the radius

or $$V \propto r^2.$$

To summarize, for a right circular cylinder,

if r is constant, $V \propto h$,
and if h is constant, $V \propto r^2$.

In experimental work, if one variable depends on two or more others, it is most convenient to see how the first depends on each of the others in turn while the remainder are held constant. As an illustration of this, consider the discharge of water through a circular hole. The volume of water V will depend in some way on:

  (i) the radius r of the hole,
  (ii) the velocity v of the water,
  (iii) the time t over which the discharge takes place.

It is found that:

  (1) if v, t are constant, $V \propto r^2$,
  (2) if t, r are constant, $V \propto v$,
  (3) if r, v are constant, $V \propto t$.

It will be seen that the equation

$$V = kr^2vt \quad (k \text{ constant})$$

satisfies the conditions (1), (2), (3) and hence it is natural to write

$$V \propto r^2vt.$$

Qu. 9.   Express the statement "If z is constant, y varies as x; if x is constant, y varies as the cube of z", as a single equation.

Qu. 10.   Write the statement, "If h, t are constant, W varies as the square of r; if r, t are constant, W varies as h; if r, h are constant, W varies inversely as t", as a single statement using the sign "$\propto$".

When one variable varies as two or more others, the word *jointly* is sometimes used. For example, with the data of the last paragraph, we might say that V varies jointly as v, t and the square of r.

Qu. 11.   "The kinetic energy T of a flywheel varies jointly as its mass m and as the square of its radius r." Express this statement (i) as an equation with a constant k, (ii) as a statement using the sign "$\propto$".

Qu. 12.   "F varies jointly as m and the square of v, and inversely as r." Express this statement as an equation.

For purposes of calculation, we can rewrite statements in the form

$$A = k\frac{x^3}{t}, \text{ where } k \text{ is some constant,}$$

in terms of the ratios of corresponding values A_1, x_1, t_1 and A_2, x_2, t_2 of the variables.   We have

$$A_1 = k\frac{x_1{}^3}{t_1},$$

$$A_2 = k \, \frac{x_2{}^3}{t_2}.$$

$$\therefore \frac{A_1}{A_2} = \frac{\dfrac{x_1{}^3}{t_1}}{\dfrac{x_2{}^3}{t_2}}.$$

Multiplying numerator and denominator by $t_1 t_2$,

$$\frac{A_1}{A_2} = \frac{x_1{}^3 t_2}{x_2{}^3 t_1}.$$

Note that A varies inversely as t, and that the ratio $\dfrac{t_1}{t_2}$ is "upside down".

Qu. 13. If x_1, y_1, z_1 and x_2, y_2, z_2 are corresponding values of x, y, z, write down equations connecting x_1, y_1, z_1 and x_2, y_2, z_2 when:

 (i) z varies jointly as x and the square of y,
 (ii) z varies as y and inversely as the square of x,
 (iii) z varies as the cube of x and as the square of y,
 (iv) z varies as x when y is constant and z varies as y when x is constant,
 (v) z varies as the square of x when y is constant and z varies as the square of y when x is constant,
 (vi) z varies as the square root of x when y is constant and inversely as y when x is constant.

Example 4. *The total sideways force experienced by a given car rounding a circular bend at a constant speed varies as the square of the speed of the car and inversely as the radius of the circle. A certain car goes round a bend of radius 50 m at 72 km/h and experiences a total sideways force of 12 kN. What sideways force will it experience on going round a bend of radius 30 m at 54 km/h?*

Let the sideways force be F kN, the speed be v km/h, and the radius r m, then

$$F \propto \frac{v^2}{r}.$$

Therefore, if F_1, v_1, r_1 and F_2, v_2, r_2 are corresponding values of F, v, r,

$$\frac{F_1}{F_2} = \frac{\dfrac{v_1{}^2}{r_1}}{\dfrac{v_2{}^2}{r_2}} = \frac{v_1{}^2 r_2}{v_2{}^2 r_1}.$$

	F (kN)	v (km/h)	r (m)
(1)	F	54	30
(2)	12	72	50

$$\therefore \frac{F}{12} = \frac{54^2 \times 50}{72^2 \times 30}.$$

$$\therefore F = \frac{12 \times 3^2 \times 5}{4^2 \times 3},$$

$$= \frac{45}{4} = 11{\cdot}25.$$

Therefore the sideways force on the car will be approximately 11 kN.

Variation in parts*

20.3. As an example of variation in parts, consider the cost of having a floor covered with lino tiles. First of all, a

* The reader is advised to delay reading this section until he has worked at least some of Exercise 20b Nos. 1–12.

man and some materials have to be transported to the site. Here the cost of the man's time and the cost of the running of a van may be taken to vary as the distance s km from the firm's premises and so we may write this part of the cost as £ks, where k is some constant to be found. Second, there is the cost of materials and the man's time doing the job, which may be taken to vary as the area A m^2 of the floor, and so this part of the cost may be written £KA, where K is another constant to be determined. Hence, if the total cost is £C,

$$C = ks + KA.$$

Let us suppose that the cost of two contracts is as given in the following table. How much would it cost to lay 40 m^2 of lino tiles at a distance of 75 km from the firm's premises?

Cost £C	Distance s km	Area A m^2
C	75	40
53	45	50
31	60	27

Substituting from the bottom two lines of the table into

$$C = ks + KA,$$

$$53 = 45k + 50K, \qquad (1)$$

$$31 = 60k + 27K. \qquad (2)$$

$4 \times (1) - 3 \times (2)$: $212 - 93 = (200 - 81)K.$

$$\therefore K = \frac{119}{119} = 1.$$

From (2), $31 = 60k + 27.$

$$\therefore 4 = 60k.$$

$$\therefore k = \frac{1}{15}.$$

Substituting $K = 1$, $k = \frac{1}{15}$,

$$C = \frac{1}{15} s + A.$$

When $s = 75$, $A = 40$,

$$C = \frac{1}{15} \times 75 + 40,$$

$$= 45.$$

Therefore the cost of laying 40 m² of lino tiles at a distance of 75 km would be £45.

Qu. 14.   The cost £C of manufacturing a certain number of wooden cubes for children is made up of two parts, one of which is constant and the other of which varies as the cube of the side x cm of a brick.

   (i) Express the above statement in symbols.

  (ii) Find the cost of making 1000 1¼-cm cubes if the same number of 2-cm and 1-cm cubes cost respectively £18 and £11.

Exercise 20b

1. The area of a sector of a circle varies jointly as the angle at the centre and the square of the radius.   Given that the area of a sector containing an angle of 36° in a circle of radius 10 cm is 31·4 cm², find the area of a sector containing an angle of 72° in a circle of radius 5 cm.

2. The number of revolutions per minute of a bicycle wheel varies as the speed of the bicycle and inversely as the diameter of the wheel. A wheel of diameter 63 cm makes 151·5 revolutions per minute when the bicycle is moving at 18 km/h. A new type of bicycle has wheels of 35 cm diameter; how many revolutions per minute will one of its wheels make when the bicycle is moving at 30 km/h?

3. The flow of water through a circular orifice varies as the square of the diameter of the orifice and as the square root of the head of water. Given that 200 litres of water per second flow through an orifice of diameter 25 mm when the head of water if 4 m, find the flow of water through an orifice of diameter 10 mm when the head of water is 9 m.

4. The kinetic energy of a car (including passengers) varies jointly as the total mass and the square of the speed. A car of total mass 1000 kg travelling at 72 km/h has a kinetic energy of 200 kJ. What is the kinetic energy of a car of total mass 1500 kg travelling at 108 km/h?

5. The volume of a given mass of gas varies directly as its absolute temperature and inversely as its pressure. At an absolute temperature of 283 K and a pressure of 73 cm of mercury, a certain mass of gas has volume 200 cm³. What will its volume be at "standard temperature and pressure", i.e. absolute temperature 273 K and pressure 76 cm of mercury? Also find an equation which expresses the volume V cm³ of the gas in terms of its absolute temperature T K and its pressure p cm of mercury.

6. The rate at which an electric fire gives out heat varies as the square of the voltage and inversely as the resistance. If a fire with resistance 57·6 ohms gives out approximately 1 kW when the voltage is 240, at what rate will heat be given out by an electric fire with resistance 69 ohms when the voltage is 220? Also find an expression which gives (approximately) the output in kW of an electric fire of resistance R ohms when the voltage is V.

7. The frequency of the note emitted by a plucked wire of a certain type varies as the square root of the tension of the

wire and inversely as its length. A wire of length 0·61 m
under a tension of 31 N emits a note of frequency 130 s⁻¹.
What will be the frequency of the note emitted by a similar
wire of length 0·25 m under a tension of 100 N. Find an
equation which gives the number of oscillations per
second f in terms of the length l m and the tension F N.

8. When a note is produced by blowing across the top of a
bottle with a circular mouth, the frequency of the note
varies as the internal diameter of the mouth and inversely
as the square root of the volume of the bottle. Blowing
across a certain bottle, I obtain a note whose frequency is
approximately 203 s⁻¹. What is the frequency of the
note I should obtain by blowing across the top of a bottle
with four times the capacity, and with three-quarters the
mouth diameter of the first?

9. The period of a simple pendulum varies as the square root
of its length and inversely as the square root of the
acceleration due to gravity. On the Earth, the period of
a pendulum 99·4 cm long is 2 s. Assuming that the
acceleration due to gravity on the surface of the Moon is
one-sixth of that on the Earth, what would be the period
of a pendulum 1 m long on the Moon?

10. The effectiveness of a spin drier is measured by the central
acceleration at a point on the internal surface of the
rotating drum. This acceleration varies as the internal
diameter of the drum and as the square of its angular
speed. Which would be the more effective: a spin drier
with internal diameter 0·5 m running at an angular speed
of 1600 rev/min, or one with internal diameter 0·3 m
running at 2000 rev/min?

11. The rate at which heat is conducted through a metal plate
varies jointly as the area of the plate and the temperature
difference between the two sides, and inversely as the
thickness of the metal. For quick heating of the contents,
which saucepan would be better: one with a diameter
15 cm and thickness 2 mm, or another with diameter 20 cm
and thickness 3 mm?

12. The light received at a point varies as the power of the source and inversely as the square of its distance from the point. Assuming that each bulb converts an equal proportion of its power into light, which gives better illumination: a 60 W bulb at $1\frac{1}{2}$ m, or a 100 W bulb at 2 m?

13. The annual cost of running a certain car is made up of two parts, one of which is fixed and the other of which varies as the distance run by the car in the year. In one year the car ran 6000 km at a total cost of £180; in the next year it ran 7200 km at a total cost of £190. How much would it cost to run the car in a year during which it ran 12 000 km? To what extent is the assumption about the cost justified?

14. The cost of printing a circular on octavo paper is partly fixed and partly varies as the number of copies printed. If 100 and 500 copies cost £1·65 and £2·85 respectively, how much will 200 copies cost? Find an equation which gives the cost £C of n copies.

15. When a body is being uniformly accelerated, the distance travelled is the sum of two parts: one part varies as the time, the other varies as the square of the time. The distances travelled by a body in 2 s and 3 s from its original position are respectively 32 m and 57 m. How far will it travel from its original position in 4 s? Find an equation which gives the distance s m in terms of the time t s from its original position.

16. In good road conditions, the driver of a car can stop the car moving at 30 km/h in 11·4 m, and if the car is moving at 60 km/h he can stop it in 33·6 m. This stopping distance is made up of two parts, one of which varies as the speed of the car, and the other of which varies as the square of the speed. In what distance can the driver stop the car if it is moving at 80 km/h? Find an equation which gives the stopping distance s m in terms of the speed v km/h.

   If the car can just be stopped in 25 m, how fast is it moving?

17. Basic slag is advertised in 5 kg packs at £0·25, 10 kg packs at £0·45 and 20 kg packs at £0·85. It is suggested

that the cost £C of these packs is partly constant and partly varies as the mass m kg of basic slag. If this is so, what is the equation which gives C in terms of m?

18. The price of a ticket to a dance is made up of two parts, one of which is fixed and the other of which varies inversely as the number of people expected at the dance. For a certain dance, it is found that the price of a ticket would need to be 60p if 100 people were to attend, but if 150 people attended the price of a ticket would need to be 50p in order to cover the cost. What would be the price of a ticket in order to cover the cost if only 75 people attended? If the price of a ticket was fixed at 54p, how many people would have to buy tickets for the cost to be covered?

19. When a certain volume of wax is cast into a square prism, the surface area of the prism may be expressed as the sum of two parts, one of which varies as the square of the side of the cross-section and the other of which varies inversely as the side of the cross-section. If the side of the cross-section is 2 cm, the surface area of the prism is 28 cm². When the side of the cross-section is 1 cm, the surface area of the prism is 42 cm². What will be the surface area of the prism when the side of the cross-section is $2\frac{1}{2}$ cm?

Also find a formula which gives the surface area S cm² of the prism in terms of the side x cm of the cross-section.

20. The volume of a cap of height h cut off from a sphere of radius r (by a plane at distance $r - h$ from the centre) is the sum of two parts, one of which varies as the square of h and the other of which varies as the cube of h. Use the formulae for the volumes of a hemisphere and a sphere (i.e. the volume of the cap when $h = r$ and when $h = 2r$) to find a formula for the volume V of the cap in terms of h, r.

21. The sum of the cubes of the first n integers may be expressed as the sum of three terms varying respectively as the fourth power of n, the cube of n, and the square of n. Use this information to find a formula for the sum of the cubes of the first n integers.

Graphical determination of laws

20.4.   A simple experiment is performed to investigate the relationship between the tension in an elastic band and its extension, by fixing the upper end and suspending bodies of different masses in turn from the lower end. The tension (y N) in the band (given by the weight of each body) is tabulated against the corresponding extension (x cm) measured to the nearest mm.

x	0	1	1·8	2·5	3·3	4·3	5·3
y	0	1	2	3	4	5	6

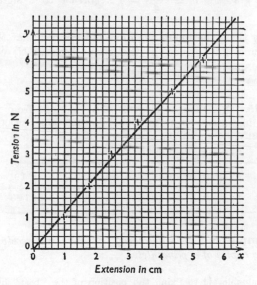

Fig. 20.2

When these results are illustrated graphically (see Fig. 20.2) we see that it is possible to draw a straight line about which the points are closely scattered; such a line is then drawn,* and we make it pass through the origin since we know that $y = 0$ when $x = 0$.

A straight line through the origin of gradient m has the equation

$$y = mx$$

and, allowing for experimental error and the limited accuracy in measuring x, we may reasonably deduce this to be the relationship between the x and y of our experiment. Referring to the straight line drawn, when $x = 4$, $y \simeq 4\cdot61$, and its gradient $m \simeq \dfrac{4\cdot61}{4} = 1\cdot2$ correct to 2 significant figures.

So by this experiment we have determined that the law connecting the tension in the given band (y N) and its extension (x cm) is

$$y \simeq 1\cdot2x.$$

Qu. 15. A trolley accelerates down a slope from rest to v km/h in t s as shown by the following table. Determine graphically the law giving v in terms of t.

v	0	10	20	30	40	50	60
t	0	2·5	4·7	7·1	9·7	11·9	14·5

Example 5. *The following estimate is received for printing copies of a pamphlet.*

---

* A valuable aid to fixing the position of the "best" line is a sheet of perspex with a straight line scored on it.

No. of copies	50	100	200	500
Cost in £	2·30	2·50	2·90	4·10

(i) *Obtain a law giving the cost, £y, of x copies.*

(ii) *Estimate the cost of 350 copies.*

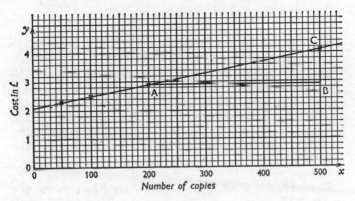

Fig. 20.3

(i) Fig. 20.3 shows a straight line graph, so we assume that the printer has used a *linear* law connecting x and y to make his estimate, i.e. there is an equation connecting the variables of the form

$$y = mx + c.$$

Now c is the intercept on the y-axis (see p. 15) and so we can refer to the graph to find that $c = 2·1$, and (from the triangle ABC) that the gradient

$$m = \frac{4·1 - 2·9}{500 - 200} = \frac{1·2}{300} = 0·004.$$

Therefore the law is

$$y = 0{\cdot}004x + 2{\cdot}1.$$

(ii) When $x = 350$, $y = 0{\cdot}004 \times 350 + 2{\cdot}1$,

$$= 1{\cdot}4 + 2{\cdot}1,$$

$$= 3{\cdot}5.$$

Therefore the cost of 350 copies is £3·50.

Qu. 16.   From the solution of Example 5 (i), when $x = 0$, $y = 2{\cdot}1$.   What interpretation may be given to this result?

Note that in Fig. 20.3 we have included the origin of the coordinates (that is to say each axis is calibrated *from zero*) and thus we were able to utilize the y-intercept to find c. This advantage must often be sacrificed in favour of the increased accuracy obtainable by using a larger scale; Example 6 demonstrates how the equation of a straight line is determined in these circumstances.

Example 6.   *Find the equation of the line $y = mx + c$ in Fig. 20.4.*

The gradient m is found from the triangle PQR (chosen so that the length of PQ is a whole number of units).

$$m = \frac{31}{5} = 6{\cdot}2.$$

Substituting in $y = mx + c$,

$$y = 6{\cdot}2x + c.$$

To find c, substitute the coordinates of a convenient point on the line, e.g. when $x = 10$ $y = 78$.

$$\therefore 78 = 6{\cdot}2 \times 10 + c,$$

$$\therefore \ c = 16.$$

Therefore the required equation is

$$y = 6 \cdot 2x + 16.$$

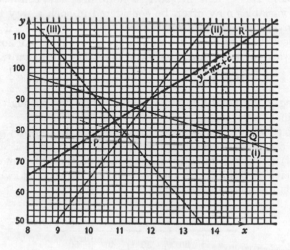

Fig. 20.4

Qu. 17. Find the equations of the lines (i), (ii), (iii) in Fig. 20.4, working to one decimal place. (Note that two of these lines have negative gradients.)

Qu. 18. The upper end of a coiled spring was fixed and bodies were hung in turn from the lower end. The mass of the bodies (y g) and the corresponding lengths of the spring (x cm) were recorded as follows:

x	8·4	9·5	10·1	11·0	11·7	12·6	13·5	14·3
y	30	40	50	60	70	80	90	100

Find a law giving y in terms of x over this range, and estimate the unstretched length of the spring.

Linear check of non-linear laws

20.5. As we saw in §20.1, a non-linear law connecting two variables may often be considered in such a way that it involves a linear relationship. For example, if we suspect that two variables x and y are inversely proportional, we wish to show that $xy = k$, where k is a constant, i.e. $y = k \times \dfrac{1}{x}$; this may be done by plotting y against $\dfrac{1}{x}$ and seeing if the points lie close to a straight line through the origin.

To take another example, let us suppose that the designer of a car windscreen wishes to find out if the air resistance (R N) is proportional to the square, or the cube, of the velocity (v km/h); he carries out an experiment which yields the following results:

v	20	30	40	50
R	4	14	33	60

The reader may check from a rough sketch that the graph of v against R does no more than indicate that R might vary as some power of v, which is of no real assistance. This problem is dealt with in the following question.

Qu. 19. With the data of the preceding paragraph, plot the following graphs, letting 1 cm represent 5 N:

    (i) R against v^2 (on the v^2-axis let 1 cm represent 200),

    (ii) R against v^3 (on the v^3-axis let 1 cm represent 10 000).

Deduce an approximate relationship giving R in terms of v.

Qu. 20. A marble was allowed to run down a sloping sheet of glass and the time (t s) taken to roll s m from rest was measured by a stop watch. The results were as follows:

s	1	2	3	4	5
t	1·4	2	2·5	2·8	3·2

Confirm that the law relating s and t is $s = kt^2$, and determine the value of the constant k to two significant figures.

Reduction of a law to linear form using logarithms*

20.6. The method of Qu. 19 is severely limited, since we assume a relationship $R = kv^n$, then we guess some integral value of n and test for it. It would be better to employ a method which tests for any rational value of n, and this is possible if we use logarithms. (At this point it may help some readers to refer back to Qu. 5 and Qu. 6 on p. 175.)

Suppose that we wish to test the law

$$R = kv^n \qquad (1)$$

where k and n are constants. If it is valid,

$$\log_{10} R = \log_{10} (kv^n).$$
$$\log_{10} R = \log_{10} v^n + \log_{10} k,$$
$$\log_{10} R = n \log_{10} v + \log_{10} k. \qquad (2)$$

Writing $\log_{10} R$ as y, $\log_{10} v$ as x and $\log_{10} k$ as c, (2) becomes

$$y = nx + c$$

which represents a straight line of gradient n.

* The reader should work some of Nos. 1 to 12 in Exercise 20c before proceeding with this section.

Thus if we plot $\log_{10} R$ against $\log_{10} v$ and we obtain a set of nearly collinear points, this means that we have established the linear relationship (2) and confirmed the law (1); we then draw the "best" straight line. Its gradient determines the value of the constant n, and the constant k is found from the y-intercept c, or by the method of Example 6.

Qu. 21. From the data of Qu. 19 the following table has been prepared:

$x = \log_{10}v$	1·30	1·48	1·60	1·70
$y = \log_{10}R$	0·60	1·15	1·52	1·78

Using a scale of 0·1 to 1 cm, plot $\log_{10} R$ against $\log_{10} v$ and deduce that $R \simeq 0{\cdot}0005v^3$ (see Example 6, p. 444).

When a given mass of gas is compressed or allowed to expand slowly, so that there is time for the transfer of heat between the gas and its surroundings, its temperature remaining constant, the pressure (p) and the volume (V) are said to undergo an *isothermal* change and obey Boyle's law $pV = k$, a constant. If however the compression or expansion takes place suddenly, and there is no appreciable exchange of heat between the gas and its surroundings, then there is a change in the temperature of the gas, and the pressure and volume undergo an *adiabatic* change which does not conform to Boyle's law.

Boyle's law may be written $p = kV^{-1}$; the experimental data from an adiabatic change suggest that in this case we have the same form of relationship, $p = kV^n$, but that n has some value other than -1.

Example 7. *A given mass of air expands adiabatically*

and the following measurements are taken of the pressure (p cm of mercury) and volume (V cm³):

V	100	125	150	175	200
p	58·6	42·4	32·8	27·0	22·3

Confirm that $p = kV^n$ and determine the values of the constants k and n.

Assuming that $p = kV^n$, and taking logarithms to the base 10 of each side,

$$\log_{10} p = \log_{10} V^n + \log_{10} k,$$
$$\log_{10} p = n \log_{10} V + \log_{10} k.$$

Writing $\log_{10} p$ as y, $\log_{10} V$ as x, $\log_{10} k$ as c,

$$y = nx + c.$$

Since this is a linear relationship between x and y, we hope to find that $\log_{10} V$ plotted against $\log_{10} p$ will yield points lying nearly on a straight line. From the following table the points have been plotted in Fig. 20.5, p. 450, and the "best" straight line has been drawn.*

$x = \log_{10} V$	2·000	2·097	2·176	2·243	2·301
$y = \log_{10} p$	1·768	1·627	1·516	1·431	1·348

* Provided that the experimental errors are random, then a reliable aid to drawing the "best" straight line is to make it pass through the point whose coordinates are the averages of the co-ordinates of the plotted points; this point is arrowed in Fig. 20.5. Sometimes there is also a point whose exact coordinates are known; such a point is (0, 0) in Fig. 20.2.

The gradient n is found from triangle PQR

$$n = -\frac{0 \cdot 282}{0 \cdot 2} = -1 \cdot 41.$$

Therefore the equation of the straight line is

$$y = -1 \cdot 41x + c.$$

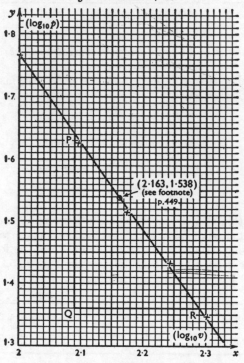

Fig. 20.5

But from the graph, when $x = 2$, $y = 1 \cdot 769$,

$$\therefore \ 1 \cdot 769 = -1 \cdot 41 \times 2 + c.$$

$$\therefore \ \log_{10} k = c = 4 \cdot 589,$$

∴ $k = 38\ 820 = 39\ 000$ to two significant figures.

Hence the experimental data confirms the relationship given between p and V, namely

$$p = 39\ 000 V^{-1\cdot41}.$$

There are other types of variation which may be confirmed by using logarithms to reduce them to a linear relationship; in laws of growth, for example, one of the variables is often in an index.

If $P = ka^x$, where k, a, are constants,

$$\log_{10} P = \log_{10} a^x + \log_{10} k,$$

$$∴ \log_{10} P = x \log_{10} a + \log_{10} k,$$

and writing $\log_{10} P$ as y, $\log_{10} a$ as m, $\log_{10} k$ as c,

$$y = mx + c.$$

This is a straight line equation which reveals a linear relationship between x and $\log_{10} P$.

Example 8. *The frequency (f oscillations per second) and the interval (x semitones) of each note of a C major scale are given in the table below; show that f, x are related by a law in the form $f = ka^x$ and determine the constants k, a.*

Note:	C	D	E	F	G	A	B	C
x	0	2	4	5	7	9	11	12
f	256	287	323	342	384	431	483	512

Assuming that $f = ka^x$, and taking logarithms to the base 10 of each side,

$$\log_{10} f = \log_{10} a^x + \log_{10} k.$$

$$\therefore \log_{10} f = x \log_{10} a + \log_{10} k.$$

Writing $\log_{10} f$ as y, $\log_{10} a$ as m, $\log_{10} k$ as c,

$$y = mx + c.$$

This shows that we must, from the data, establish a linear relationship between x and $\log_{10} f$. From the following table the points have been plotted in Fig. 20.6 and the "best" straight line has been drawn; we have confirmed that the law is of the form $f = ka^x$.

x	0	2	4	5	7	9	11	12
$y = \log_{10} f$	2·408	2·458	2·509	2·534	2·584	2·634	2·684	2·709

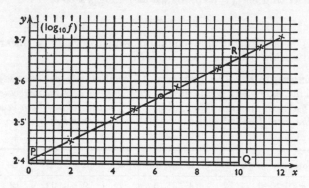

Fig. 20.6

If we now consider the straight line in Fig. 20.6 to have the equation $y = mx + c$, we see from the triangle PQR that its gradient $m = \log_{10} a = \dfrac{0·252}{10} = 0·0252$.

From antilogarithm tables, $a = 1·059$.

We may now write $y = 0 \cdot 0252x + c$.

From the graph when $x = 0$, $y = 2 \cdot 408$.

$$\therefore c = \log_{10} k = 2 \cdot 408.$$

From antilogarithm tables, $k = 256$.

Hence from the data we deduce the required law to be

$$f = 256 \, (1 \cdot 059)^x.*$$

Qu. 22. A remote and isolated tribe came under the influence of medical missionaries in 1935, when a very careful count of the population was made. Less reliable counts were made in later years, as shown in the following table:

Year:	1935	1940	1950	1955
Lapse of years (t)	0	5	15	20
Population (P)	2070	2500	4200	5100

Show that the data points to the operation of a law of the form $P = ka^t$, and determine the constants k, a. Also estimate the population in 1948.

Two final points deserve mention, starting with a word of warning. A graphical method may confirm that a certain law is obeyed but *only within the given ranges of values of the variables*; guard against false deductions. For example, remember that an elastic band may be stretched beyond its elastic limit; or a gas undergoing changes of pressure and volume may also be approaching a change of state.

* In fact standard musical pitch has been set slightly higher than that used in this example, with $f = 440$ for A above middle C, giving $f = 261 \cdot 6$ for middle C. Also, since the ratio of the frequency of a note to that of an octave below is 2:1, $a^{12} = 2$ and calculation gives a better value of a as $1 \cdot 05946$. Thus the corresponding law for a correctly tuned piano is $f = 261 \cdot 6(1 \cdot 059)^x$.

Secondly, the use of logarithmic graph paper has not been mentioned in this chapter. It can be a time-saver in repetitive work, and the reader who has mastered the idea of this last section will have no difficulty in using it should the need arise.

Exercise 20c

1. A round bolt with nominal diameter D mm has a counter-sunk head of diameter A mm. D and A are found to be as follows:

D	6·4	7·9	9·5	11·1	12·7	15·9	19·0	22·2	25·4
A	11·7	14·6	17·5	20·4	23·4	29·2	35·0	40·9	46·7

Find the linear equation giving A in terms of D. Does A vary as D?

2. The mass m kg of a 300 mm square of lead sheeting of thickness t mm is given as follows:

t	1·25	1·80	2·24	2·50	3·15	3·55
m	1·275	1·835	2·285	2·550	3·215	3·625

Obtain a linear relation giving m in terms of t. What is the connection between the gradient of the graph of m against t and the relative density of lead?

3. A marble was dropped from a height h_1 cm and observed to rise to a height h_2 cm. Four such observations are given in the table below:

h_1	4	9	16	22
h_2	$1\frac{1}{2}$	3	$5\frac{1}{2}$	$7\frac{1}{2}$

Does it appear that there is a law connecting h_1, h_2? If so, what is it?

4. A letter in a daily paper gave the following table relating the deaths in a certain group due to lung cancer with the number of cigarettes smoked per day.

No. of cigarettes per day n	0	1 to 14	15 to 24	over 25
Deaths per 100 000 per annum d	7	57	139	227

Investigate the justification for assuming from these figures that a linear relationship exists.

5. Some printers quoted the price of a small book as follows:

No. of copies	500	1000	2000	5000	6000
Cost in £	130	173	260	520	607

Does this bear out the idea that one gets a reduction for ordering in quantity? Can you estimate the cost of (i) 3500 copies, (ii) getting the type set up ready to print, without running off any copies?

6. A man bought a car when the distance travelled registered as 71 km, the fuel tank containing an unknown amount of petrol. According to his log book, he bought 20 litres of petrol at the following kilometre readings:

$$241, \quad 432, \quad 685, \quad 907, \quad 1123.$$

Estimate the average number of km travelled per litre of petrol up to the last distance.

Given that the car ran out of petrol at 1378 km, estimate the quantity of petrol originally in the fuel tank.

7. While some water was cooling, the temperature was recorded at minute intervals as follows:

Time t minutes	0	1	2	3	4	5	6
Temperature θ°C	62	61·5	61	60·5	60	59·5	59

Find an equation giving θ in terms of t. Can you expect this equation to hold over a wider range of values? Give reasons for your answer.

8. The flow of water through a circular hole is thought to vary as the square root of the head of water. For a certain hole, the following results were obtained:

Head of water, h m	1·5	3	4·5	6
Flow of water, x litres/min	119	170	205	240

Do they confirm the conjecture? Estimate the flow of water through the hole when the head is 5 m.

9. A crane on a building site displayed the following figures:

Load in tonnes	2	1·5	1	0·75
Radius in metres	7·5	10	15	20

Do these figures confirm the expectation that the radius is inversely proportional to the load? Is there an equation giving the load l tonnes in terms of the radius r metres?

10. The mass m kg of 100 m lengths of a certain type of steel wire rope is given for nominal diameters d mm as follows:

d	8	9·5	11	13	16	19
m	21·6	30·5	40·9	57·2	86·6	122

Examine the suggestion that the mass varies as the square of the nominal diameter of the rope.

11. A hose squirts a stream of water horizontally and the height of the stream y m at distance x m along level ground is estimated to be as follows:

Distance x m	0	2	4	5	6	7	8
Height y m	3·50	3·40	3·10	2·88	2·60	2·28	1·90

Obtain an equation in the form $y = a + bx^2$ connecting these values approximately.

12. For purposes connected with a survey, the digits 0, 1, 2, ..., 9 were required in a random order. However, when they were taken from a list of random numbers, it was noticed that the intervals between new digits tended to increase. Noting the intervals on a number of occasions the following averages were obtained:

Position of digit p	1	2	3	4	5	6	7	8	9	10
Average interval i	1·0	1·2	1·4	1·4	2·1	1·7	2·3	4·0	3·0	16·2

Find a law in the form $p = a - \dfrac{b}{i}$. [In finding a, use the fact that $i = 1$ when $p = 1$.] Hence express i in terms of p.

13. The periods and mean distances of some of the planets are given in the table below:

Period P days	87·97	224·7	365·3	687·0	4333	10 760
Mean distance s in millions of km	58	108	150	228	778	1426

Find a law in the form $P = ks^n$.

14. For a certain survey in which n housewives are to be interviewed, a market research organisation calculates that it has an even chance of obtaining correct within $p\%$ the percentage in favour of the product concerned in the survey. n and p are related as below:

n	500	1000	2000	5000	10 000
p	1·51	1·07	0·75	0·48	0·34

Find how p varies with n.

15. Some molecules are made out of two atoms. The moment of inertia and the distance between the nuclei of the atoms is given for four such molecules in the table below:

Moment of inertia I (10^{-40} g cm²)	1·34	2·66	3·31	4·31
Distance between nuclei r (10^{-8} cm)	0·92	1·28	1·42	1·62

Find a law in the form $I = kr^n$. (Source of data: S. Glasstone, *Theoretical chemistry*.)

16. The width of successive whorls of a shell of *Turbo duplicatus* have been measured:

Position of whorl n	1	2	3	4	5	6	7	8
Width of whorl w cm	3·33	2·84	2·39	2·03	1·70	1·45	1·22	1·04

Find a law in the form $w = ab^n$. (Source of data: H. Moseley, *Phil. Trans.* 1838, 356.)

17. Two substances in a chemical reaction have the same initial concentration a moles per litre, and after t min the concentration of each is $(a - x)$ moles per litre. The following experimental results were obtained:

t	5	15	25	35	55	120
$a - x$	10·24	6·13	4·32	3·41	2·31	1·10

In order to establish that this is a second order reaction (i.e. the rate of reaction $\dfrac{dx}{dt}$ is a quadratic function of x) show graphically that a linear relationship exists between t and the reciprocal of $(a - x)$; deduce that $\dfrac{dx}{dt} = k(a - x)^2$, and determine the value of k, the reaction velocity constant.

18. A given mass of ozone is subjected to an adiabatic change and the pressure p 10^{-10} N/m² and volume V cm³ are observed as follows:

Volume V cm³	100	90	80	70	60	50
Pressure p (10^{-10}N/m²)	1·18	1·35	1·57	1·82	2·27	2·87

Verify graphically that $pv^\gamma = k$, where γ, k are constants, and determine the value of γ.

19. Steinmetz's law, $E = \eta B^{1.6}$, gives an approximation for the energy lost per cycle of magnetisation in a transformer core, where the energy lost is E ergs/cm³, the maximum magnetic flux density is B gauss, and η is the Steinmetz coefficient for the given material. Values of B and E are tabulated below:

B	1000	2000	3000	4000	5000	6000
$\dfrac{E}{10^3}$	0·316	0·956	1·83	2·90	4·14	5·55

Use a graphical method to show that these values agree with the given law, and determine the value of η for this material.

20. In the Ehrenfest game, n balls numbered from 1 to n are placed in a container A and another container B is left empty. Numbers in the range 1 to n are drawn at random. When a number is drawn, the corresponding ball is transferred from the container it is in to the other container. In such a game with $n = 100$, the total T balls left in container A after x numbers had been drawn was as follows:

x	0	10	20	30	40	50
T	100	92	84	78	72	68

Find a law in the form $T = ab^{-x}$, where a, b are constants, to fit these data as well as possible. As x becomes large, can T be expected to obey this law?

21. Rutherford and Soddy in "The cause and nature of radioactivity" (*Phil. Mag.* IV, 370) give the following table of the radioactivity of a sample of what they describe as "Thorium X". The initial activity of ThX was taken to be 100.

Time in days t	0	1	2	3	4	6	9	10	13	15
Activity of ThX	100	117	100	88	72	53	29·5	25·2	15·2	11·1

Show that, apart from the first reading, the activity after t days is approximately given by ab^t, where a, b are constants. Give the value of b and find the initial activity given by this formula.

22. The population of England and Wales in the census years from 1841 to 1901 was as follows:

Year t	1841	1851	1861	1871	1881	1891	1901
Population (1000's) P	15 914	17 928	20 066	22 712	25 974	29 003	32 528

Find the values of a, b so that the equation $P = ab^x$, where $x = \dfrac{t - 1841}{10}$, shall fit the data as closely as possible.

If the population had obeyed this law exactly, by what factor would the population have increased in each year? (Source of data: The Registrar General's *Statistical review of England and Wales for the year* 1962.)

ANSWERS

CHAPTER 1

2. **Qu. 1.** $(-3, 2)$, $(2, -3)$, $(0, 0)$.

3. **Qu. 3.** (i) 13, (ii) $\sqrt{41}$, (iii) $\sqrt{\{(r - p)^2 + (s - q)^2\}}$.

4. **Qu. 4.** (i) $(5, 6)$, (ii) $(-1, 4)$, (iii) $(-\frac{3}{2}, -\frac{5}{2})$,

    (iv) $\left(\dfrac{p + r}{2}, \dfrac{q + s}{2}\right)$.

Exercise 1a

5. **1.** (i) 4, (ii) 5, (iii) 6, (iv) 13, (v) $\sqrt{74}$, (vi) 10.

    2. (i) $(3, 2)$, (ii) $(5, \frac{5}{2})$, (iii) $(1, 3)$, (iv) $(0, \frac{7}{2})$,

       (v) $(-\frac{1}{2}, -\frac{9}{2})$, (vi) $(-6, -7)$.

    3. 17.   **4.** $(-\frac{5}{2}, \frac{9}{2})$.   **5.** $(-\frac{3}{2}, -\frac{3}{2})$.   **6.** P, R, S.

    7. A, B, D; $\sqrt{50}$.   **8.** 13, $6\frac{1}{2}$.

7. **Qu. 5.** (i) $\frac{9}{4}$, (ii) $\frac{3}{2}$, (iii) $-\frac{4}{3}$, (iv) $-\frac{10}{11}$, (v) 0, (vi) ∞,

       (vii) $(s - q)/(r - p)$, (viii) -1, (ix) b/a.

    Qu. 6. $\frac{4}{3}$, $-\frac{3}{4}$, -1.   **Qu. 7.** $\frac{12}{5}$, $-\frac{5}{12}$, -1.

10. **Qu. 8.** $-\frac{1}{3}$, -4, $\frac{1}{6}$, $\frac{3}{2}$, $-1/2m$, a/b, $2/m$.

    Qu. 9. (i) parallel, (ii) perpendicular, (iii) neither.

12. **Qu. 10.** 5, 20, -1.   **Qu. 11.** 2, 0, $-\frac{5}{2}$.

    Qu. 12. (i) $(-\frac{1}{2}, 0)$, $(1, 0)$, (ii) $(0, -1)$.

    Qu. 13. (i) yes, (ii) no, (iii) no, (iv) no, (v) yes, (vi) yes.

13. **Qu. 14.** (i) 0, (ii) 2, (iii) 3, (iv) $\frac{1}{2}$, (v) -1.

14. **Qu. 15.** (i) $y = \frac{1}{3}x$, (ii) $y = -2x$, (iii) $y = mx$.

    Qu. 16. (i) $\frac{1}{4}$, (ii) $-\frac{5}{4}$, (iii) $\frac{3}{4}$, (iv) $\frac{7}{4}$, (v) q/p.

    Qu. 17. (i) $y = 3x + 2$, (ii) $y = 3x + 4$, (iii) $y = 3x - 1$,

       (iv) $y = \frac{1}{5}x + 2$, (v) $y = \frac{1}{5}x + 4$.

17. **Qu. 18.** (i) $\frac{2}{3}$, 2, (ii) $\frac{1}{4}$, $\frac{1}{2}$, (iii) -3, -6, (iv) $\frac{7}{3}$, $-\frac{5}{3}$,

       (v) 0, -4, (vi) $-l/m$, $-n/m$.

    Qu. 19. (i) $y = 0$, (ii) $x = 0$, (iii) $x = 4$, (iv) $y = -7$.

Exercise 1b

1. (i) $\frac{9}{5}$, (ii) $-\frac{7}{5}$, (iii) $-\frac{1}{14}$, (iv) $\frac{1}{5}$.

3. (i) 1, (ii) -1, (iii) $\sqrt{3}$, (iv) $-1/\sqrt{3}$.

4. (i) perpendicular, (ii) parallel, (iii) perpendicular,

   (iv) parallel, (v) perpendicular, (vi) neither.

page
18. 6. $\sqrt{50}$; $(3\frac{1}{2}, 4\frac{1}{2})$.   7. $\frac{1}{2}\sqrt{34}$; $(\frac{3}{4}, -1\frac{3}{4})$.

  8. 10, 1, 2, 26; ± 2, ± 4.

  9. -27, -1, 1, 27; -2, 0, 2.

  10. (i) yes, (ii) no, (iii) no, (iv) yes.   11. $-\frac{10}{3}$, $+5$; $\frac{25}{3}$.

  12. (i) (4, 0), $(-3, 0)$, $(0, -12)$, (ii) $(\frac{2}{3}, 0)$, $(\frac{1}{2}, 0)$, (0, 2),
    (iii) (0, 9), and touches x-axis at (3, 0),
    (iv) (9, 0), and cuts y-axis, touches x-axis at (0, 0),
    (v) $(-1, 0)$, (0, 25) and touches x-axis at (5, 0),
    (vi) (1 , 0), $(-1, 0)$, (3, 0), $(-3, 0)$, (0, 9).

  13. (i) $y = x$, (ii) $y = -x$, (iii) $y = \frac{1}{2}x$,
    (iv) $y = \frac{1}{2}x - 4$, (v) $x = -5$, (vi) $y = -\frac{2}{3}x + 5$.

  14. (i) $y = 11$, (ii) $x = 4$, (iii) $y = 6x - 10$,
    (iv) $y = -8x + 2$, (v) $y = -\frac{2}{3}x - 1$.

19. 15. $y = \frac{1}{8}x$.   16. M$(0, -\frac{9}{2})$; S$(5, -1)$.

  17. (i) $(b - q)/(a - p)$, 7, (ii) $-\frac{3}{2}$.

22. Qu. 20. (i) $5x - 2y - 26 = 0$, (ii) $5x + 2y - 1 = 0$.

  Qu. 21. (i) $3x - 2y - 19 = 0$, (ii) $12x + 5y - 1 = 0$.

  Qu. 22. $y - y_1 = m(x - x_1)$.

  Qu. 23. (i) $(4\frac{1}{2}, 1)$, (ii) (1, 5), (iii) (0, c), (iv) $(-a, c - a)$.

  Qu. 24. No.   They are parallel.

23. Qu. 25. $(-\frac{3}{4}, 0)$, $(\frac{2}{3}, 0)$.

Exercise 1c

  1.   (i) $4x - y - 1 = 0$,   (ii) $3x - y + 11 = 0$,
    (iii) $x - 3y - 17 = 0$,   (iv) $3x + 4y - 41 = 0$,
    (v) $3x - 6y - 4 = 0$,   (vi) $20x + 12y + 31 = 0$.

  2.   (i) $3x - 4y + 21 = 0$,   (ii) $5x + 4y - 23 = 0$,
    (iii) $3x + 11y - 38 = 0$, (iv) $x - 5y - 19 = 0$,
    (v) $2x + 3y - 7 = 0$,   (vi) $2x - y + 1 = 0$.

  3.   (i) $(7, -7)$,   (ii) $(-\frac{3}{2}, -1\frac{1}{2})$,
    (iii) $(\frac{11}{7}, -\frac{13}{7})$,   (iv) $(4, -7)$.

  4.   (i) $3x - 4y + 1 = 0$, (ii) $5x - 2y + 16 = 0$,
    (iii) $7x - y - 28 = 0$, (iv) $3x - 4y - 6 = 0$.

  5. $2x - 5y + 19 = 0$.   6. $26x + 4y - 21 = 0$.

  7. $7x - 10y - 70 = 0$; $7x + 10y = 0$.

  8. $2x - 7y - 3 = 0$.   9. $\frac{8}{21}$.

24 10. (2, -5).   11. $4x - 3y - 13 = 0$; 5.

  12. $x + 4y - 15 = 0$.   13. $7x - 4y - 43 = 0$.

  14. (0, 0), (16, 64).   15. $\sqrt{512} = 16\sqrt{2}$.

Exercise 1d

24.   **1.** $5x + y - 33 = 0.$

     2. $2x + 7y - 14 = 0;\ 2x - 7y - 14 = 0.$

     3. (i) $3x - 5y + 14 = 0$, (ii) $3x + 5y - 14 = 0$,
        (iii) $2x + 5y + 14 = 0.$

     4. $m_1 m_2 = -1$, (i) $5x + 2y - 11 = 0$,
        (ii) $2x - 5y - 16 = 0.$

     5. $(0, 0), (2, \frac{5}{2}), (5, -\frac{5}{2});\ y = 0,\ 9x + 2y - 21 = 0, (2\frac{1}{3}, 0).$

25.   **6.** (i) $3x + 2y + 5 = 0$, (ii) $2x + 7y - 19 = 0$,
        (iii) $2x + 5y + 11 = 0.$

     7. (i) $2x - 3y - 14 = 0$, (ii) $3x + 2y - 8 = 0$, $(\frac{32}{13}, \frac{4}{13})$.

     8. $\sqrt{85},\ 6x + 7y - 85 = 0.$ **9.** $(1, 8), 52.$

    11. (i) $AB = AC = 13$, (ii) $12, 78.$

    12. $2x + y - 17 = 0,\ 72\frac{1}{4}.$

    13. $x - y = 0,\ 2x + 2y - 9 = 0, (\frac{9}{4}, 3), (3, \frac{3}{2}), \frac{3}{2}\sqrt{5}.$

    14. $x + 3y + 2 = 0,\ x - 3y - 4 = 0.$

26. **15.** $x + 4y - 9 = 0.$

    16. $3x + 4y + 1 = 0,\ 4x - 3y - 7 = 0.$

    17. $13x - 8y = 0,\ 4x + y - 30 = 0,\ x - 2y + 12 = 0,$
        $(\frac{16}{3}, \frac{26}{3}).$

    18. $x + 2y + 4 = 0,\ 8x - y + 15 = 0,$
        $10x + 3y - 45 = 0.$

    19. $(\frac{1}{3}\sqrt{3}, \frac{1}{3}\sqrt{3}).$ **20.** $(-11, 3), 174.$

    21. $3x + 4y - 15 = 0,\ 4x - 3y - 1 = 0.$

    22. $3x + 2y - 2 = 0,\ 4x + y + 1 = 0, (-\frac{4}{5}, \frac{11}{5}).$

    23. $13x + y - 22 = 0.$ **24.** $9, \frac{2}{13}\sqrt{13}, 2\frac{1}{13}.$

CHAPTER 2

29.   **Qu. 1.** The circle.   **Qu. 2.** $90°.$

     Qu. 3. $PQ \to 0,\ QO \to 0,\ PQ/QO = \frac{1}{2}.$

32.   **Qu. 4.** $3, 2\frac{1}{2}, 2 \cdot 1, 2 \cdot 01; 2.$

     Qu. 5. $Q \to P$; gradient of $PQ \to$ gradient of tangent at
        $P; 2.$

     Qu. 6. $4.$

34.   **Qu. 7.** $-3, -2, 1, 4.$

     Qu. 8. (i) $6x$, (ii) $10x$, (iii) x, (iv) $2cx$, (v) $2x$, (vi) $2x.$

35.   **Qu. 9.** $4x^3.$ **Qu. 10.** $6x^2.$

page

37. **Qu. 11.** (i) $12x^2$, (ii) $20x^3$, (iii) $2ax$, (iv) $4nx^{n-1}$,
(v) $k(n+1)x^n$.

39. **Qu. 12.** (i) $3x^2 + 4x + 3$, (ii) $16x^3 - 6x$, (iii) $2ax + b$.

40. **Qu. 13.** (i) $12x^2 - 4x$, (ii) $2x - 1$, (iii) 5.

Exercise 2a

1. $12x^{11}$. **2.** $21x^6$. **3.** 5. **4.** 5. **5.** 0. **6.** $10x - 3$.

7. $12x^3 - 6x^2 + 2x - 1$. **8.** $8x^3 + x^2 - \frac{1}{2}x$.

9. $3ax^2 + 2bx + c$. **10.** $18x^2 - 8$.

11. $15x^2 + 3x$. **12.** -1. **13.** 0.

14. $12x^2 - 3$. **15.** $ax - 2b$. **16.** $4x + 2$.

17. $6x - 3$. **18.** $x^2 - 1$. **19.** $2x - 1$.

20. $6x$. **21.** $x + \frac{7}{4}$. **22.** $\frac{4}{3}x - \frac{1}{3}$.

23. x. **24.** $1; 2$. **25.** $1; 1$. **26.** $3; -4$.

27. $-5; 4$. **28.** $28; -36$. **29.** $9; -24$.

40. **30.** $(4, 16)$: **31.** $(-2, -8), (2, 8)$. **32.** $(0, 0)$.

41. **33.** $(\frac{3}{2}, -\frac{5}{4})$. **34.** $(-1, 8), (1, 6)$. **35.** $(2, -12)$.

36. $(0, 1), (\frac{3}{2}, -\frac{11}{16})$. **37.** $(-\frac{1}{3}, \frac{4}{27}), (1, 0)$. **38.** $(1, 4), (3, 0)$.

Exercise 2b

42. **1.** (i) $4x - y - 4 = 0$, (ii) $24x - y - 40 = 0$,
(iii) $x + y - 1 = 0$, (iv) $8x + y - 5 = 0$,
(v) $18x + y + 54 = 0$.

2. (i) $x + 4y - 18 = 0$, (ii) $x + 24y - 1204 = 0$,
(iii) $x - y + 1 = 0$, (iv) $x - 8y - 25 = 0$,
(v) $x - 18y + 3 = 0$.

3. $9x - y - 27 = 0; x + 9y - 3 = 0$.

4. $16x - y = 0; x + 16y = 0$.

43. **5.** $2x - y - 10 = 0$. **6.** $4x + y - 3 = 0$.

7. $y + 4 = 0; y - 23 = 0$.

8. $y - 10 = 0; y + 17 = 0$.

Exercise 2c

1. $6, 6x - y + 2 = 0, (-\frac{2}{5}, -\frac{2}{5})$.

2. $10x - y - 16 = 0, (-4, -56)$.

3. $5x - y - 1 = 0, (2, 4), (4, -8)$.

4. $4x - 2y + 5 = 0, (\frac{2}{3}, \frac{41}{27})$.

5. $(0, 0), (1, 0), (2, 0), y = 2x, x + y - 1 = 0$,
$y = 2x - 4$.

page

43. **6.** $x - y + 3 = 0$, $x - 2y + 12 = 0$, $(6, 9)$.

   7. $y = -5x + 4$, $(2, -6)$, $y = -4x + 2$.

44. **8.** $(-2, 0)$, $x - 2y + 2 = 0$, $x + 2y - 2 = 0$, $(0, 1)$.

   9. $y = 3x + 2$, $y = 3x + 6$.

   10. $(-\sqrt{\frac{2}{3}}, -\frac{1}{3}\sqrt{\frac{2}{3}})$, $(0, 0)$, $(\sqrt{\frac{2}{3}}, \frac{1}{3}\sqrt{\frac{2}{3}})$.

   11. $3x - y - 9 = 0$, $x + 3y - 33 = 0$, 20.

   12. $(0, -3)$. **13.** $6x - y - 9 = 0$, $(1, 1)$.

   14. $2x + y + 16 = 0$, $(-16, 16)$.

   15. $2x - y - 1 = 0$, $6x - 3y - 8 = 0$, $\frac{1}{3}\sqrt{5}$.

45. **16.** $4h$, $y - k = 4h(x - h)$, $y = \pm 12x$.

   17. $(1, 4)$, $(3, 12)$.

CHAPTER 3

49. **Qu. 1.** $6 \cdot 1$ m, $12 \cdot 2$ m/s.

   Qu. 2. (i) $1 \cdot 0$ m, 10 m/s, (ii) $4 \cdot 9$ $(2h + h^2)$ m,

   (iii) $4 \cdot 9(2 + h)$ m/s.

   Qu. 3. $9 \cdot 8$ m/s.

   Qu. 4. (i) $24 \cdot 5$ m, $24 \cdot 5$ m/s, (ii) 11 m, 22 m/s,

   (iii) $2 \cdot 0$ m, 20 m/s, (iv) $4 \cdot 9$ $(4h + h^2)$ m,

   $4 \cdot 9$ $(4 + h)$ m/s; $19 \cdot 6$ m/s.

50. **Qu. 5.** (i) $6 \cdot 9$, $23 \cdot 6$, $50 \cdot 1$, $86 \cdot 4$ m below top,

   (ii) $11 \cdot 8$, $21 \cdot 6$, $31 \cdot 4$, $41 \cdot 2$ m/s, (iii) $26 \cdot 5$ m/s.

Exercise 3a

51. **1.** (i) $10 \cdot 5$ m, $10 \cdot 5$ m/s, (ii) 13, 15, $(15 \cdot 4 - 4 \cdot 9h)$ m/s,

   (iii) $15 \cdot 4$ m/s.

   2. $v = 24 \cdot 5 - 9 \cdot 8t$, (i) $t = 0$, 5 seconds,

   (ii) $19 \cdot 6$, $29 \cdot 4$, $29 \cdot 4$, $-29 \cdot 4$ m; $14 \cdot 7$, $4 \cdot 9$, $-4 \cdot 9$,

   $-34 \cdot 3$ m/s, (iii) below ledge; falling,

   (iv) $t = 2 \cdot 5$; $30 \cdot 6$ m, (v) $2 \cdot 4$ m.

   3. $v = 3 + 2t$; (i) At O, 3 m/s, (ii) $t = 0$, or -3,

   (iii) $t = -\frac{3}{2}$, $\frac{9}{4}$ m from O on the negative side,

   (iv) -3 m/s.

   4. (i) 0, 8, 9, 8, 0, -7 m; on AO produced.

   (ii) 6, 2, -2, -6 m/s; moving in direction $\overrightarrow{AO}$,

   (iii) $t = 3$; 9 m from O, on OA.

ANSWERS   467

page
53. **5.** (i) 11.59 a.m.; 12.03 p.m., (ii) $\frac{5}{27}$, 1 km,
(iii) $\frac{8}{27}$ km/min $= 17\frac{7}{9}$ km/h,
(iv) $\frac{1}{3}$ km/min $= 20$ km/h.

6. (i) 11.57 a.m.; 12.02 p.m., (ii) $\frac{9}{8}$, $\frac{11}{18}$ km,
(iii) $\frac{25}{72}$ km/min $= 20\frac{5}{6}$ km/h,
(iv) $\frac{1}{2}$ km/min $= 30$ km/h.

7. 29·4 m/s.

54. **Qu. 6.** (i) 19·8, 29·6, 39·4, $10 + 9·8t$, m/s,
(ii) straight line through (0, 10) of gradient 9·8.

Exercise 3b

1. 2·5 m/s^2. **2.** 3 m/s^2.
3. (i) 18 km/h per s, (ii) 64 800 km/h^2.
4. (i) 3·6 km/h per s, (ii) 1 m/s^2, (iii) 12 960 km/h^2.
5. 6·25 s. **6.** $- 1·5$ m/s^2; $- 5$.
55. **7.** 130 km/h.

Exercise 3c

56. **1.** (i) $+ 5·6$ m, $+ 0·7$ m/s (up), $- 9·8$ m/s^2 (decreasing speed),
(ii) $+ 1·4$ m, $- 9·1$ m/s (down), $- 9·8$ m/s^2 (increasing speed),
(iii) $- 12·6$ m, $- 18·9$ m/s (down), $- 9·8$ m/s^2 (increasing speed).

2. 24·9 m, 29·8 m/s, 9·8 m/s^2.
3. (i) 31·5 m, $- 4·2$ m/s, (ii) $t = 2\frac{4}{7}$, (iii) 32·4 m,
(iv) 2·5 m, (v) $- 9·8$ m/s^2 (constant).
4. (i) 18, 54, 114 m/s^2, (ii) 58 m/s^2.
59. **5.** (i) $t = 2$, (ii) $t = \frac{2}{3}$, $\frac{32}{27}$ m from O on OA; $t = 2$, at O,
(iii) $\frac{32}{27}$, $\frac{64}{27}$ m, (iv) 3 m/s,
(v) 1 m from O, on OA; towards O $(- 1$ m/s);
increasing $(a = - 2$ m/s^2).

6. 9 m from O on AO produced $(s = - 9)$; towards O $(+ 15$ m/s); decreasing $(a = - 14$ m/s^2).

7. (i) After 0, 1, 2 s,
(ii) 2, $- 1$, 2 m/s; $- 6$, 0, $+ 6$ m/s^2.
(iii) 0 m/s, (iv) 0 m/s^2.

Exercise 3d

59. **1.** 0·7 m/s. **2.** 16 m/s, 14 m/s².

60. **3.** 84 m/s, 4 m/s².

4. 0 cm/s, — 4 cm/s², $1\frac{1}{2}$ s, 18 cm.

5. (i) 1, 3 s, 4, 0 cm, (ii) — 6, + 6 cm/s²,
(iii) — 3 cm/s.

6. 0 cm/s, 16 cm, (— 24 cm/s).

7. $(24n — 5)$ cm, $(24n + 7)$ cm/s, 24 cm/s².

CHAPTER 4

62. **Qu. 1.** (i) $2x — 4$, (ii) $6x$, (iii) $6x^2 — 10x$, (iv) $2x — 2$,
(v) $3x^2 — 4x — 3$.

65. **Qu. 2.** (i) $(1, 2)$, (ii) $(— \frac{1}{3}, — 5\frac{1}{3})$, (iii) $(\frac{3}{4}, — \frac{1}{4})$.

Qu. 3. (i) $\frac{5}{3}$, highest, (ii) $\frac{7}{6}$, lowest, (iii) $— \frac{3}{2}$, lowest,
(iv) $— \frac{4}{5}$, highest.

Exercise 4a

67. **1.** (i) $6x — 2$, (ii) $10x + 4$, (iii) $2 — 4x$, (iv) $6x + 1$,
(v) $48x + 6$.

2. (i) $(— 2\frac{1}{2}, — 8\frac{1}{4})$, (ii) $(\frac{9}{14}, 7\frac{25}{28})$, (iii) $(\frac{1}{3}, — \frac{1}{3})$,
(iv) $(— \frac{5}{8}, 7\frac{9}{16})$.

3. (i) $2\frac{1}{2}$, lowest, (ii) — 6, lowest, (iii) $\frac{4}{3}$, highest,
(iv) — 25, highest.

68. **4.** (i) $— 2\frac{1}{4}$, least, (ii) 4, greatest, (iii) 16, greatest,
(iv) $— 6\frac{1}{8}$, least.

6. 12·1 m, $1\frac{4}{7}$ s. **7.** 50 m by 50 m. **8.** 10 cm.

9. 250 m, 500 m, 125 000 m². **10.** 50 m, 5 s.

11. 2 cm, 3 cm.

70. **Qu. 4.** (i) $— 4x^{-5}$, (ii) $— 6x^{-3}$, (iii) $— 6x^{-4}$, (iv) $— \frac{3}{2}x^{-4}$,
(v) $— mx^{-m-1}$, (vi) $4x — 3 — 5x^{-2}$,
(vii) $1 — 3x^{-2} + 8x^{-3}$.

71. **Qu. 5.** (i) A, E min., D, F max.; B, C infl.,
(ii) G max., I min., H infl.
(iii) K max., J, L, infl.

Exercise 4b

74. **1.** (i) 0, infl. (ii) 0, y max. (iii) 2, y max.; 3, y min.

(iv) $-$ 3, y max.; 5, y min. (v) $-$ 6, y max.; $-$ 1, y min.

(vi) 1, y min.; 3, y max.· (vii) $-$ 3, y min.; 4, y max.

(viii) $-$ 6, y min.; 1, y max. (ix) $-$ 5, y min.; 3, y max.

(x) $-\sqrt{\frac{27}{5}}$, y max.; 0, infl.; $\sqrt{\frac{27}{5}}$, y min.

(xi) $-$ 2, y max.; 2, y min.

75. **2.** (i) $-\frac{16}{9}$, min.; $\frac{16}{9}$, max. (ii) 0, max.; $-$ 27, min.

(iii) 0, max.; $-\frac{256}{27}$, min. (iv) $-$ 2, max.; $+$ 2, min.

(v) $\frac{100}{27}$, max., $-$ 9, min.

3. (i) $(-$ 2, 16) max.; (2, $-$ 16) min.

(ii) $(\frac{1}{3}, 2\frac{13}{27})$ max.; (3, $-$ 7) min.

(iii) (0, 0) min.; (2, 4) max. (iv) $(\frac{1}{2}, 3)$ min.

(v) $(3\frac{1}{3}, 181\frac{13}{27})$ max.; (12, $-$ 144) min.

4. (i) 0, min. (ii) 3, infl. (iii) 0, infl.; $\frac{27}{16}$, max.

(iv) 19, infl.; 3, min.

5. 18 cm³; $x = 1$. **6.** $7\frac{11}{27}$ cm³; $x = \frac{2}{3}$.

7. 2 cubic feet.

76. **8.** $8/\pi$ cubic feet. **9.** $\sqrt[3]{(5/\pi)}$ cm; $2\sqrt[3]{(5/\pi)}$ cm.

11. 4 cm. **12.** 6, 6, 3 cm.

Exercise 4c

77. **1.** (0, 0), (3, 0); (0, 0) min., (2, 4) max.

2. (0, 0,) (6, 0); (0, 0) max., (4, $-$ 32) min.

78. **3.** (0, 0), (1, 0); $(\frac{1}{3}, \frac{4}{27})$ max., (1, 0) min.

4. $(-$ 1, 0), (2, 0), (0, 2); $(-$ 1, 0) min., (1, 4) max.

5. (0, 0), (2, 0); (0, 0) min., (1, 1) max., (2, 0) min.

6. (0, 0), (8, 0); (0, 0) infl., (6, $-$ 432) min.

7. $(\pm$ 1, 0), $(\pm$ 3, 0), (0, 9); (0, 9) max., $(\pm \sqrt{5}, -$ 16) min.

8. (0, 0), $(- \sqrt[3]{32}, 0)$; $(-$ 2, $-$ 48) min.

9. (0, 0), $(1\frac{1}{4}, 0)$; (0, 0) max., (1, $-$ 1) min.

10. (0, 0), $(\pm \sqrt{\frac{2}{3}}, 0)$; $(-$ 1, 2) max., (0, 0) infl., (1, $-$ 2) min.

11. (0, 0), $(- \sqrt[3]{\frac{5}{2}}, 0)$; $(-$ 1, 3) max., (0, 0) min.

81. **Qu. 6.** (i) neg., pos., decreasing, (ii) pos., pos., increasing, (iii) neg., zero, neither.

Exercise 4d

81.   **2.** $v = 6t^2 - 22t + 12$, $a = 12t - 22$;
     (i) 4 m from O on BO produced $(s = -4)$,
     (ii) away, (iii) 8 m/s $(v = -8)$, (iv) decreasing,
     (v) 2 m/s² $(a = +2)$.

82.   **3.** (i) 3 m from O on OB $(s = +3)$, (ii) away,
     (iii) 4 m/s $(v = -4)$, (iv) increasing,
     (v) 10 m/s² $(a = -10)$.
     4. After $\frac{11}{6}$ s; $s = -\frac{143}{54}$.
     5. (i) 100 m from O on OA $(s = +100)$; approaching A
     at 40 m/s $(v = +40)$; retarding at 14 m/s²
     $(a = -14)$,
     (ii) $t = 3\frac{1}{3}$ to $t = 12$, (iii) $t = 7\frac{2}{3}$.

Exercise 4e

1. (i) $(1, 6)$ max., $(3, 2)$ min.;
     (ii) $(-1, 15)$ max.; $(2, -12)$ min.;
     (iii) $(-1, 2)$ max., $(1, -2)$ min.;
     (iv) $(1, -1)$ min., $(-\frac{1}{2}, 5\frac{3}{4})$ max.;
     (v) $(0, 0)$ max., $(-2, -16)$, $(2, -16)$ min.;
     (vi) $(-\frac{1}{3}, \frac{91}{27})$ max., $(1, 1)$ min.
2. $(-2, 27)$ max., $(1, 0)$ min., $(-3\frac{1}{2}, 0)$.   **3.** ± 1.
4. (i) max., (ii) infl.

83.   **5.** $y = 1$, $y = 1$, $y = 0$.
     6. $2\frac{4}{5}$, $x + 3y - 7 = 0$.   **7.** $a = 2$, $b = -4$, $c = -1$.
     8. $x + t^2y - 4t = 0$.   **9.** $4x - y \pm 15 = 0$.
     10. 256 cm³.   **11.** 32 m.   **12.** (i) $4\frac{1}{2}$ cm². (ii) 4 cm².
     13. $\frac{1}{144}(17x^2 - 16xl + 8l^2)$, min.

84.   **16.** 48 m².
     17. $2\pi r(r + h)$; (i) $(12/r) - r$, $\pi r(12 - r^2)$; (ii) $r = 2$.
     18. (i) 5 cm, (ii) 6 cm.
     19. $V = \pi r^2(5 - 2\pi r)$, $125/27\pi \simeq 1{\cdot}47$ m³.
     20. 10 m by 10 m by 5 m.   **21.** 6 cm by 3 cm by 4 cm.
85.   **24.** 20.   **25.** $AP^2 = x^4 - x^2 + 1$, $(\frac{1}{2}\sqrt{2}, \frac{1}{2})$, $(-\frac{1}{2}\sqrt{2}, \frac{1}{2})$.
     27. $x = \frac{1}{3}$, $\frac{8}{27}$ of the sphere.

CHAPTER 5

88. **Qu. 1.** (i) $2x + c$, (ii) $mx + c$, (iii) $x^3 + c$, (iv) $\frac{3}{2}x^2 + c$,
(v) $\frac{3}{5}x^5 + c$, (vi) $3x + x^2 + c$,
(vii) $\frac{1}{2}x^2 - \frac{1}{3}x^3 + c$, (viii) $\frac{1}{2}ax^2 + bx + c$.

Qu. 2. (i) $-\frac{1}{2}x^{-2} + c = \dfrac{-1}{2x^2} + c$,

(ii) $-\frac{1}{3}x^{-3} + c = \dfrac{-1}{3x^3} + c$,

(iii) $-2x^{-1} + c$, (iv) $\dfrac{-x^{-(n-1)}}{n-1} + c$.

Qu. 3. $\dfrac{x^0}{0} + c$ is meaningless.

89. **Qu. 4.** $y = 4x + 18$. A straight line of gradient 4 through $(-2, 10)$.

Exercise 5a

1. (i) $\frac{1}{2}x + c$, $\frac{1}{6}x^3 + c$, $\frac{1}{3}x^3 + \frac{3}{2}x^2 + c$,
$\frac{3}{4}x^3 + 6x^2 + 9x + c$, $-\frac{1}{4}x^{-4} + c$, $\frac{2}{3}x^{-3} + c$;
(ii) $\frac{1}{2}at^2 + c$, $\frac{1}{12}t^4 + c$, $\frac{1}{3}t^3 - \frac{1}{2}t^2 - 2t + c$,
$-\dfrac{1}{n}t^{-n} + c$, $-t^{-1} + 3t + t^2 + c$;
(iii) $ay^{-1} + c$, $-ky^{-1} + c$, $\frac{1}{3}y^3 - y + 6y^{-1} + c$.

2. (i) $y = ax^3 + c$, (ii) $s = \frac{3}{4}t^4 + c$,
(iii) $s = ut + \frac{1}{2}at^2 + c$, (iv) $x = t + t^{-1} + c$,
(v) $y = t + 3t^{-1} - 2t^{-2} + c$,
(vi) $A = -x^{-1} - w - \frac{2}{3}x^3 + c$.

3. $x - 6y + 34 = 0$. **4.** $y = x^2 + 5x - 25$.

5. $y = x^3 + \dfrac{1}{x} - 8\frac{1}{2}$.

90. **6.** $(1, 0)$, $(3, 0)$. **7.** $(4, 0)$; $y = 9\frac{13}{27} = \frac{256}{27}$.

8. $s = \frac{3}{2}t^2 + \dfrac{8}{t} - 8$.

9. $A = c - 3x^{-1} - \frac{1}{2}x^{-2} + x^{-3} + \frac{1}{4}x^{-4}$; $\frac{45}{64}$.

91. **Qu. 5.** $v = 15 + 9\cdot81t$, $s = 15t + 4\cdot90(5)t^2$.

Qu. 6. $+9\cdot8$ m/s (rising), $-9\cdot8$, $-29\cdot4$ m/s (falling); $14\cdot7$, $14\cdot7$ m (above start), $-24\cdot5$ m (below).

93. **Qu. 7.** (i) 9, (ii) 42, (iii) $-$ 6, (iv) 35.

Qu. 8. $12\frac{2}{3}$ m.

Qu. 9. (i) 13 m past O, (ii) 5 m past O, (iii) 7 m past O, (iv) 100 m short of O.

Exercise 5b

1. $v = 20 + 9{\cdot}81t$; $s = 20t + 4{\cdot}90t^2$.

94. **2.** $v = -12 + 9{\cdot}8t$; $s = -12t + 4{\cdot}9t^2$;
$-2{\cdot}2$ m/s (rising), $+ 7{\cdot}6$, $+ 17{\cdot}4$ m/s (falling), $- 7{\cdot}1$, $- 4{\cdot}4$ m (above ground level), $+ 8{\cdot}1$ m (below).

3. (i) $s = 3t + 3$, (ii) $s = 2t^2 - t - 6$,
(iii) $s = t^3 + \frac{5}{2}t^2 - 2t - 13$,
(iv) $s = \frac{1}{3}t^3 + 5t + 2t^{-1} - 7$.

4. (i) 32, (ii) 328, (iii) $-$ 21, (iv) 16.

5. (i) $s = 2t^2 + 3t + c$, 14 m,
(ii) $s = \frac{1}{3}t^3 - 3t + c$, $3\frac{1}{3}$ m;
(iii) $s = \frac{1}{3}t^3 - \frac{3}{2}t^2 + 2t + c$, $3\frac{5}{6}$ m;
(iv) $s = \frac{1}{2}t^2 + 3t + \dfrac{1}{t} + c$, $179\frac{19}{20}$ m.

6. $v = \frac{1}{2}At^2 + B$; $s = \frac{1}{6}At^3 + Bt + c$.

7. (i) $v = \frac{3}{2}t^2 + 3$, $s = \frac{1}{2}t^3 + 3t$;
(ii) $v = 2t + \frac{1}{2}t^2$, $s = -3 + t^2 + \frac{1}{6}t^3$;
(iii) $v = -7\frac{1}{2} + 10t - \frac{1}{2}t^2$, $s = -7\frac{1}{2}t + 5t^2 - \frac{1}{6}t^3$;
(iv) $v = \frac{1}{4}t^2 + 5$, $s = \frac{1}{12}t^3 + 5t + c$;
(v) $v = \frac{1}{3}t^3 + c$, $s = \frac{1}{12}t^4 + ct + 9\frac{11}{12} - c$.

95. **8.** (i) $13\frac{1}{2}$ m past O, (ii) $2\frac{1}{2}$ m past O, (iii) 8 m past O, (iv) $7\frac{1}{2}$ m.

9. (i) $s = -5 + 6t - t^2$, 5 m; (ii) 13 m.

10. (i) $1\frac{2}{7}$ s; (ii) $8{\cdot}1$ m; (iii) $7{\cdot}7$, $2{\cdot}9$ m.

11. (i) 40 km, (ii) 20 km/h, (iii) 30 km/h.

96. **12.** (i) $13\frac{1}{3}$ km/h, (ii) 20 km/h.

13. 35 m/s, 28 m/s². **14.** $k = 6$; 4800 m.

15. (i) After 4 s, $- 64$ m/s; (ii) 27 m;
(iii) 16 m/s, 64 m/s.

16. (i) $11.59\frac{1}{2}$ a.m., $12.01\frac{1}{2}$ p.m.;
(ii) $s = \frac{1}{48}(5 + 18t + 12t^2 - 8t^3)$;
(iii) 20 km/h; (iv) 30 km/h.

page

99. **Qu. 10.** (i) 72, (ii) 9, (iii) 36, (iv) 21.

101. **Qu. 11.** (i) $3\frac{1}{4}$, (ii) 9, (iii) 2, (iv) -8, (v) -38, (vi) $9\frac{1}{4}$.
　　Qu. 12. 25.

102. **Qu. 13.** (i) 9, (ii) 81.

103. **Qu. 14.** $2\frac{2}{3}$. 　**Qu. 15.** $\frac{1}{2}$.

Exercise 5c

105. **1.** (i) $3\frac{63}{64}$, (ii) -2, (iii) $10\frac{2}{3}$, (iv) $36\frac{137}{144}$.
　　2. 50. 　**3.** (i) 26, (ii) $58\frac{1}{3}$, (iii) $22\frac{7}{60}$, (iv) $2\frac{1}{2}$.
　　4. $5\frac{1}{3}$. 　**5.** $-\frac{1}{3}$. 　**6.** $-\frac{5}{12}$, $2\frac{2}{3}$.
　　7. (i) $-2\frac{1}{3}$, (ii) 4, (iii) $2\frac{9}{20}$, (iv) $-\frac{1}{2}$.
　　8. $1\frac{1}{3}$. 　**9.** $4\frac{1}{2}$.
　　10. (i) $(0, 0)$, $(4, 8)$, $5\frac{1}{3}$; (ii) $(-2, 12)$, $(1, 3)$, $13\frac{1}{2}$;
　　(iii) $(-1, 0)$, $(3, 4)$, $10\frac{2}{3}$.

106. **11.** (i) 96, (ii) 60, (iii) $1\frac{1}{2}$. 　**12.** $833\frac{1}{3}$.

Exercise 5d

　　1. $y = 3 - \dfrac{3}{x} + \dfrac{1}{x^2}$. 　**2.** $y = x^2 - 1 + \dfrac{1}{x}$. 　**3.** $20\frac{5}{6}$.

　　4. $-4\frac{1}{2}$. 　**5.** $6\frac{3}{4}$. 　**6.** 36.
　　7. $\frac{1}{6}(4t^3 - 27t^2 + 60t)$, $4t - 9$, $7\frac{1}{6}$, $7\frac{7}{24}$.
　　8. $4\frac{5}{6}$ m, 1 m/s².

107. **9.** 14 m/s², 44 m. 　**10.** $y = x^3 - 3x^2 + 4x + 8$, $10\frac{3}{4}$.
　　11. $10\frac{2}{3}$. 　**14.** $(t^2 - 4t + 3)$ m/s, 1, $\frac{4}{3}$ m.
　　15. 20 m/s, 467 m. 　**16.** 3, -1 m/s², $11\frac{1}{3}$ m, 1, 6.
　　17. 16 m/s, $42\frac{2}{3}$ m, $85\frac{1}{3}$ m.

108. **18.** 9 m/s, 3 m/s². 　**19.** 9 m/s, 6 m/s², $2\frac{2}{9}$ m.

CHAPTER 6

110. **Qu. 1. D.** (i) $-4x^{-5}$, (ii) $-6x^{-4}$, (iii) $-\dfrac{2}{x^3}$, (iv) $-\dfrac{4}{x^2}$,

　　　(v) $\dfrac{4}{x^3}$, (vi) $-\dfrac{1}{x^4}$, (vii) $\dfrac{4}{x^5}$, (viii) $-\dfrac{3}{x^6}$.

　　I. (i) $-\frac{1}{2}x^{-2} + c$, (ii) $-2x^{-1} + c$, (iii) $-\dfrac{1}{x} + c$,

　　　(iv) $-\dfrac{1}{x^2} + c$, (v) $-\dfrac{1}{6x^2} + c$, (vi) $-\dfrac{2}{15x^3} + c$.

110. **Qu. 2. D.** (i) $\frac{1}{2}x^{-\frac{1}{2}}$, (ii) $-\frac{2}{3}x^{-\frac{4}{3}}$, (iii) $\dfrac{1}{2\sqrt{x}}$, (iv) $\dfrac{1}{3\sqrt[3]{x^2}}$,

(v) $-\dfrac{1}{3\sqrt[3]{x^4}}$, (vi) $\dfrac{2}{3\sqrt[3]{x^4}}$, (vii) $3\sqrt{x}$,

(viii) $-\dfrac{1}{3\sqrt{x^3}}$.

I. (i) $\frac{4}{5}x^{\frac{3}{4}} + c$, (ii) $\frac{4}{5}x^{\frac{5}{2}} + c$, (iii) $\frac{2}{3}\sqrt{x^3} + c$,

(iv) $\frac{3}{4}\sqrt[3]{x^4} + c$, (v) $2\sqrt{x} + c$, (vi) $-\dfrac{2}{\sqrt{x}} + c$.

Exercise 6a

113. **1.** (i) $4(2x + 3)$, (ii) $24(3x + 4)^3$, (iii) $\dfrac{-2}{(2x + 5)^2}$,

(iv) $2(3x - 1)^{-\frac{2}{3}}$, (v) $(3 - 2x)^{-\frac{3}{2}}$, (vi) $12(3 - 4x)^{-4}$.

2. (i) $\frac{1}{12}(3x + 2)^4 + c$, (ii) $\frac{1}{6}(2x + 3)^3 + c$,

(iii) $-\frac{1}{3}(3x - 4)^{-1} + c$, (iv) $\frac{1}{3}(2x + 3)^{\frac{3}{2}} + c$.

3. (i) $\dfrac{-3}{(3x + 2)^2}$, (ii) $\dfrac{-4}{(2x + 3)^3}$,

(iii) $\dfrac{-3}{2\sqrt{(3x + 1)^3}}$, (iv) $\dfrac{-4}{3(2x - 1)^{\frac{5}{3}}}$.

4. (i) $\dfrac{-1}{2(2x - 3)} + c$, (ii) $\frac{2}{3}\sqrt{(3x + 2)} + c$,

(iii) $2(2x - 1)^{\frac{1}{2}} + c$.

5. (i) $18x(3x^2 + 5)^2$, (ii) $(18x^2 + 10)(3x^3 + 5x)$,

(iii) $\dfrac{14x}{3}(7x^2 - 4)^{-\frac{2}{3}}$, (iv) $-(36x^2 - 8)(6x^3 - 4x)^{-3}$,

(v) $-\frac{2}{3}(6x - 5)(3x^2 - 5x)^{-\frac{5}{3}}$.

6. (i) $\dfrac{-6x}{(3x^2 + 2)^2}$ (ii) $\dfrac{-3x}{\sqrt{(2 + x^2)^3}}$,

(iii) $\dfrac{1}{\sqrt{x}(1 + \sqrt{x})^3}$, (iv) $\dfrac{3}{x^2}\left(1 - \dfrac{1}{x}\right)^2$,

(v) $\dfrac{-2x}{3(x^2 - 1)^{\frac{4}{3}}}$.

page

113. **7.** (i) $3(3\sqrt{x} - 2x)^2 \left(\dfrac{3}{2\sqrt{x}} - 2 \right)$, (ii) $\dfrac{1}{\sqrt{x}(2 - \sqrt{x})^2}$,

(iii) $\dfrac{1}{3} \left(2x^2 - \dfrac{3}{x^2} \right)^{-\frac{2}{3}} \left(4x + \dfrac{6}{x^3} \right)$,

(iv) $\dfrac{1}{2} \left(x - \dfrac{1}{x} \right)^{-\frac{1}{2}} \left(1 + \dfrac{1}{x^2} \right)$.

8. (i) $\dfrac{-\frac{3}{2}\sqrt{x}}{(x^{\frac{3}{2}} - 1)^2}$, (ii) $\dfrac{1}{2x^{\frac{3}{2}}\sqrt{(x - 1)}}$,

(iii) $\dfrac{-1}{6\sqrt{x}\sqrt[3]{(1 - \sqrt{x})^2}}$, (iv) $\dfrac{x^2 - 1}{x^2}$.

114. **9.** (i) $\dfrac{-3(2x - 7)}{(x^2 - 7x)^4}$, (ii) $\dfrac{1 - 4x^{\frac{3}{2}}}{\sqrt{x}(x^2 - \sqrt{x})^3}$,

(iii) $\dfrac{x}{(1 - x^2)^{\frac{3}{2}}}$, (iv) $\dfrac{1}{\sqrt{x}(1 - \sqrt{x})^3}$,

10. (i) $\dfrac{x^4 + 1}{x^2\sqrt{(x^4 - 1)}}$, (ii) $\dfrac{-2(\sqrt{x} + 1)}{\sqrt{x}(x + 2\sqrt{x})^2}$,

(iii) $\dfrac{1}{3x^{\frac{3}{2}} \left(1 - \dfrac{2}{\sqrt{x}} \right)^{\frac{2}{3}}}$, (iv) $\dfrac{1}{4x^{\frac{3}{2}}\sqrt{\left(1 - \dfrac{1}{\sqrt{x}} \right)}}$.

Exercise 6b

116. **1.** $1458 \text{ cm}^3/\text{s}$. **2.** $10\pi \text{ cm}^2/\text{s}$. **3.** $\frac{5}{17} \text{ cm/s}$.
4. $\frac{3}{2}$ cm/s. **5.** Decreasing, $8\pi \text{ cm}^2/\text{s}$.

117. **6.** 24. **7.** $\dfrac{1}{8\pi}$ cm/s. **8.** $\dfrac{1}{2\pi}$ cm/s. **9.** $\frac{4}{45}$ cm/s.
10. (i) 6 cm, (ii) $\frac{1}{8}$ cm/min. **11.** 30. **12.** $\frac{4}{15}$.
13. $0 \cdot 27$ cm/s. **14.** $4 \cdot 8$ litres/min.

119. **Qu. 3.** (i) $2x + 3$, (i) $2x(2x^2 + 1)$,
(iii) $4(x - 2)(x^2 - x - 1)$,
(iv) $2(x + 1)(x + 2)(2x + 3)$.

121. **Qu. 5.** (i) and (ii) $\dfrac{2(3x - 1)}{(x + 3)^3}$.

Exercise 6c

1. $x(5x + 2)(x + 1)^2$.

2. $(9x^2 + 1)(x^2 + 1)^3$.

3. $2(2x - 1)(x + 1)^2$.

4. $\dfrac{1}{(x + 1)^2}$.

5. $\dfrac{-4x}{(1 + x^2)^2}$.

6. $\dfrac{2(x - 1)}{(x + 1)^3}$.

7. $2x(1 + x^2)(1 - 3x^2)$.

8. $2x - \frac{3}{2}\sqrt{x}$.

9. $-x(x + 1)(x - 1)^2(4 + 7x + 7x^2)$.

10. $\dfrac{2x^2 - x + 1}{\sqrt{(x^2 + 1)}}$.

11. $\dfrac{x(2 + 3x^2)}{\sqrt{(1 + x^2)}}$.

12. $\dfrac{x(2 + x^2)}{\sqrt{(1 + x^2)^3}}$.

13. $\dfrac{(x - 1)(3x + 1)}{2\sqrt{x^3}}$.

14. $\dfrac{x(-2x^3 + 2x^2 + 3x - 4)}{\sqrt{(x^2 - 1)^3}}$.

15. $\dfrac{2x + 5}{2\sqrt{(x + 3)}\sqrt{(x + 2)}}$.

16. $\dfrac{1}{2\sqrt{\{x(x + 1)^3\}}}$.

17. $\dfrac{-1}{\sqrt{x}(1 + \sqrt{x})^2}$.

18. $\dfrac{1}{2\sqrt{\{(1 + x)(2 + x)^3\}}}$.

19. $\dfrac{(4x + 5)\sqrt{(x + 2)}}{2\sqrt{(x + 1)}}$.

20. $\dfrac{(2x + 5)\sqrt{(x + 1)}}{2\sqrt{(x + 2)^3}}$.

21. $\dfrac{x(2x^2 + 5)\sqrt{(1 + x^2)}}{\sqrt{(2 + x^2)^3}}$.

124. Qu. 6. (a) (i) 1, (ii) $\dfrac{dy}{dx}$, (iii) $2x$, (iv) $2y\dfrac{dy}{dx}$,

(v) $y + x\dfrac{dy}{dx}$, (vi) $2xy + x^2\dfrac{dy}{dx}$, (vii) $y^2 + 2xy\dfrac{dy}{dx}$.

(b) $\dfrac{2x - 6y + 3}{6x - 2y + 2}$.

125. Qu. 7. $\dfrac{2x + y}{3y^2 - x}$.

Exercise 6d

1. $\pm \frac{1}{3}$. **2.** $-1, \frac{11}{3}$. **3.** $\frac{3}{7}$. **4.** $-\frac{3}{2}$. **5.** (i) $\dfrac{3t}{2}$, (ii) $\frac{3}{2}\sqrt{x}$.

6. $(9, 3), (-1, 3)$. **7.** (i) $\dfrac{-2y}{3x}$, (ii) $\dfrac{y(2x - y)}{x(2y - x)}$.

8. (i) $\dfrac{1}{t}$, (ii) $\dfrac{t}{t + 1}$. **9.** $2t - t^2$. **10.** $\dfrac{2(x - y - 1)}{2x - 2y - 3}$.

11. $\dfrac{9(t + 2)^2}{4(t + 3)^2}$. **12.** $\dfrac{4y - 3x}{3y - 4x}$.

Exercise 6e

page

128. **1.** 8π cm². **2.** 9%. **3.** (i) 2·000 83, (ii) 5·01.

4. $\frac{1}{2}x$. **5.** $\Delta p/p = -\Delta v/v$. **6.** (i) $1\frac{1}{4}$, (ii) $1\frac{1}{4}$.

7. $1·6\pi$ cm³. **8.** 4%. **9.** 2%. **10.** (i) 25·04, (ii) 10·0166.

11. $1\frac{1}{3}$%.

129. **13.** $70·4\pi$ cm³.

Qu. 8. (i) $2x + \dfrac{2}{x^3}$, $2 - \dfrac{6}{x^4}$; (iii) $-\dfrac{1}{(x-1)^2}$, $\dfrac{2}{(x-1)^3}$.

130. Qu. 9. $\dfrac{1}{t}$, $-\dfrac{1}{2at^3}$.

Exercise 6f

1. (i) $-\dfrac{2n}{x^{2n+1}}$, (ii) $(n+1)x^n$, (iii) $\frac{1}{2}(2a-1)x^{2a-2}$,

(iv) $2mx^{2m-1}$, (v) $\frac{1}{2}nx^{\frac{1}{2}n-1}$.

2. (i) $\dfrac{1}{2k-1}x^{2k-1} + c$, (ii) $\dfrac{1}{(1-2n)}x^{1-2n} + c$,

(iii) $-n^2x^{-1/n}$, (iv) $\dfrac{1}{k}x^k + c$.

3. (i) $\dfrac{1}{n}x^{(1/n)-1} + c$, (ii) $\frac{1}{2}nx^{\frac{1}{2}n-1}$, (iii) $2(1-n)/x^n$,

(iv) $-\frac{1}{2}nx^{-\frac{1}{2}n-1}$, (v) $-\frac{2}{3}nx^{-\frac{2}{3}n-1}$.

4. (i) $-\dfrac{6}{n}x^{-(3/n)-1}$, (ii) $-\frac{1}{3}x^{-\frac{4}{3}}$, (iii) $-\dfrac{3n}{2}x^{-(3n/2)-1}$,

(iv) $\dfrac{3\sqrt{2}}{2}\sqrt{x}$, (v) $\dfrac{n}{n+1}x^{1/(n+1)}$.

131. **5.** (i) $8x(x^2+3)^3$, (ii) $\dfrac{3x^2}{\sqrt{(2x^3-3)}}$,

(iii) $\dfrac{3}{2\sqrt{x}}(\sqrt{x}+1)^2$, (iv) $n\left(\dfrac{1}{2}+\dfrac{2}{x^2}\right)\left(\dfrac{x}{2}-\dfrac{2}{x}\right)^{n-1}$.

6. (i) $-\dfrac{(2\sqrt{x}+1)}{2\sqrt{x}(x+\sqrt{x})^2}$, (ii) $-\dfrac{2x}{(x^2-1)^2}$,

(iii) $-\dfrac{1}{\sqrt{x}(\sqrt{x}-1)^3}$, (iv) $\dfrac{2x-2}{3(2x-x^2)^{\frac{4}{3}}}$.

131. **7.**   (i) $x(5x - 2)(x - 1)^2,$

      (ii) $(4x + 1)(x + 1)^{\frac{1}{2}}(x - 1)^{\frac{3}{2}},$

      (iii) $\dfrac{(5x - 3)(x - 1)}{2\sqrt{(x - 2)}}.$

  8.   (i) $\dfrac{4x + 1}{2}\sqrt{\left(\dfrac{x - 2}{x + 1}\right)},$ (ii) $\dfrac{14x^2 - 6x - 2}{3\sqrt[3]{(1 - 2x)^2}},$

      (iii) $\dfrac{2x^2 - 1}{\sqrt{(x^2 - 1)}}.$

  9.   (i) $-\dfrac{1 + x^2}{(x^2 - 1)^2},$ (ii) $\dfrac{x - 2}{2(x - 1)^{\frac{3}{2}}},$

      (iii) $-\dfrac{1 + x}{2\sqrt{x}(x - 1)^2},$ (iv) $-\dfrac{1}{2\sqrt{x}\sqrt{(x - 1)}(\sqrt{x} - 1)}.$

 10.   (i) $\dfrac{4(x - 1)}{(x + 2)^3},$ (ii) $\dfrac{3(x - 1)^3(x + 1)}{(x^3 - 1)^2}.$

 11.   (i) $\dfrac{1}{2\sqrt{(x + 1)}\sqrt{(x + 2)^3}},$ (ii) $\dfrac{(2x - 5)\sqrt{(x + 2)}}{2(x - 1)^{\frac{3}{2}}}.$

 12.   (i) $-\dfrac{2x}{\sqrt{\{(x^2 - 1)^3(x^2 + 1)\}}},$

      (ii) $\dfrac{(1 - \sqrt{x})(1 - x\sqrt{x})}{\sqrt{x}(x^2 - 1)^{\frac{3}{2}}}.$

 13. $-1.$ **14.** $\dfrac{3y - 2x - 4}{2y - 3x - 2}.$ **15.** $\dfrac{3x - 2y}{2x}.$

 16. $-\dfrac{2t}{1 - t^2}$ **17.** $-\dfrac{1}{t}.$ **18.** $-1.$

132. **19.** $1\%.$ **20.** $3\%.$ **21.** $2\%.$ **22.** (i) $4 \cdot 021,$ (ii) $6 \cdot 083.$

 23. $12x^2 - 12x - 9,\ -24,\ 24.$ **24.** $\dfrac{1}{t},\ -\dfrac{1}{2at^3}$

 25. $\frac{5}{3},\ -\frac{3}{4}.$ **26.** $5x - 4y = 9.$

 28. $\dfrac{3x^2 + 4x}{2(x + 1)^{\frac{3}{2}}},\ \dfrac{3x^2 + 8x + 8}{4(x + 1)^{\frac{5}{2}}}.$ **29.** $1\frac{1}{2},\ 1\frac{7}{10},\ 2,\ 2.$

page

132. 30. (i) $2 \sec 2x \tan 2x$, (ii) $2 \sin x \cos x$, (iii) $\cos x - x \sin x$,

(iv) $3 \tan^2 x \sec^2 x$, (v) $\dfrac{1}{2\sqrt{x}} \cos \sqrt{x}$.

31. $\dfrac{3t}{4(t^2 - 1)}$, $- \dfrac{3(1 + t^2)}{16t(t^2 - 1)^3}$. **32.** 1%. **33.** $- \frac{8}{5}\sqrt{5}$.

133. 35. $y = 3x + \frac{1}{3}$, $(\frac{1}{9}, \frac{2}{3})$, $3y + x = 2\frac{1}{9}$.

36. $\dfrac{bt}{s}$ m/s, $\dfrac{b}{s^3}(s^2 - bt^2)$ m/s².

37. $x \sin t + y \cos t = a \sin t \cos t$,
$x \cos t - y \sin t = a \cos 2t$.

39. $0, 0$, min.; $\dfrac{64\sqrt{2}}{25\sqrt{5}}$, max.

40. $x \sin \theta - y \cos \theta = 2a\theta \sin \theta$,
$x \cos \theta + y \sin \theta = 2a\theta \cos \theta + 2a \sin \theta$.

41. $0 \cdot 08\pi$ cm³.

43. (i) $6x \sin (6x^2 + 8)$,

(ii) $\dfrac{1}{2\sqrt{(x + 1)}} \sec^2 \sqrt{(x + 1)}$, (iii) $\dfrac{1}{1 + \cos x}$,

(iv) $\dfrac{\sec^4 x}{2\sqrt{\tan x}\,(1 - \tan^2 x)^{\frac{3}{2}}}$.

44. $- 5, 78$. **45.** $\dfrac{x}{\sqrt{(x^2 + 1)}}$. **46.** $- \frac{1}{2}\%$.

134. 48. At $x = 4, 8$. **49.** Decreasing $\frac{1}{2}\%$. **51.** $\dfrac{ah \sin \theta}{h - a \cos \theta}$.

53. (i) $0 \cdot 15\pi$ cm³/s, (ii) $0 \cdot 16\pi$ cm²/s.

54. $t(1 - t^2)\dfrac{d^2y}{dt^2} + (1 - 3t^2)\dfrac{dy}{dt} = ty$.

Exercise 7a

138. 4. (i) $\frac{3}{4}x^{\frac{4}{3}} + c$, (ii) $\frac{4}{5}x^{\frac{5}{4}} + c$, (iii) $\frac{5}{3}x^{\frac{3}{5}} + c$, (iv) $\frac{3}{4}kx^{\frac{4}{3}} + c$,
(v) $2x^{\frac{1}{2}} + c$, (vi) $\frac{3}{2}x^{\frac{2}{3}} + c$, (vii) $\frac{6}{5}x^{\frac{5}{6}} + c$, (viii) $\frac{5}{2}x^{\frac{2}{5}} + c$,
(ix) $\frac{3}{5}x^{\frac{5}{3}} + c$, (x) $\frac{3}{10}x^{\frac{10}{3}} + c$, (xi) $\frac{2}{5}x^{\frac{5}{2}} + c$,

page

138. $\quad$ (xii) $- 3x^{\frac{1}{3}} + c$, (xiii) $\dfrac{a}{a+1}x^{(a+1)/a} + c$,

$\quad$ (xiv) $\dfrac{n}{n-1}x^{(n-1)/n} + c$, (xv) $\frac{2}{7}x^{\frac{7}{2}} + \frac{4}{5}x^{\frac{5}{2}} - 2x^{\frac{3}{2}} + c$,

$\quad$ (xvi) $\frac{2}{5}x^{\frac{5}{2}} + 4x^{\frac{1}{2}} + c$, (xvii) $\frac{1}{2}x^2 - \frac{2}{3}x^{\frac{3}{2}} - 6x + c$,

$\quad$ (xviii) $\frac{2}{3}(x+2)^{\frac{3}{2}} + c$, (xix) $\frac{1}{3}(x^2-3)^{\frac{3}{2}} + c$.

$\quad$ **5.** (i) $-\frac{1}{2}$, (ii) 21, (iii) $12\frac{2}{3}$.

142. Qu. 1. $60 + 50/n$; 60.

143. Qu. 2. (i) $\frac{3}{4}x^2 - 4x + c$, (ii) $\frac{2}{3}x^3 + 3x^{-1} + c$,

$\quad$ (iii) $\frac{7}{8}x^{\frac{8}{7}} + c$, (iv) $- t^2 + \frac{10}{3}t^{\frac{3}{2}} - 3t + c$.

$\quad$ Qu. 3. (i) $17\frac{5}{6}$, (ii) $\frac{7}{48}$, (iii) $-5\frac{3}{5}$.

Exercise 7b

147. $\quad$ **1.** (i) $\frac{1}{3}x^3 - \frac{3}{2}x^2 + c$, $\qquad$ (ii) $- 2x^{-1} + x^{-2} + c$,

$\quad$ (iii) $\frac{1}{3}at^3 + bt - ct^{-1} + k$,

$\quad$ (iv) $\frac{1}{5}x^5 - \frac{3}{5}x^{\frac{5}{3}} + 2x + x^{-1} + c$,

$\quad$ (v) $\frac{1}{3}y^3 + \frac{2}{3}y^{\frac{3}{2}} + y - 2y^{-\frac{1}{2}} + c$,

$\quad$ (vi) $\frac{3}{8}s^{\frac{8}{3}} + \frac{6}{5}s^{\frac{5}{3}} + \frac{3}{2}s^{\frac{2}{3}} + c$.

$\quad$ **2.** (i) $26\frac{2}{3}$, (ii) $25\frac{2}{3}$, (iii) $1\frac{11}{15}$, (iv) $3\frac{19}{24}$, (v) $21\frac{1}{3}$, (vi) $24\frac{2}{3}$.

$\quad$ **3.** (i) 12, (ii) $- 31\frac{1}{4}$, (iii) $1\frac{7}{24}$.

$\quad$ **4.** (i) 9, (ii) $11\frac{1}{4}$, (iii) 12, (iv) $2(\sqrt{3} - \sqrt{2})$.

$\quad$ **5.** (i) $4\frac{1}{2}$, on the negative side of the y-axis, (ii) $4\frac{1}{2}$,

$\quad$ (iii) $1\frac{1}{3}$.

148. $\quad$ **6.** $18\frac{2}{3}$. $\quad$ **7.** $- 36$. $\quad$ **8.** $28\frac{4}{5}$. $\quad$ **9.** (i) 4, (ii) $\frac{3}{4}$.

$\quad$ **10.** (i) $10\frac{2}{3}$, (ii) $1\frac{1}{3}$, (iii) $4\frac{1}{2}$, (iv) $4\frac{1}{2}$, (v) $20\frac{5}{6}$, (vi) $20\frac{5}{6}$.

$\quad$ **11.** (i) $\frac{9}{5}$, (ii) $\frac{1}{3}$, (iii) $\frac{1}{24}$, (iv) $\frac{16}{3}$, (v) $41\frac{2}{3}$, (vi) $13\frac{1}{2}$.

$\quad$ **12.** Other points of intersection $(2, \frac{1}{4})$ and $(\frac{1}{2}, 4)$; $1\frac{11}{16}$.

149. Qu. 4. (i) (a) A cone, vertex **C**, (b) two cones with common

$\quad$ base, vertices **A** and **C**,

$\quad$ (ii) sphere, (iii) hemisphere,

$\quad$ (iv) ring internal dia. 4, external dia. 8,

$\quad$ (v) cylinder.

151. Qu. 5. (i) $31\pi/5$, (ii) $56\pi/15$.

Exercise 7c

page
154. 1. (i) 144π, (ii) $28\pi/15$, (iii) 2π, (iv) $16\pi/15$, (v) $\pi/105$, (vi) $3\pi/4$.

2. (i) 18π, (ii) $9\pi/2$, (iii) $96\frac{2}{5}\pi$, (iv) $34\frac{2}{15}\pi$, (v) $3\pi/5$, (vi) $3\pi/10$.

3. (i) $8\pi/3$, (ii) $17\frac{1}{15}\pi$, (iii) $8\pi/3$, (iv) $16\pi/15$, (v) $128\pi/105$, (vi) $7\pi/3$.

4. (i) 5π, (ii) $64\pi/15$, (iii) $32\pi/3$, (iv) $\frac{1}{2}\pi$.

155. 5. $\frac{1}{3}\pi r^2 h$. 6. $\frac{4}{3}\pi r^3$. 7. $661\frac{1}{3}\pi$ cm³. 8. 1296π cm³.

9. $57\frac{5}{7}\pi$ cm³. 10. $27\frac{1}{2}\pi$ cm³. 11. $16\pi/15$.

12. $37\pi/10$. 13. $37\frac{1}{3}\pi$. 14. 8π. 15. $45\pi/2$.

159. Qu. 6. (i) $(\frac{8}{5}, 0)$, (ii) $(\frac{6}{5}, 3\sqrt{2}/8)$.

Exercise 7d

160. 1. (i) $(\frac{12}{5}, 0)$, (ii) $(0, \frac{3}{5})$, (iii) $(\frac{8}{5}, 0)$, (iv) $(0, \frac{7}{3})$.

2. (i) $(\frac{2}{5}, 0)$, (ii) $(0, \frac{83}{70})$.

3. (i) $(8, \frac{4}{3})$, (ii) $(\frac{3}{4}, \frac{3}{5})$, (iii) $(\frac{9}{4}, \frac{27}{10})$, (iv) $(\frac{8}{5}, \frac{16}{7})$.

161. 4. (i) $(1\frac{1}{3}, 0)$, (ii) $(\frac{8}{5}, 0)$, (iii) $(0, \frac{10}{9})$, (iv) $(\frac{5}{4}, 0)$, (v) $(\frac{9}{7}, 0)$, (vi) $(0, \frac{1275}{248})$.

5. $\frac{1}{4}h$ above the base. 6. $\frac{3}{8}r$. 7. $9\frac{1}{7}$ cm.

8. $(4r/3\pi, 4r/3\pi)$.

Exercise 7e

1. 0, (i) $\frac{1}{4}$, (ii) $-\frac{1}{4}$. 2. $\frac{2}{15}$.

3. A$(1, 1\frac{1}{2})$, B$(4, 3)$, $\frac{3}{2}\sqrt{5}$.

162. 4. 134π. 5. $\frac{14\pi}{5}$, 10π. 6. 120π. 7. 16π.

9. $y = 2x - 2$. 10. $10\frac{2}{3}$. 11. 255π.

163. 13. $6\frac{3}{4}, 1\frac{4}{5}$. 14. $5\frac{3}{5}, 2\frac{4}{21}$. 15. $3\frac{1}{4}, \frac{73}{195}$.

16. (i) $\frac{7}{3}$, (ii) $(\frac{45}{28}, \frac{93}{70})$. 17. $(\frac{9}{5}, \frac{54}{35})$. 18. $6\frac{3}{4}, \frac{1}{5}$.

19. $(1\frac{17}{28}, 6\frac{9}{14})$. 20. $\frac{32}{5}\pi, \frac{31}{24}$. 21. $8\pi, \frac{9}{8}$. 22. $4\pi, \frac{5}{4}$.

164. 23. $\frac{4}{3}\pi, \frac{4}{5}$. 24. $\frac{4}{3}\pi, \frac{6}{5}$.

CHAPTER 8

166. Qu. 1. (1) 2, (ii) 6, (iii) a, (iv) ab, (v) 18, (vi) 80, (vii) $4a$, (viii) 6, (ix) 35, (x) 16, (xi) 36, (xii) ab.

Exercise 8a

167

1. (i) 5, (ii) $\frac{1}{2}$, (iii) 48, (iv) $\frac{1}{2}$, (v) a/b, (vi) 15, (vii) 21,
(viii) p/q, (ix) $1/(4p)$, (x) $9a/(2b)$.

2. (i) $2\sqrt{2}$, (ii) $2\sqrt{3}$, (iii) $3\sqrt{3}$, (iv) $5\sqrt{2}$, (v) $3\sqrt{5}$,
(vi) $11\sqrt{10}$, (vii) $5\sqrt{3}$, (viii) $4\sqrt{2}$, (ix) $6\sqrt{2}$,
(x) $7\sqrt{2}$, (xi) $2\sqrt{15}$, (xii) $16\sqrt{2}$.

3. (i) $\sqrt{18}$, (ii) $\sqrt{12}$, (iii) $\sqrt{80}$, (iv) $\sqrt{24}$, (v) $\sqrt{72}$,
(vi) $\sqrt{216}$, (vii) $\sqrt{128}$, (viii) $\sqrt{1000}$, (ix) $\sqrt{\frac{1}{2}}$,
(x) $\sqrt{\frac{1}{3}}$, (xi) $\sqrt{\frac{1}{6}}$, (xii) $\sqrt{\frac{2}{3}}$.

4. (i) $\sqrt{5}/5$, (ii) $\sqrt{7}/7$, (iii) $-\sqrt{2}/2$, (iv) $2\sqrt{3}/3$,
(v) $\sqrt{6}/2$, (vi) $\sqrt{2}/4$, (vii) $-\sqrt{3}/2$, (viii) $3\sqrt{6}/8$,
(ix) $\sqrt{2}-1$, (x) $2+\sqrt{3}$, (xi) $(4+\sqrt{10})/6$,
(xii) $\sqrt{6}-2$, (xiii) $(\sqrt{5}+\sqrt{3})/2$, (xiv) $3\sqrt{6}+3\sqrt{5}$,
(xv) $3+2\sqrt{2}$, (xvi) $(3\sqrt{2}+2\sqrt{3})/6$.

Exercise 8b

168

1. (i) $3\sqrt{2}$, (ii) $6\sqrt{3}$, (iii) $4\sqrt{7}$, (iv) $5\sqrt{10}$, (v) $28\sqrt{2}$,
(vi) 0.

2. (i) 25·5, (ii) 2·26, (iii) 3·15, (iv) 19·5, (v) 0·354,
(vi) 0·260.

3. (i) $\frac{6}{7}+\frac{2}{7}\sqrt{2}$, (ii) $9+4\sqrt{5}$, (iii) $-1+\sqrt{2}$,
(iv) $4-2\sqrt{3}$, (v) $-1-\sqrt{2}$, (vi) $\frac{1}{2}+\frac{3}{4}\sqrt{2}$,
(vii) $\frac{2}{9}\sqrt{3}$, (viii) $\frac{6}{25}\sqrt{5}$, (ix) $\frac{8}{11}+\frac{5}{11}\sqrt{3}$, (x) $\frac{3}{2}+\frac{1}{2}\sqrt{5}$,
(xi) $\frac{3}{7}+\frac{5}{14}\sqrt{2}$, (xii) 0.

4. (i) $5+2\sqrt{6}$, (ii) $\frac{1}{2}(5+\sqrt{3}+\sqrt{5}+\sqrt{15})$,
(iii) $-7+3\sqrt{6}$, (iv) $4+\sqrt{10}$, (v) $3+2\sqrt{2}$,
(vi) $\sqrt{2}$.

5. (i) $2-\sqrt{2}$, (ii) $4(2+\sqrt{3})$, (iii) $-(2+\sqrt{3})$,
(iv) $2+\sqrt{3}$, (v) $3+2\sqrt{2}$, (vi) $6+4\sqrt{2}$.

Qu. 2. (i) 3, (ii) 3, (iii) 9, (iv) 2, (v) 8, (vi) 243,
(vii) 16, (viii) 8.

Exercise 8c

1. (i) 5, (ii) 3, (iii) 2, (iv) 7, (v) $\frac{1}{2}$, (vi) 1, (vii) -2,
(viii) -1, (ix) 16, (x) 9, (xi) 125, (xii) 343, (xiii) $\frac{1}{8}$,
(xiv) $\frac{2}{3}$, (xv) $1\frac{1}{2}$, (xvi) $\frac{2}{3}$.

2. (i) 1, (ii) $\frac{1}{3}$, (iii) 1, (iv) $\frac{1}{4}$, (v) $\frac{1}{3}$, (vi) 2, (vii) 9,
(viii) 1, (ix) $\frac{1}{27}$, (x) $-\frac{1}{6}$, (xi) 1, (xii) $\frac{9}{4}$.

page
172.
    (xiii) 4, (xiv) 3, (xv) $4\frac{1}{2}$, (xvi) $\frac{5}{9}$.

3. (i) $\frac{1}{2}$, (ii) $\frac{1}{4}$, (iii) $\frac{1}{2}$, (iv) $\frac{1}{8}$, (v) $\frac{1}{9}$, (vi) 2, (vii) 2, (viii) 9, (ix) $1\frac{1}{2}$, (x) $1\frac{1}{2}$, (xi) $1\frac{1}{2}$, (xii) $\frac{16}{81}$.

Exercise 8d

1. (i) 16, (ii) 36, (iii) 4, (iv) 6, (v) $1\frac{1}{2}$, (vi) $1\frac{1}{3}$, (vii) $\frac{1}{2}$, (viii) $\frac{1}{5}$, (ix) $\frac{1}{16}$, (x) $\frac{1}{27}$, (xi) $2\frac{3}{4}$, (xii) 64, (xiii) $\frac{4}{3}$, (xiv) 1·1, (xv) 125, (xvi) $\frac{1}{2}$.

2. (i) $\frac{1}{2}$, (ii) 1, (iii) $\frac{1}{3}$, (iv) 2, (v) 2, (vi) 1.

3. (i) 2^{-n}, (ii) 3^{n+1}, (iii) 4, (iv) 3, (v) 12, (vi) $10^{\frac{1}{2}n}$.

173. **4.** (i) $x^{-\frac{7}{12}}$, (ii) 2, (iii) $x^{\frac{1}{2}n+1\frac{1}{2}}$, (iv) 1, (v) y^{-q}, (vi) 1.

5. (i) $-\dfrac{1}{x^2(x^2+1)^{\frac{1}{2}}}$, (ii) $\dfrac{x-2}{2x^2(1-x)^{\frac{1}{2}}}$,

    (iii) $-\dfrac{1}{2x^{\frac{3}{2}}(1+x)^{\frac{1}{2}}}$, (iv) $\dfrac{3+2x}{3(1+x)^{\frac{4}{3}}}$,

    (v) $\dfrac{1}{(1-x)\sqrt{(1-x^2)}}$.

174. **Qu. 4.** Bases: (i) 10, (ii) 10, (iii) 3, (iv) 4, (v) 2, (vi) $\frac{1}{2}$, (vii) a.

    Logarithms: (i) 2, (ii) 1·6021, (iii) 2, (iv) 3, (v) 0, (vi) -3, (vii) b.

Exercise 8e

1. (I) $\log_2 16 = 4$, (II) $\log_3 27 = 3$, (iii) $\log_5 125 = 3$, (iv) $\log_{10} 1\,000\,000 = 6$, (v) $\log_{12} 1728 = 3$, (vi) $\log_{16} 64 = \frac{3}{2}$, (vii) $\log_{10} 10\,000 = 4$, (viii) $\log_4 1 = 0$, (ix) $\log_{10} 0 \cdot 01 = -2$, (x) $\log_2 \frac{1}{2} = -1$, (xi) $\log_9 27 = \frac{3}{2}$, (xii) $\log_8 \frac{1}{4} = -\frac{2}{3}$, (xiii) $\log_{\frac{1}{3}} 81 = -4$, (xiv) $\log_e 1 = 0$, (xv) $\log_{16} \frac{1}{2} = -\frac{1}{4}$, (xvi) $\log_{\frac{1}{3}} 1 = 0$, (xvii) $\log_{81} 27 = \frac{3}{4}$, (xviii) $\log_{\frac{1}{16}} 4 = -\frac{1}{2}$, (xix) $\log_{\frac{2}{3}} \frac{4}{9} = 2$, (xx) $\log_{-3}(-\frac{1}{3}) = -1$, (xxi) $\log_a c = 5$, (xxii) $\log_a b = 3$, (xxiii) $\log_p r = q$, (xxiv) $\log_b a = c$.

page

174.   **2.** (i) $2^5 = 32$, (ii) $3^2 = 9$, (iii) $5^2 = 25$,
         (iv) $10^5 = 100\,000$.

175.   (v) $2^7 = 128$, (vi) $9^0 = 1$, (vii) $3^{-2} = \frac{1}{9}$,
         (viii) $4^{\frac{1}{2}} = 2$, (ix) $e^0 = 1$, (x) $27^{\frac{1}{3}} = 3$, (xi) $a^2 = x$,
         (xii) $3^b = a$, (xiii) $a^c = 8$, (xiv) $x^y = z$, (xv) $q^p = r$.

     3. (i) 6, (ii) 2, (iii) 7, (iv) 2, (v) $\frac{1}{3}$, (vi) 0, (vii) $\frac{1}{3}$,
         (viii) 2, (ix) 3, (x) -1, (xi) 3, (xii) -1.

     Qu. 5. $x = \log_c a$, $y = \log_c b$, $x + y = \log_c (ab)$,
         $x - y = \log_c (a/b)$.

     Qu. 6. $x = \log_c a$, $nx = \log_c a^n$.

Exercise 8f

177.   **1.** (i) $\log a + \log b$, (ii) $\log a - \log c$, (iii) $- \log b$,
         (iv) $2 \log a + \frac{3}{2} \log b$, (v) $-4 \log b$,
         (vi) $\frac{1}{2} \log a + 4 \log b - 3 \log c$, (vii) $\frac{1}{2} \log a$,
         (viii) $\frac{1}{3} \log b$, (ix) $\frac{1}{2} \log a + \frac{1}{2} \log b$, (x) $1 + \log a$,
         (xi) $-2 - 2 \log b$, (xii) $\frac{1}{2} \log a - \frac{1}{2} \log b$,
         (xiii) $\frac{1}{2} \log a + \frac{3}{2} \log b - \frac{1}{2} \log c$,
         (xiv) $\log b + \frac{1}{2} \log a - \frac{1}{3} \log c$,
         (xv) $\frac{1}{2} + \frac{1}{2} \log a - \frac{5}{2} \log b - \frac{1}{2} \log c$.

     2. (i) $\log 6$, (ii) $\log 2$, (iii) $\log 6$, (iv) $\log 2$,
         (v) $\log (ac)$, (vi) $\log (xy/z)$, (vii) $\log (a^2/b)$,
         (viii) $\log (a^2 b^3/c)$, (ix) $\log \sqrt{(x/y)}$, (x) $\log (p/\sqrt[3]{q})$,
         (xi) $\log (100\,a^2)$, (xii) $\log (10a/\sqrt{b})$,
         (xiii) $\log (a^2/2000c)$, (xiv) $\log (10x^2/\sqrt{y})$.

     3. (i) 3, (ii) 2, (iii) 2, (iv) 1, (v) $\log 2$, (vi) $\log 7$,
         (vii) $\log \frac{1}{2}$, (viii) 0, (ix) 0, (x) 3, (xi) 2, (xii) $\frac{2}{3}$.

     4. (i) $2 \cdot 322$, (ii) $0 \cdot 6309$, (iii) $0 \cdot 3155$, (iv) $1 \cdot 161$,
         (v) $-2 \cdot 585$, (vi) $6 \cdot 838$.

178.   **5.** (i) $3 \cdot 170$, (ii) $0 \cdot 7211$, (iii) $1 \cdot 042$, (iv) $2 \cdot 303$,
         (v) $1 \cdot 145$, (vi) $-0 \cdot 6309$.

     7. (i) $3 \cdot 119$, (ii) $1 \cdot 297$, (iii) $23 \cdot 14$,
         (iv) $0 \cdot 7936$, (v) $0 \cdot 3674$, (vi) $0 \cdot 000\,759\,7$.

179.   Qu. 7. (i) $\frac{2}{3}$, $-\frac{7}{3}$; (ii) $-\frac{11}{3}$, $\frac{2}{3}$; (iii) $-\frac{5}{2}$, $-\frac{1}{2}$;
         (iv) $-\frac{1}{2}$, $-\frac{7}{2}$.

     Qu. 8. (i) $x^2 - 7x + 12 = 0$, (ii) $x^2 - 3x - 2 = 0$,
         (iii) $8x^2 + 4x - 3 = 0$, (iv) $3x^2 - 2x = 0$.

     Qu. 9. -3, $\frac{7}{2}$.

Exercise 8g

182. **1.** (i) $\frac{11}{2}$, $\frac{3}{2}$; (ii) $-\frac{1}{2}$, $-\frac{1}{2}$; (iii) $\frac{7}{3}$, -2; (iv) -1, -1;
(v) 1, -3; (vi) 1, -5; (vii) 4, 1; (viii) 3, -2.

2. (i) $x^2 - 3x + 4 = 0$, (ii) $x^2 + 5x + 6 = 0$,
(iii) $2x^2 - 3x - 5 = 0$, (iv) $3x^2 + 7x = 0$,
(v) $x^2 - 7 = 0$, (vi) $5x^2 - 6x + 4 = 0$,
(vii) $36x^2 + 12x + 1 = 0$,
(viii) $10x^2 + 25x - 16 = 0$.

3. (i) $\frac{25}{4}$, (ii) $\frac{3}{4}$, (iii) $-\frac{5}{2}$, (iv) $-\frac{25}{8}$.

4. (i) $\frac{5}{9}$, (ii) $\frac{16}{9}$, (iii) $\frac{80}{27}$, (iv) $\frac{80}{9}$.

5. (i) 72, (ii) 5, (iii) $-\frac{9}{8}$, (iv) -32.

6. (i) $x^2 - 39x + 49 = 0$, (ii) $x^2 - 7x - 1 = 0$,
(iii) $x^2 + 35x - 343 = 0$.

183. **7.** (i) $2x^2 + 4x + 1 = 0$, (ii) $x^2 - 4x + 2 = 0$,
(iii) $x^2 - 6x + 1 = 0$. **8.** $4x^2 - 49x + 36 = 0$.

9. $\frac{25}{4}$. **10.** ± 6.

12. (i) $- bc/a^2$, (ii) $(b^2 - 2ac)/a^2$, (iii) $b(3ac - b^2)/a^3$,
(iv) $- b/c$, (v) $(b^2 - 2ac)/ac$,
(vi) $(b^4 - 4ab^2c + 2a^2c^2)/a^4$.

13. (i) $ax^2 - bx + c = 0$,
(ii) $ax^2 + (b - 2a)x + a - b + c = 0$,
(iii) $a^2x^2 + (2ac - b^2)x + c^2 = 0$,
(iv) $cx^2 - bx + a = 0$, (v) $a^2x^2 - (b^2 - 4ac) = 0$,
(vi) $a^2x^2 + 3abx + (2b^2 + ac) = 0$.

17. 2, -9, 9; 3, $\frac{3}{2}$.

184. **18.** (i) $ay^2 + y(b - 2a) + a - b + c = 0$, $\alpha + 1$, $\beta + 1$;
(ii) $ay^4 + by^2 + c = 0$, $\pm \sqrt{\alpha}$, $\pm \sqrt{\beta}$;
(iii) $a^2y^2 + (2ac - b^2)y + c^2 = 0$, α^2, β^2.

19. (i) $ay^2 + (b - 4a)y + 4a - 2b + c = 0$,
(ii) $cy^2 + by + a = 0$,
(iii) $ay^4 - 4ay^3 + (6a + b)y^2 - 2(2a + b)y + a + b + c = 0$.

20. $\frac{5}{2}$, $\frac{11}{4}$; $\frac{11}{4}$; $\frac{11}{4}$. **21.** 1. **22.** $\frac{5}{4}$. **23.** (i) 6, (ii) 4.
24. (i) 1, (ii) $3\frac{7}{8}$.

185. **Qu. 10.** Polynomials.

186. **Qu. 11.** (i) 0, (ii) -4, (iii) -8, (iv) -18,
(v) $a^3 + 3a - 4$.

Exercise 8h

187. **1.** (i) $-12, -12, -6, 0, 0$; $(x-2)$, or $(x+2)$;
(ii) $-1, 0, -2, 19, -21, (x-1)$.

188. (iii) $0, 6, -2, 88, -24, x$.
(iv) $3, 0, 0, 3, 3, (x-1)$, or $(x+1)$.
2. (i) 2, (ii) 18, (iii) -11, (iv) -1, (v) 2, (vi) $-2\frac{1}{2}$.
3. (i) -3, (ii) -10, (iii) 2, (iv) 4, (v) 4, (vi) 2.
4. $(x+3)(2x-1)$. **5.** $(2x-1)(2x+3)(3x+1)$.
6. (i) $(x-1)(x+2)(x-3)$, (ii) $(x+1)(x-2)(x-3)$.
(iii) $(2x+1)(x-2)(x+2)$.
(iv) $(x+1)(x+2)(2x-1)$.
(v) $(x+2)(x+3)(2x+1)$, (vi) $(x^2+1)(2x-1)$.
7. $a=3, b=2$. **8.** $p=1, q=-3$.
9. $a=3, b=-1, c=-2$.

189. **10.** $a=2, b=-1, c=-2$.

Exercise 8i

1. (i) $5\sqrt{5}$, (ii) $\sqrt{2}$, (iii) $18\sqrt{3}$.
2. (i) $18\cdot9$, (ii) $6\cdot29$, (iii) $0\cdot642$.
3. (i) $(11+6\sqrt{2})/7$, (ii) $13+2\sqrt{2}$.
4. (i) $\frac{1}{16}, \frac{8}{27}, \frac{1}{4}$, (ii) 8, 27.
5. (i) $x-3x^{\frac{1}{3}}-2$, 0; (ii) $x-3+x^{-1}, \frac{5}{4}$.
6. (i) 8, (ii) 2.
7. (i) $1\cdot079\,18$, (ii) $0\cdot653\,21$, (iii) $0\cdot592\,72$.

190. **8.** (i) $2+2\log a-3\log b-\frac{1}{2}\log c$, (ii) $1\cdot602\,060$.
9. (i) $0\cdot698\,970$, (ii) $1\cdot255\,273$, (iii) $0\cdot176\,091$.
10. (i) $1\frac{1}{3}$, (ii) (a) $0\cdot153$, (b) $3\cdot17$.
11. (i) $7\cdot525$ cm, (ii) $4\cdot402$ cm.
12. $-\frac{5}{3}, -\frac{1}{3}$; $9x^2-31x+1=0$. **13.** 1, 9; -3.
14. (i) $7\frac{1}{4}$, (ii) $x^2+5x-2=0$.
15. $k \leqslant 1, k \geqslant 5$. **17.** $a=2, b=-\frac{3}{2}, c=\frac{1}{2}$, (ii) $\frac{1}{2}$.

191. **18.** $10\frac{1}{8}-2(x+\frac{5}{4})^2$. **19.** $a=3, b=-63$.
20. $(x-1)(x+2)(3x-2)$.
21. 3. **22.** $(x+1)(x-5)(3x+1)$.
23. $a=1, b=-1$. **24.** $p=12, q=4$.

Exercise 9a

page
194. **1.** 720. **2.** 360. **3.** 24, 120. **4.** 243. **5.** 72. **6.** 24.
7. 27 000.

195. **8.** 120. **9.** 900. **10.** 120. **11.** 719. **12.** 72.
13. 5040. **14.** 168. **15.** 336, 144.
16. 3 628 800, 3 628 800. **17.** 78. **18.** 80.
19. 10 368 000.

196. **20.** 40 320, 384.

Exercise 9b

197. **1.** (i) 6, (ii) 24, (iii) 120, (iv) 90, (v) 210, (vi) 1320,
(vii) 330, (viii) $2\frac{1}{8}$, (ix) 4, (x) 20, (xi) 120, (xii) 2520.

2. (i) $\frac{6!}{3!}$, (ii) $\frac{10!}{8!}$, (iii) $\frac{12!}{8!}$, (iv) $\frac{n!}{(n-3)!}$, (v) $\frac{(n+2)!}{(n-1)!}$,

(vi) $\frac{10!}{8!2!}$, (vii) $\frac{7!}{4!3!}$, (viii) $\frac{52!}{49!3!}$,

198. (ix) $\frac{n!}{(n-2)!2!}$, (x) $\frac{(n+1)!}{(n-2)!3!}$, (xi) $\frac{(2n)!}{(2n-2)!2!}$,

(xii) $\frac{n!}{(n-r)!}$.

3. (i) 20! . 22, (ii) 25! . 25, (iii) 13! . 12, (iv) 14! . 19,
(v) $n!(n+2)$, (vi) $(n-2)!(n-2)$,
(vii) $(n-1)!(n+2)$, (viii) $n!(n+2)^2$.

4. (i) $\frac{16!}{12!4!}$, (ii) $\frac{22!}{14!8!}$, (iii) $\frac{18!}{7!11!}$, (iv) $\frac{37!}{19!18!}$,

(v) $\frac{(n+1)!}{r!(n-r+1)!}$, (vi) $\frac{(n+2)!}{r!(n-r+2)!}$.

Exercise 9c

202. **1.** 282 240. **2.** 362 880, 40 320.
203. **3.** 6720, 1680. **4.** 24 × 17!, 48 × 16! **5.** $\frac{1}{60}$ × 13!.
6. 181 440. **7.** 20 × 10! **8.** 768. **9.** 16. **10.** 144.
11. 30 240. **12.** 60 480. **13.** 528.

page
204. **14.** 1 404 000. **15.** 2400. **16.** 11 520, 276 480.
17. 23 520. **18.** 100. **19.** 138 600. **20.** 34 560, 31 680.
206. Qu. **1.** (i) 56; (ii) 210.

Qu. **2.** $\dfrac{n!}{(n-r)!r!}$.

Exercise 9d

1. (i) 45, (ii) 15, (iii) 35, (iv) 126, (v) 70; (vi) $\frac{1}{2}n(n-1)$,
(vii) $\frac{1}{6}n(n-1)(n-2)$, (viii) $\frac{1}{2}n(n-1)$,
(ix) $\frac{1}{2}n(n+1)$, (x) $\frac{1}{2}n(n+1)$.
2. 78. **3.** 70.
207. **4.** 252. **5.** 126. **6.** 30. **7.** 252. **8.** 286. **9.** 792.
10. 200. **11.** 495. **12.** 840. **13.** 182. **14.** 420.
15. 11 550.
208. **16.** 34 650. **17.** 25 200. **18.** 2142. **19.** 31 733.

Exercise 9e

1. 2160. **2.** 1960. **4.** 15 120. **5.** 5040, 240.
6. 360, 240. **7.** 728. **8.** $\frac{1}{2}n(n-3)$. **9.** 48.
209. **10.** 120 960. **11.** 2520. **12.** 240, 15 552. **13.** 277 200.
14. 4200. **15.** 5120. **16.** 504. **17.** 876. **18.** 1013.
19. 1 693 440. **20.** 300. **21.** 319. **22.** 646.
23. 28 732.
210. **24.** 6006. **25.** 240. **26.** (i) 917, (ii) 296.

CHAPTER 10

211. Qu. **1.** (i) 9, 11; (ii) 14, 17; (iii) 16, 32; (iv) $\frac{1}{48}$, $\frac{1}{96}$,
(v) 5^3, 6^3; (vi) $\frac{5}{6}$, $\frac{6}{7}$; (vii) 25, 36; (viii) 720, 5040;
(ix) $\frac{5}{81}$, $\frac{6}{243}$; (x) -4, -6; (xi) 1, -1;
(xii) $\frac{1}{16}$, $-\frac{1}{32}$.

Exercise 10a

213. **1.** (i) $1\frac{1}{2}$, (ii) -3, (iii) $0\cdot1$, (v) $\frac{1}{3}$, (vii) n, (ix) $1\frac{1}{4}$, (x) -7,
(xii) $-0\cdot2$.
2. (i) 75, 147; (ii) -34, -82; (iii) $7\frac{1}{3}$, $\frac{1}{3}(5n-3)$.

page
214.
(iv) $- 148, 52 - 2n$; (v) $- 13\frac{1}{2}, \frac{1}{2}(15 - n)$;
(vi) $799, 3 + 4n$.

3. (i) 23, (ii) 13, (iii) 31, (iv) 21, (v) 91, (vi) 13, (vii) $2n$,
(viii) n, (ix) n, (x) $(l - a)/d + 1$.

4. (i) 2601, (ii) 632, (iii) 420, (iv) 288, (v) 250·5,
(vi) $60\frac{1}{2}$, (vii) $121x$, (viii) $\frac{1}{2}n(2a + n - 1)$,
(ix) $\frac{1}{2}n\{2a + (n - 1)d\}$.

5. (i) 444, (ii) $- 80$, (iii) 20 100, (iv) $- 520$,
(v) $n(2n + 4)$, (vi) $\frac{1}{2}n(11 - n)$.

6. 2, 13, 220.   **7.** 33, $- 72$.

215. **8.** 5.   **9.** 14, 4.   **10.** 7500.   **11.** 7650.

12. $3\frac{1}{2}, \frac{1}{10}, 148\frac{1}{2}$.   **14.** 60.   **15.** £32 960.

Qu. 2. (i) (a) 6, 8; (b) 8, 16. (ii) (a) 0, $- 6$; (b) 3, $1\frac{1}{2}$.

Exercise 10b

216. **1.** (i) 3, (ii) $\frac{1}{4}$, (iii) $- 2$, (iv) $- 1$.
217. (vi) a, (vii) $1·1$, (x) 6.

2. (i) $5.2^{10}, 5.2^{19}$, (ii) $10(\frac{5}{2})^6, 10(\frac{5}{2})^{18}$, (iii) $\frac{2}{3}(\frac{9}{8})^{11}, \frac{2}{3}(\frac{9}{8})^{n-1}$,
(iv) $3(- \frac{2}{3})^7, 3(- \frac{2}{3})^{n-1}$, (v) $\frac{2}{7}(- \frac{3}{2})^8, \frac{2}{7}(- \frac{3}{2})^{n-1}$,
(vi) $3(\frac{1}{2})^{18}, 3(\frac{1}{2})^{2n-1}$.

3. (i) 9, (ii) 8, (iii) 7, (iv) 8, (v) $n + 1$, (vi) n.

4. (i) $2^{10} - 2$, (ii) $\frac{1}{2}(3^5 - \frac{1}{27})$, (iii) $0·03(2^7 - 1)$,
(iv) $- \frac{16}{405}\{(\frac{3}{2})^8 - 1\}$, (v) $5(2^{n+1} - 1)$,
(vi) $a\left(\dfrac{1 - r^n}{1 - r}\right)$.

5. (i) $2(3^{10} - 1)$, (ii) $\frac{45}{2}\{1 - (\frac{1}{3})^{20}\}$, (iii) $- \frac{1}{3}(2^{50} - 1)$,
(iv) $16\{1 + (\frac{1}{2})^{17}\}$, (v) $11(1·1^{23} - 1)$, (vi) $1 - (\frac{1}{2})^{13}$,
(vii) $3(2^n - 1)$, (viii) $\frac{3}{4}\{1 - (- \frac{1}{3})^n\}$.

6. 2, $2\frac{1}{2}$, $157\frac{1}{2}$.   **7.** ± 3, $\pm \frac{2}{3}$.

218. **8.** 6, $13\frac{1}{2}$.   **9.** £10 700 000.   **10.** $6\frac{3}{4}$.   **12.** $\frac{5}{2}$, $- \frac{1}{3}$.

13. $\sqrt{2} - 1, 5\sqrt{2} - 7$.   **14.** 1023.

224. **Qu. 3.** (i) 34, (ii) 16.   **Qu. 4.** 8, $12\frac{1}{2}$, 10.

Qu. 5. $2ac/(a + c)$.

Exercise 10c

1. 2550.   **2.** 8.   **3.** 98.   **4.** $\frac{3}{4}$, $- \frac{3}{2}$, 3.   **5.** 16 400.

6. 432.   **7.** $\frac{7}{2}$, 2.   **8.** 17, $- 2$, 10th.

page
225. **9.** $1\frac{1}{2}$, 2, 24.   **10.** 3, 4; 3, 7, 11, 15, 19.
 11. -2, 1, 4, 7, 10.   **12.** -3, -2.   **13.** 18.
 14. 18th, 655 360.   **15.** 14.   **16.** -9, 5.
 17. 2, 4, 6, 8, 10.   **18.** 5808.   **19.** 6, 8, 10.
 20. $2\frac{1}{2}$, 5, $7\frac{1}{2}$, 10.   **21.** £1840, to 3 sig. fig.
226. **22.** £1910, to 3 sig. fig.
228. **Qu. 6.** (i) $n(2n + 1)$, (ii) $\frac{1}{6}(n + 1)(n + 2)(2n + 3)$,
229. (iii) $\frac{1}{4}(n - 1)^2 n^2$, (iv) $n(2n - 1)$,
 (v) $\frac{1}{3}n(2n + 1)(4n + 1)$, (vi) $n^2(2n - 1)^2$.

Exercise 10e

231. **1.** (i) $1^3 + 2^3 + 3^3 + 4^3$, (ii) $2^2 + 3^2 + \ldots + n^2$,

 (iii) $2 + 6 + \ldots + (n^2 + n)$, (iv) $\frac{1}{1.2} + \frac{1}{2.3} + \frac{1}{3.4}$,

 (v) $2^2 + 2^3 + 2^4 + 2^5$, (vi) $-1 + 4 - 9 + 16$,
 (vii) $1 + 2^2 + \ldots + n^n$, (viii) $-\frac{1}{3} + \frac{1}{4} - \frac{1}{5} + \frac{1}{6}$,
 (ix) $n(n - 1) + (n + 1)n + (n + 2)(n + 1)$,

 (x) $\dfrac{n - 2}{n - 1} + \dfrac{n - 1}{n} + \dfrac{n}{n + 1}$.

 2. (i) $\displaystyle\sum_{1}^{n} m$, (ii) $\displaystyle\sum_{1}^{n+1} m^4$, (iii) $\displaystyle\sum_{1}^{5} \frac{1}{m}$, (iv) $\displaystyle\sum_{2}^{5} 3^m$,

 (v) $\displaystyle\sum_{2}^{6} m(m + 5)$, (vi) $\displaystyle\sum_{1}^{5} \frac{m}{3^{m-1}}$, (vii) $\displaystyle\sum_{1}^{5} \frac{m(2m + 1)}{2(m + 1)}$,

 (viii) $\displaystyle\sum_{1}^{6} (-1)^m m$, (ix) $\displaystyle\sum_{0}^{5} (-2)^m$,

 (x) $\displaystyle\sum (-1)^{m+1} m(2m + 1)$.

 3. (i) $(n + 1)(2n + 1)$, (ii) $\frac{1}{6}n(n - 1)(2n - 1)$,
 (iii) $n^2(2n + 1)^2$, (iv) $n(n + 2)$, (v) $\frac{1}{2}n(3n + 1)$,
 (vi) $n(2n + 3)$, (vii) $\frac{1}{6}n(2n^2 + 3n + 7)$,
 (viii) $\frac{1}{3}n(n + 1)(n + 2)$, (ix) $\frac{1}{6}n(n + 1)(2n + 7)$,
 (x) $\frac{2}{3}n(n + 1)(2n + 1)$.
232. (xi) $\frac{1}{3}n(2n - 1)(2n + 1)$, (xii) $\frac{1}{4}n(n + 1)(n^2 + n + 2)$,
 (xiii) $\frac{1}{12}n(n + 1)(n + 2)(3n + 1)$.

Exercise 10f

page
233.
234.

1. (i) $1\frac{1}{2}$, (ii) 24, (iii) $\frac{1}{3}$, (iv) $\frac{13}{99}$, (v) $\frac{5}{9}$,
(vi) $\frac{6}{11}$, (vii) $\frac{2}{3}$, (viii) $40\frac{1}{2}$.

2. (i) $\frac{8}{9}$, (ii) $\frac{4}{33}$, (iii) $3\frac{5}{9}$, (iv) $2\frac{23}{33}$, (v) $1\frac{4}{999}$, (vi) $2\frac{317}{330}$.

3. $\frac{2}{3}$. **4.** 2, $\frac{1}{2}$, $\frac{1}{4}$. **5.** $\frac{2}{5}$, 60; $\frac{3}{5}$, 40.

Exercise 10g

1. 1683. **2.** 20. **5.** 17. **6.** 2. **7.** $6n + 7$.

8. $\frac{1}{3}n(n - 1)(n + 1)$. **9.** $27 + 29 + \ldots + 113$.

10. 4234.

235. **11.** $\frac{5}{2}(3^n - 1)$, 16. **12.** 3, 2, $\frac{4}{3}$, $\frac{8}{9}$. **14.** $\frac{9}{4}$, 3, $\frac{15}{4}$. **15.** 35.

16. 4, $- 12$, $15\frac{7}{8}$, $57\frac{7}{8}$. **18.** $\frac{1}{6}n(2n^2 + 3n + 13)$.

19. $(ar + b)/(r + 1)$, $(br + a)/(r + 1)$.

20. 3, 12, 48, 3.4^{n-1}. **21.** 13, 9. **23.** 1, $\frac{1}{2}$, 2.

236. **26.** $\frac{1}{2}n(n + 1)$, $\dfrac{1 - (n + 1)x^n + nx^{n+1}}{(1 - x)^2}$.

Exercise 11a

241. **1.**

(i) $a^5 + 5a^4b + 10a^3b^2 + 10a^2b^3 + 5ab^4 + b^5$,

(ii) $x^3 + 3x^2y + 3xy^2 + y^3$,

(iii) $x^4 + 8x^3y + 24x^2y^2 + 32xy^3 + 16y^4$,

(iv) $1 - 4z + 6z^2 - 4z^3 + z^4$,

(v) $16x^4 + 96x^3y + 216x^2y^2 + 216xy^3 + 81y^4$,

(vi) $64z^3 + 48z^2 + 12z + 1$,

(vii) $a^6 - 6a^5b + 15a^4b^2 - 20a^3b^3 + 15a^2b^4 - 6ab^5 + b^6$,

(viii) $a^3 - 6a^2b + 12ab^2 - 8b^3$,

(ix) $81x^4 - 108x^3y + 54x^2y^2 - 12xy^3 + y^4$,

(x) $8x^3 + 4x^2 + \frac{2}{3}x + \frac{1}{27}$,

(xi) $x^5 - 5x^3 + 10x - 10x^{-1} + 5x^{-3} - x^{-5}$,

(xii) $\frac{1}{16}x^4 + x^2 + 6 + 16x^{-2} + 16x^{-4}$,

(xiii) $a^7 + 7a^6b + 21a^5b^2 + 35a^4b^3 + 35a^3b^4 + 21a^2b^5 + 7ab^6 + b^7$,

(xiv) $a^{10} - 5a^8b^2 + 10a^6b^4 - 10a^4b^6 + 5a^2b^8 - b^{10}$,

(xv) $a^6 - 3a^4b^2 + 3a^2b^4 - b^6$.

2. (i) 14, (ii) 194, (iii) 14·14, (iv) 391·8, (v) 98,
(vi) 56·56.

page

241. 3. $32 + 80x + 80x^2 + 40x^3 + 10x^4 \cdot \because x^5$, 32·080 08.

4. $1 + x + \frac{3}{8}x^2 + \frac{1}{16}x^3 \div \frac{1}{256}x^4$, 1·104.

5. $64 - 192x + 240x^2 - 160x^3 + 60x^4 - 12x^5 + x^6$,
63·616 96, 5.

6. $1 + x + \frac{7}{16}x^2 + \frac{7}{64}x^3$, 1·104.

242. 7. 65·944.　8. 0·904.

Exercise 11b

246. 1. (i) $448x^5$, (ii) $1080u^3$, (iii) $- 3168t^7$, (iv) $1320x^2y^3$.

2. (i) $84x^3$, (ii) $- 14\,080x^3$, (iii) $945x^4$, (iv) $190x^2$.

3. (i) $\frac{105}{512}$, (ii) 540, (iii) 6048, (iv) 1386.

4. (i) 120, (ii) $- 9120$, (iii) 4320, (iv) 5670.

247. 5. (i) $15x^2$, (ii) 20.　6. (i) 70, (ii) $3\frac{3}{4}$.

7. (i) 6, (ii) 14, (iii) $- 16$.　8. $3/5x$.　9. $8/45x$.

10. $b(r + 1)/a(n - r)$.

11.　(i) $1 + 10x + 45x^2 + 120x^3$,

(ii) $1 + \frac{9}{2}x + 9x^2 + \frac{21}{2}x^3$,

(iii) $1 - 11x + 55x^2 - 165x^3$,

(iv) $1 + 12x + 66x^2 + 220x^3$,

(v) $256 + 512x + 448x^2 + 224x^3$,

(vi) $128 - 224x + 168x^2 - 70x^3$.

12. (i) 1·105, (ii) 1029·13, (iii) 0·965, (iv) 253·96.

13.　(i) $1 + 3x + 6x^2 + 7x^3$,

(ii) $1 + 12x + 54x^2 + 100x^3$,

(iii) $1 - 4x + 2x^2 + 8x^3$,

(iv) $32 + 80x + 160x^2 + 200x^3$,

(v) $1 - 8x + 36x^2 - 112x^3$,

(vi) $128 + 448x - 224x^2 - 2128x^3$,

(vii) $81 - 216x + 324x^2 - 312x^3$,

(viii) $81 + 108x + 54x^2 + 120x^3$.

Exercise 11c

252. 1.　(i) $1 - 2x + 3x^2 - 4x^3$, $-1 < x < 1$;

(ii) $1 + \frac{1}{3}x - \frac{1}{9}x^2 + \frac{5}{81}x^3$, $-1 < x < 1$;

(iii) $1 + \frac{3}{4}x + \frac{3}{8}x^2 - \frac{1}{16}x^3$, $-1 < x < 1$;

(iv) $1 - x - \frac{1}{2}x^2 - \frac{1}{2}x^3$, $-\frac{1}{2} < x < \frac{1}{2}$;

(v) $1 - \frac{3}{2}x + \frac{3}{2}x^2 - \frac{5}{4}x^3$, $-2 < x < 2$;

page
252.

(vi) $1 + \frac{2}{3}x + \frac{17}{6}x^2 + \frac{125}{16}x^3$, $-\frac{1}{3} < x < \frac{1}{3}$;

(vii) $1 - 3x + 9x^2 - 27x^3$, $-\frac{1}{3} < x < \frac{1}{3}$;

(viii) $1 - \frac{1}{2}x^2$, $-1 < x < 1$;

(ix) $1 - \frac{1}{3}x - \frac{1}{9}x^2 - \frac{5}{81}x^3$, $-1 < x < 1$;

(x) $1 - x + \frac{3}{2}x^2 - \frac{5}{2}x^3$, $-\frac{1}{2} < x < \frac{1}{2}$;

(xi) $1 - x + \frac{3}{4}x^2 - \frac{1}{2}x^3$, $-2 < x < 2$;

(xii) $1 - 3x + \frac{9}{2}x^2 + \frac{9}{2}x^3$, $-\frac{1}{3} < x < \frac{1}{3}$;

(xiii) $\frac{1}{2} - \frac{1}{4}x + \frac{1}{8}x^2 - \frac{1}{16}x^3$, $-2 < x < 2$;

(xiv) $\sqrt{2}(1 - \frac{1}{4}x - \frac{1}{32}x^2 - \frac{1}{128}x^3)$, $-2 < x < 2$;

(xv) $\sqrt[3]{3}(1 + \frac{1}{9}x - \frac{1}{81}x^2 + \frac{5}{2187}x^3)$, $-3 < x < 3$;

(xvi) $\frac{1}{2}\sqrt{2}(1 - \frac{1}{4}x^2)$, $-\sqrt{2} < x < \sqrt{2}$;

(xvii) $\frac{1}{9} + \frac{2}{27}x + \frac{1}{27}x^2 + \frac{4}{243}x^3$, $-3 < x < 3$;

(xviii) $\sqrt[3]{9}(1 + \frac{1}{9}x^3)$, $-\sqrt[3]{3} < x < \sqrt[3]{3}$.

2. (i) $1 \cdot 000\ 500$, (ii) $0 \cdot 9612$, (iii) $0 \cdot 998\ 999$, (iv) $1 \cdot 0000$, (v) $1 \cdot 0102$.

3. (i) $1 + 2x + 2x^2 + 2x^3$, (ii) $2 - 3x + 4x^2 - 5x^3$, (iii) $1 - \frac{3}{2}x + \frac{1}{8}x^2 - \frac{11}{16}x^3$, (iv) $1 + x + \frac{1}{2}x^2 + \frac{1}{2}x^3$, (v) $-\frac{3}{2} + \frac{7}{4}x - \frac{7}{8}x^2 + \frac{7}{16}x^3$, (vi) $1 - 2x + \frac{3}{2}x^2 - x^3$, (vii) $3 + 4x + 7x^2 + 16x^3$.

253. **4.** $1 - 4x - 8x^2 - 32x^3$, $4 \cdot 7958$.

5. $1 - \frac{1}{3}x - \frac{1}{9}x^2 - \frac{5}{81}x^3$, $3 \cdot 332\ 22$.

6. $1 - 4v - 24v^2 - 224v^3$, $2 \cdot 499\ 00$, six.

Exercise 11d

1. $252(3x)^5(2y)^5$, 252.

2. (i) $32x^5 + 40x^3 + 20x + \frac{5}{x} + \frac{5}{8x^3} + \frac{1}{32x^5}$

(ii) $40\sqrt{6}$.

3. $a^5 - 5a^4b + 10a^3b^2 - 10a^2b^3 + 5ab^4 - b^5$, $77\ 400$.

4. $1024 - 640x + 180x^2$.

5. (i) $a^{11} + 11a^{10}b + 55a^9b^2 + 165a^8b^3$, (ii) $8064x^6y^5$, (iii) 5376.

6. $x^5 + 10x^4 + 40x^3 + 80x^2 + 80x + 32$, $x^4 - 8x^3 + 24x^2 - 32x + 16$, 96.

7. $4 - 28x + 85x^2 - 146x^3 + 155x^4$.

8. (i) $16 + 96x + 216x^2 + 216x^3 + 81x^4$, (ii) $1 + 12x + 78x^2 + 340x^3$.

page
254. **9.** (i) $1 - 5x + 20x^2 - 50x^3$,
    (ii) $1 - 4x + 10x^2 - 20x^3$.

10. (i) $1 - 3x + 6x^2 - 10x^3 + 15x^4$,
    (ii) $1024 + 1280x + 720x^2 + 240x^3$, 1159.

11. (i) $70 \ (2x)^4 3^4$, $\frac{3.5}{8}$, (ii) $1 + 4x + 12x^2 + 32x^3$.

12. (i) 2520, (ii) $1 + \frac{1}{3}x - \frac{1}{9}x^2$, 2·080.

13. $1 + 4x - 8x^2 + 32x^3$, 1·732 05.

14. $1 + 8x - 64x^2$, 1·259 92.

15. $1 + \frac{1}{3}x - \frac{1}{9}x^2 + \frac{5}{81}x^3 - \frac{10}{243}x^4$.

16. $1 + x - 2x^2 + 6x^3$. **17.** $a = 2, b = \frac{3}{16}$.

255. 18. $3 - 5x + 7x^2 - 9x^3$.

CHAPTER 12

259. **Qu. 1.** (i) $\sin 10°$, (ii) $-\tan 60°$, (iii) $-\cos 20°$,
    (iv) $-\sin 50°$, (v) $\cos 20°$, (vi) $-\sin 35°$,
    (vii) $\tan 40°$, (viii) $-\cos 16°$, (ix) $-\csc 50°$,
    (x) $-\tan 37°$, (xi) $-\cos 50°$, (xii) $-\sin 70°$,
    (xiii) $-\tan 50°$, (xiv) $\cot 20°$, (xv) $\cos 67°$,
    (xvi) $\sin 50°$, (xvii) $-\sec 38°$, (xviii) $-\cot 24°$,
    (xix) $-\csc 53°$, (xx) $-\sec 8°$.

261. **Qu. 3.** 360°, 180°.
    Qu. 4. (i) $\frac{1}{2}$, (ii) $\sqrt{3}/2$, (iii) $1/\sqrt{2}$, (iv) $1/\sqrt{3}$, (v) 2,
    (vi) $2/\sqrt{3}$, (vii) 1, (viii) $\sqrt{2}$.

Exercise 12a

264. **1.** (i) 0, (ii) 0, (iii) -1, (iv) -1, (v) $\frac{1}{2}$, (vi) $-\sqrt{3}/2$,
    (vii) $-\sqrt{3}$, (viii) $\sqrt{3}/2$, (ix) $-\sqrt{3}/2$, (x) $1/\sqrt{2}$,
    (xi) $-1/\sqrt{2}$, (xii) $-1/\sqrt{2}$, (xiii) $-\sqrt{3}$, (xiv) 1,
    (xv) $1/\sqrt{3}$.

3. 360°. **4.** 180°.

5. (i) 180°, (ii) 720°, (iii) 240°, (iv) 360°, (v) 360°.

6. (i) $\pm 120°$; (ii) 45°, $-135°$; (iii) 30°, 150°;
    (iv) $-50°$, $-130°$; (v) $\pm 53° 8'$; (vi) 120°, $-60°$;
    (vii) $-60°$, $-120°$; (viii) $\pm 180°$; (ix) $\pm 111° 21'$;
    (x) $\pm 50°$; (xi) 0°, $-120°$; (xii) 90°, 150°;
    (xiii) 6° 34', $-173° 26'$; (xiv) 118° 28', $-38° 28'$.

264. **7.**   (i) 30°, 150°, 210°, 330°;

(ii) 30°, 150°, 210°, 330°;

(iii) 15°, 75°, 195°, 255°;

(iv) $67\frac{1}{2}$°, $157\frac{1}{2}$°, $247\frac{1}{2}$°, $337\frac{1}{2}$°;

(v) 10°, 110°, 130°, 230°, 250°, 350°;

(vi) 90°, 210°, 330°;

(vii) 45°, 135°, 225°, 315°;

(viii) 35° 16′, 144° 44′, 215° 16′, 324° 44′;

(ix) 15°, 45°, 75°, . . . 345°;

(x) 37° 46′, 142° 14′, 217° 46′, 322° 14′.

265.   (xi) 11° 34′, 48° 26′, 191° 34′, 228° 26′;

(xii) 23° 51′, 83° 51′, 143° 51′, 203° 51′, 263° 51′, 323° 51′;

(xiii) 131° 40′;

(xiv) 84° 44′, 155° 16′, 264° 44′, 335° 16′.

8. (i) — 180°, — 45°, 0°, 135°, 180°; (ii) ± 60°, ± 90°;

(iii) 0°, ± 180°, — 19° 28′, — 160° 32′;

(iv) — 150°, — 30°, 90°; (v) ± 120°, ± 180°;

(vi) ± 60°, ± 90°, ± 120°; (vii) 0°, ± 180°;

(viii) ± 45°, ± 135°; (ix) ± 90°, 11° 32′, 168° 28′;

(x) + 40° 54′, ± 139° 16′;

(xi) ± 90°, 41° 49′, 138° 11′;

(xii) — 104° 02′, — 45°, 75° 58′, 135°;

(xiii) 23° 35′, 156° 25′; (xiv) ± 109° 28′.

9. (Maxima first), (i) 1, 90°; — 1, 270°.

(ii) 3, 0°; — 3, 180°.   (iii) 2, 0°; — 2, 360°.

(iv) $\frac{1}{2}$, 135°; $\frac{1}{2}$, 45°.   (v) 3, 270°;   1, 90°.

(vi) 5, 0°; 1, 60°.   (vii) 1, 270°; $\frac{1}{3}$, 90°

(viii) 1, 0°; $\frac{1}{7}$, 180°.   (ix) — 1, 120°; 1, 0°.

(x) No max.; 0, 0°.   (xi) $\frac{1}{2}$, 90°; no min.

(xii) None.   (xiii) None.

10. (i), (iii), (iv), (v), (vii).

267. Qu. 5. (i) cot θ, (ii) cosec θ, (iii) — cosec θ, (iv) — tan θ,

(v) sec θ, (vi) — cosec θ, (vii) — sin θ, (viii) sin θ,

(ix) — tan θ, (x) — cos θ, (xi) — cos θ,

(xii) cosec θ.

Exercise 12b

page
270. 1. (i) $\cos \theta$, (ii) $\tan \theta$, (iii) $\cos \theta \cot \theta$.

2. (i) $\sin \theta$, (ii) $\tan \theta$, (iii) $\operatorname{cosec} \theta \cot \theta$.

271. 3. (i) $\sec \theta$, (ii) $\sec^2 \theta \tan \theta$, (iii) $\sin \theta$.

4. (i) $\cot \theta$, (ii) $\cos \theta$, (iii) $\operatorname{cosec} \theta \tan^2 \theta$.

5. (i) $a^2 \cos^2 \theta$, (ii) $\dfrac{1}{a} \sec \theta$, (iii) $a \cos \theta \cot \theta$.

6. (i) $b^2 \operatorname{cosec}^2 \theta$, (ii) $b^2 \cot \theta \operatorname{cosec} \theta$, (iii) $\dfrac{1}{b} \sin \theta \cos \theta$.

7. (i) $a^2 \tan^2 \theta$, (ii) $\dfrac{1}{a} \cot \theta$, (iii) $\sin \theta$.

8. $0°, 60°, 300°, 360°$.　9. $270°$.

10. $45°, 63° 26', 225°, 243° 26'$.

11. $26° 34', 135°, 206° 34', 315°$.　12. $60°, 300°$.

13. $30°, 41° 49', 138° 11', 150°$　14. (i) $\pm \frac{4}{5}$, (ii) $\pm \frac{3}{4}$.

15. $\frac{15}{17}, -\frac{8}{15}$.　16. (i) $-\frac{25}{24}, -\frac{7}{25}$.

272. 37. $\cos \theta = \dfrac{2u}{u^2 + 1}$, $\sin \theta = \dfrac{u^2 - 1}{u^2 + 1}$, $\tan \theta = \dfrac{u^2 - 1}{2u}$.

38. $\dfrac{x^2}{a^2} + \dfrac{y^2}{b^2} = 1$.　　39. $\dfrac{y^2}{b^2} - \dfrac{x^2}{a^2} = 1$.

40. $\dfrac{b^2}{y^2} - \dfrac{x^2}{a^2} = 1$.　　41. $(x - 1)^2 + (y - 1)^2 = 1$.

42. $x(y - b) = ac$.　　43. $\dfrac{a^2}{x^2} + \dfrac{b^2}{y^2} = 1$.

44. $y^2(x - 1)^2 + y^2 = 1$.　45. $x^2 + y^2 = 2$.

46. $xy = 1$.

273. 47. $\dfrac{4}{(x + y)^2} - \dfrac{4}{(x - y)^2} = 1$.

Exercise 12c

1. (i) $-\cos 25°$, (ii) $-\tan 27°$, (iii) $\sec 51°$;
(iv) $\sin 35°$, (v) $\cot 46°$, (vi) $-\operatorname{cosec} 36°$.

2. (i) -1, (ii) $-\frac{1}{2}\sqrt{3}$, (iii) $\sqrt{3}$, (iv) $\frac{1}{2}\sqrt{2}$, (v) $-\frac{2}{3}\sqrt{3}$,
(vi) $-\sqrt{2}$, (vii) -1, (viii) $\frac{1}{2}$, (ix) $\frac{1}{2}$.

page
273. **3.** (i) 30°, 150°; (ii) 135°, 315°; (iii) 36° 52′, 323° 08′;
(iv) 22½°, 112½°, 202½°, 292½°;
(v) 37° 46′, 142° 14′, 217° 46′, 322° 14′;
(vi) 60°, 300°; (vii) 80° 32 ′, 299° 28′;
(viii) 14° 26′, 105° 34′; (ix) 96° 02′.

4. (i) 0°, ± 180°; − 30°, − 150°.
(ii) ± 90°; − 123° 41′, 56° 19′.   (iii) 30°, 150°; 90°.
(iv) ± 131° 49′.   (v) 30°, 150°.
(vi) ± 66° 25′, ± 120°.
(vii) 45°, − 135°; 63° 26′, − 116° 34′.
(viii) ± 60°; − 23° 35′, − 156° 25′.

5. (i) max. 5, 90°; min. 1, 270°.
(ii) max. 4, 180°; min. − 2, 0°.
(iii) max. 4, 60°; min. − 4, 180°.
(iv) max. 3, 180°; min. 0, 0°.
(v) max. − 1, 180°; min. $\frac{1}{3}$, 0°.
(vi) max. 1, 45°; min. $\frac{1}{2}$, 135°.

6. (i) $\tan \theta$, (ii) $\cos \theta$, (iii) $\sin \theta$, (iv) $- \cot \theta$.

274.　(v) $- \operatorname{cosec} \theta$, (vi) $- \sec \theta$, (vii) $- \sin \theta$,
(viii) $- \tan \theta$, (ix) $\sin \theta$.

7. (i) $\cot^2 \theta$, (ii) $\sin \theta$, (iii) $- \operatorname{cosec} \theta$, (iv) 1, (v) 1,
(vi) $\sec \theta \operatorname{cosec} \theta$.

8. (i) 90°; 210°, 330°.   (ii) 41° 25′, 318° 35′.
(iii) 0°, 360°; 131° 49′, 228° 11′.
(iv) 23° 35′, 156° 25′; 16° 36′, 163° 24′.
(v) 60°, 300°.   (vi) 56° 19′, 236° 19′.
(vii) 53° 08′, 135°, 315°, 233° 08′.

9. (i) $\frac{3}{5}$, $\frac{4}{5}$; (ii) − $\frac{12}{13}$, $\frac{5}{13}$; (iii) $\frac{8}{17}$, $\frac{5}{13}$; (iv) − $\frac{21}{29}$, − $\frac{13}{20}$.

275. **18.** $b^2x^2 - a^2y^2 = a^2b^2$.   **19.** $(x - 1)^2 + (y - 1)^2 = 1$.
20. $a^2b^2 - x^2y^2 = a^2y^2$.   **21.** $x^2y^2 - a^2b^2 = a^2y^2$.
22. $b^2x^2 - a^2y^2 = x^2y^2$.   **23.** $xy = 1$.
24. $(y + 1)^2 = x^2(1 + y^2)$. **25.** $y^2(1 + x) = 1 - x$.
26. max. $\sqrt{2}$, 45°; min. − $\sqrt{2}$, − 135°.
27. − 36° 52′, 90°.   **28.** 18°.
29. $x = 60°$, $y = 75°$, 345°; $x = 120°$, $y = 15°$, 285°;
$x = 240°$, $y = 165°$, 255°; $x = 300°$, $y = 105°$, 195°.

Exercise 13a

281. **1.** (i) $\frac{1}{4}\sqrt{2}(\sqrt{3}+1)$,   (ii) $\frac{1}{4}\sqrt{2}(\sqrt{3}+1)$,

(iii) $\frac{1}{4}\sqrt{2}(\sqrt{3}+1)$,   (iv) $\frac{1}{4}\sqrt{2}(1-\sqrt{3})$,

(v) $-\frac{1}{4}\sqrt{2}(\sqrt{3}+1)$,   (vi) $\frac{1}{4}\sqrt{2}(\sqrt{3}-1)$,

(vii) $\frac{1}{4}\sqrt{2}(\sqrt{3}-1)$,   (viii) $\frac{1}{4}\sqrt{2}(\sqrt{3}-1)$.

2. (i) $\frac{56}{65}$, (ii) $\frac{33}{65}$, (iii) $\frac{33}{56}$.   **3.** (i) $\frac{63}{65}$, (ii) $-\frac{63}{16}$, (iii) $-\frac{33}{56}$.

4. (i) $\frac{56}{65}$, (ii) $\frac{56}{33}$, (iii) $-\frac{63}{65}$.   **5.** $\frac{1}{3}$.   **6.** -2.

7. $45°$.   **8.** $135°$.

9. (i) $\cos(x+60°)=\sin(30°-x)$,

(ii) $\cos(45°-x)=\sin(45°+x)$.

282. (iii) $\tan(x+60°)$, (iv) $\sin 26°$, (v) $\sec 39°$,

(vi) $\cos 15°=\sin 105°=\sin 75°$.

10. (i) $\frac{1}{2}$, (ii) $\frac{1}{2}$, (iii) $\frac{1}{3}\sqrt{3}$, (iv) 0, (v) $\frac{1}{2}$, (vi) $\frac{1}{2}\sqrt{2}$,

(vii) $\frac{1}{3}\sqrt{3}$, (viii) $\frac{1}{2}\sqrt{6}$.

11. 2.   **12.** $\frac{12}{31}$.   **13.** (i) $\frac{1}{3}$, (ii) 1, (iii) $-\frac{7}{4}$, (iv) $2-\sqrt{3}$.

16. (i) $9°\,54'$, $189°\,54'$; (ii) $157\frac{1}{2}°$, $337\frac{1}{2}°$.

283. (iii) $49°\,06'$, $229°\,06'$; (iv) $56°\,32'$, $236°\,32'$.

284. **41.** $\dfrac{\tan\theta+2\tan\phi-\tan\theta\tan^2\phi}{1-\tan^2\phi-2\tan\theta\tan\phi}$.   **43.** $70°\,12'$, $10°\,12'$.

Exercise 13b

287. **1.** $\sin 34°$.   **2.** $\tan 60°$. ·   **3.** $\cos 84°$.

4. $\sin\theta$.   **5.** $\cos 45°$.   **6.** $\tan\theta$.

7. $\cos 30°$.   **8.** $\sin 4A$.   **9.** $\cos\theta$.

10. $\cos 6\theta$.   **11.** $\frac{1}{2}\tan 4\theta$.   **12.** $\frac{1}{2}\sin 2x$.

13. $2\cot 40°$.   **14.** $2\operatorname{cosec} 2\theta$.   **15.** $\cos\theta$.

Exercise 13c

1. (i) $\frac{1}{2}$, (ii) 1, (iii) $-\frac{1}{2}\sqrt{3}$, (iv) $-\frac{1}{2}\sqrt{2}$, (v) $\frac{1}{2}\sqrt{2}$,

(vi) $2\sqrt{3}$, (vii) 1, (viii) $2\sqrt{2}$.

2. (i) $\pm\frac{24}{25}$, $\frac{7}{25}$; (ii) $\pm\frac{120}{169}$, $\frac{119}{169}$; (iii) $\pm\frac{1}{2}\sqrt{3}$, $-\frac{1}{2}$.

3. (i) $-\frac{24}{7}$, (ii) $\frac{240}{161}$, (iii) $\pm\frac{120}{119}$.

288. **4.** (i) $\pm\frac{3}{4}$, $\pm\frac{1}{4}\sqrt{7}$; (ii) $\pm\frac{4}{5}$, $\pm\frac{3}{5}$; (iii) $\pm\frac{12}{13}$, $\pm\frac{5}{13}$.

5. (i) $\frac{1}{3}$, -3; (ii) $\frac{1}{2}$, -2; (iii) $-\frac{2}{3}$, $\frac{3}{2}$.   **6.** $\sqrt{2}-1$.

7. $90°$, $120°$, $240°$, $270°$.   **8.** $0°$, $180°$, $360°$; $60°$, $300°$.

9. $30°$, $150°$; $270°$.   **10.** $56°\,27'$, $123°\,33'$; $270°$.

page

288. 11. 30°, 150°; 90°, 270°.

12. 0°, 180°, 360°; 85° 13′, 274° 47′.

13. 0°, 180°, 360°; 120°, 240°; 36° 52′, 323° 08′.

14. 0°, 180°, 360°; 30°, 150°, 210°, 330°.

15. 45°, 225°; 120° 58′, 300° 58′.

16. 18° 26′, 161° 34′, 198° 26′, 341° 34′.

17. (i) $y = 2x^2 - 1$,   (ii) $2y = 3(2 - x^2)$,

     (iii) $y(1 - x^2) = 2x$,   (iv) $x^2y = 8 - x^2$.

290. 46. (i) $\dfrac{(1 + t)^2}{1 + t^2}$,   (ii) $\dfrac{2}{1 + t^2}$,

     (iii) $3 + 6t - 5t^2$,   (iv) $\dfrac{1 - t}{1 + t}$,

     (v) $\dfrac{2(2 - t^2)}{1 - 4t + t^2}$,   (vi) $\dfrac{1 - t^2}{2t^2}$.

47. (i) 0°, 112° 37′, 360°;   (ii) 53° 8′, 323° 08′.

     (iii) 73° 44′, 180°;   (iv) 119° 33′, 346° 43′.

Exercise 13d

292. 1. 90°, 330°.   2. 94° 52′, 219° 54′.

3. 114° 18′, 335° 42′.   4. 204° 31′, 351° 42′.

5. 72° 37′, 319° 17′.   6. 76° 43′, 209° 33′.

7. 28° 07′, 208° 07′; 159° 28′, 339° 28′.

8. 0°, 180°, 360°; 45°, 225°.

10. max. 2, 330°; min. − 2, 150°.

11. max. $\sqrt{13}$, 33° 41′; min. − $\sqrt{13}$, − 146° 19′.

12. 5, 53° 08′.

293. 13. max. $\sqrt{5}$, 63° 26′; min. − $\sqrt{5}$, − 116° 34′.

14. $\sqrt{2}$, 45°; − $\sqrt{2}$, 225°.

15. 5, 126° 52′; − 5, 306° 52′.

16. 2, 60°; − 2, 240°.

17. 17, 298° 04′; − 17, 118° 04′.

18. $\sqrt{37}$, 170° 32′; − $\sqrt{37}$, 350° 32′.

19. 1, 240°; − 1, 60°.

20. 5, 53° 08′; − 5, 233° 08′.

21. − $\frac{1}{2}\sqrt{2}$, 135°; $\frac{1}{2}\sqrt{2}$, 315°.

22. 13, 56° 19′, 236° 19′; 0, 146° 19′, 326° 19′.

23. min. $\frac{1}{5}$, 153° 26′, 333° 26′.

Exercise 13e

294. 1. (i) $\frac{140}{221}$, (ii) $-\frac{21}{221}$, (iii) $\frac{171}{140}$.

2. (i) $\frac{468}{493}$, (ii) $-\frac{475}{493}$, (iii) $\frac{475}{132}$.

3. (i) $\frac{1}{2}$, (ii) 1, (iii) 1.

4. (i) $\pm\frac{2}{3}$, $\pm\frac{1}{3}\sqrt{5}$; (ii) $\pm\frac{4}{9}$, $\pm\frac{1}{9}\sqrt{65}$.

5. (i) $\frac{5}{2}$, $-\frac{2}{5}$; (ii) $\frac{2}{9}$, $-\frac{9}{2}$. 6. (i) $\frac{840}{1369}$, (ii) $-\frac{1081}{1369}$.

8. $20°\ 06'$, $200°\ 06'$. 9. $79°\ 12'$, $259°\ 12'$.

10. $67°\ 31'$, $247°\ 31'$.

295. 11. $60°$, $300°$. 12. $0°$, $180°$, $360°$; $41°\ 25'$, $318°\ 35'$.

13. $90°$, $270°$. 14. $0°$, $180°$, $360°$; $60°$, $120°$, $240°$, $300°$.

15. $41°\ 36'$, $244°\ 40'$. 16. $79°\ 47'$, $347°\ 35'$.

17. $252°\ 54'$, $332°\ 20'$. 18. $9x = 4y^2 - 18$.

19. $y(4 - x^2) = 4x$. 20. $x^2 - 1 = 2xy$.

21. $4x^3 = 27(x + y)$. 22. $y^2 = (1 - x^2)(1 - 4x^2)^2$.

23. $2(t + 2)^2/(1 + t^2)$. 24. $(1 + t)/(1 - t)$.

25. $(3t - 7)^2$. 26. 13, $292°\ 37'$; -13, $112°\ 37'$.

27. 37, $71°\ 05'$; -37, $251°\ 05'$.

28. 73, $311°\ 07'$; -73, $131°\ 07'$.

296. 47. $\cos A = r/q$, $\cos B = r/p$.

Exercise 14a

297. 1. $\cos(x + y) - \cos(x - y)$.

2. $\cos(x + y) + \cos(x - y)$.

3. $\cos 4\theta + \cos 2\theta$. 4. $\cos 2S - \cos 2T$.

5. $\cos 2x - \cos 8x$. 6. $\cos 2x + \cos 2y$.

7. $\cos A + \cos B$. 8. $\cos B - \cos C$.

9. $\cos 2x$. 10. $\cos 4x + \cos 60°$.

Exercise 14b

298. 1. $\sin(x + y) + \sin(x - y)$.

2. $\sin(x + y) - \sin(x - y)$.

3. $\sin 4\theta + \sin 2\theta$. 4. $\sin 2S + \sin 2T$.

5. $\sin 8x - \sin 2x$. 6. $\sin 2x - \sin 2y$.

7. $\sin 2x - \sin 6x$. 8. $\sin A + \sin B$.

9. $\sin A - \sin B$. 10. $\sin R - \sin S$.

Exercise 14c

302

1. $2 \cos \frac{1}{2}(x + y) \cos \frac{1}{2}(x - y)$. 2. $2 \sin 4x \cos x$.

3. $2 \cos (y + z) \sin (y - z)$. 4. $2 \cos 6x \cos x$.

5. $- 2 \sin \frac{3}{4}A \sin \frac{1}{4}A$. 6. $2 \cos 3x \sin x$.

7. $2 \sin 4A \sin A$. 8. $2 \sin 6\theta \cos \theta$.

9. $\sqrt{3} \sin x$. 10. $\sqrt{2} \cos (y - 35°)$.

11. $- 2 \cos 4\theta \sin \theta$. 12. $- \sin x$.

13. $- 2 \sin x \sin \frac{1}{2}x$. 14. $2 \sin 2x \cos 80°$.

15. $2 \cos (45° - \frac{1}{2}x + \frac{1}{2}y) \cos (45° - \frac{1}{2}x - \frac{1}{2}y)$.

16. $2 \cos (45° - \frac{1}{2}A + \frac{1}{2}B) \cos (45° - \frac{1}{2}A - \frac{1}{2}B)$.

17. $2 \sin (\frac{3}{2}x + 45°) \cos (\frac{3}{2}x - 45°)$.

18. $2 \sin (x + 45°) \cos (x - 45°)$.

19. $2 \cos (45° - \frac{1}{2}A + \frac{1}{2}B) \sin (45° - \frac{1}{2}A - \frac{1}{2}B)$.

20. $2 \cos (30° + \theta) \cos (30° - \theta)$.

Exercise 14d

303

14. 30°, 90°, 150°, 210°, 270°, 330°; 45°, 135°, 225°, 315°.

15. 0°, 120°, 240°, 360°; 72°, 144°, 216°, 288°.

16. 0°, 180°, 360°; 45°, 135°, 225°, 315°.

17. 0°, 72°, 144°, 180°, 216°, 288°, 360°.

18. 175°, 355°. 19. 45°, 135°, 225°, 315°.

20. 25°, 205°. 21. 280°; 68°, 140°, 212°, 284°, 356°.

22. 0°, 180°, 360°. 23. 0°, 360°.

24. 54°, 126°, 198°, 270°, 342°.

25. 45°, 225°; 67½°, 157½°, 247½°, 337½°.

26. 30°, 150°. 27. 300°.

28. 112½°, 157½°, 292½°, 337½°. 29. 290°, 350°.

30. 15°, 75°, 195°, 255°. 31. 0°, 180°, 360°.

32. 8° 15′, 66° 45′, 188° 15′, 246° 45′.

33. 3° 03′, 61° 57′, 93° 03′, 151° 57′, 183° 03′,
241° 57′, 273° 03′, 331° 57′.

Exercise 14e

1. $\sin A$. 2. $- \cos (U + A)$. 3. $- \cos B$.

4. $- \tan (B + C)$. 5. $- \cot (A + B)$.

6. $\cos \frac{1}{2}(B + C)$. 7. $- \sin (2B + 2C)$. 8. $\sin \frac{1}{2}A$.

9. $\cot \frac{1}{2}(A + B)$. 10. $\cos 2C$. 11. $\sin \frac{1}{2}(A + C)$.

12. $- \sin 2A$. 13. $\operatorname{cosec} \frac{1}{2}B$. 14. $\tan \frac{1}{2}(B + C)$.

15. $\sin (3B + 3C)$.

Exercise 14f

306.
1. 60°, 120°; 30°, 90°, 150°.
2. 0°, 45°, 90°, 135°, 180°; 60°, 120°.
3. 0°, 90°, 180°.   4. 60°, 180°.
5. 45°, 135°; 30°, 150°.   6. 0°, 72°,144°; 90°; 180°.
7. $22\frac{1}{2}°$, $67\frac{1}{2}°$, $112\frac{1}{2}°$, $157\frac{1}{2}°$; 45°, 135°; 90°.

Exercise 15a

313.
1. (i) $A = 48°$, $b = 13\cdot84$, $c = 15\cdot44$.
    (ii) $B = 56°\ 08'$, $a = 6\cdot533$, $c = 5\cdot032$.
   (iii) $C = 45°\ 06'$, $a = 231\cdot1$, $b = 212\cdot6$.

314.
2. (i) $B = 95°$, $a = 1\cdot398$, $c = 1\cdot795$.
    (ii) $B = 19°\ 40'$, $b = 4\cdot625$, $c = 8\cdot297$.
   (iii) $A = 32°\ 45'$, $b = 243\cdot8$, $c = 171\cdot7$.

3. (i) $B = 59°\ 07'$, $A = 72°\ 38'$, $a = 19\cdot57$;
    or $B = 120°\ 53'$, $A = 10°\ 52'$, $a = 3\cdot866$.
    (ii) $C = 26°\ 41'$, $A = 24°\ 19'$, $a = 4\cdot181$.
   (iii) $B = 55°\ 30'$, $C = 96°\ 15'$, $c = 17\cdot85$;
    or $B = 124°\ 30'$, $C = 27°\ 15'$, $c = 8\cdot223$.
   (iv) $C = 72°\ 06'$, $B = 42°\ 40'$, $b = 109\cdot0$;
    or $C = 107°\ 54'$, $B = 6°\ 52'$, $b = 19\cdot22$.
    (v) $B = 6°\ 58'$, $C = 8°\ 51'$, $c = 3\cdot184$.

4. (i) $A = 38°\ 13'$, $B = 81°\ 47'$, $C = 60°$.
    (ii) $A = 54°\ 38'$, $B = 78°\ 08'$, $C = 47°\ 13'$.
   (iii) $A = 64°\ 09'$, $B = 43°\ 29'$, $C = 72°\ 21'$.

5. (i) $a = 13$, $B = 32°\ 12'$, $C = 87°\ 48'$.
    (ii) $b = 11\cdot74$, $A = 72°\ 16'$, $C = 54°\ 42'$.
   (iii) $c = 7\cdot596$, $A = 82°\ 35'$, $B = 54°\ 13'$.

6. (i) $A = 29°\ 32'$, $B = 38°\ 03'$, $C = 112°\ 25'$.
    (ii) $A = 17°\ 54'$, $B = 120°$, $C = 42°\ 06'$.
   (iii) $A = 35°\ 47'$, $B = 49°\ 18'$, $C = 94°\ 55'$.

7. (i) $A = 11°\ 38'$, $b = 73$, $C = 48°\ 22'$.
    (ii) $a = 17\cdot41$, $B = 33°\ 50'$, $C = 41°\ 55'$.
   (iii) $A = 31°\ 13'$, $B = 44°\ 37'$, $c = 57\cdot99$.

8. (i) $B = 80°\ 47'$, $C = 45°\ 13'$, $a = 1\cdot197$.
    (ii) $A = 59°\ 50'$, $B = 80°\ 10'$, $c = 1\cdot279$.
   (iii) $C = 31°\ 07'$, $A = 38°\ 53'$, $b = 15\cdot27$.
   (iv) $A = 26°\ 55'$, $B = 33°\ 05'$, $c = 6\cdot505$.
    (v) $B = 57°\ 15'$. $C = 63°\ 27'$, $a = 12\cdot88$.
   (vi) $A = 82°\ 18'$, $B = 62°\ 48'$, $c = 90\cdot06(5)$.

page

314. **9.** 1·434 km.   **10.** 25·8 m.

315. **11.** 1° 02½′.   **12.** N 12° 40′ W, 3·635 n.m.

   13. 200 m.   **14.** 11·42(5) cm.

   15. 5·439 n.m., N 73° 17′ W.

   16. 34·07 cm, 60° 37′.   **17.** 60°, 13.

Exercise 15b

319. **1.** (i) 2, (ii) 0·8660, (iii) 3·988.

   2. (i) $A = 46° 03′$, $B = 68° 56′$, $C = 65° 01′$;
       (ii) $A = 67° 20′$, $B = 60°$, $C = 52° 40′$;
      (iii) $A = 63° 09′$, $B = 60°$, $C = 56° 51′$;
      (iv) $A = 98° 43′$, $B = 46° 20′$, $C = 34° 57′$;
       (v) $A = 100° 14′$, $B = 27° 00′$, $C = 52° 46′$.

Exercise 15c

322. **1.** (i) 6·495, (ii) 72·36, (iii) 32·18, (iv) 43·82, (v) 76·26,
      (vi) 122·7, (vii) 32 600.

   2. 6000 m².   **3.** 4 cm, $2\sqrt{13}$ cm.

   4. 5 cm, 4 cm.   **5.** 10·95 cm².

   6. 144·3, 4·899, 36° 59′.

   7. $A = 51° 20′$, $B = 69° 51′$, $C = 58° 49′$.

323. **12.** $A = 50° 31′$, $C = 65° 17′$.

   13. $a = 13·98$, $B = 53° 29′$, $C = 62° 31′$.

   19. (i) $A = 64° 01′$, $B = 48° 51′$, $c = 23·98$;
       (ii) $C = 99° 25′$, $a = 9·537$, $b = 5·225$;
      (iii) $A = 33° 33′$, $B = 62° 11′$, $C = 84° 16′$;
      (iv) $A = 59° 55′$, $B = 54° 02′$, $C = 67° 03′$;
       (v) $B = 54° 50′$, $C = 46° 40′$, $a = 19·54$;
      (vi) $B = 38° 49′$, $C = 110° 02′$, $c = 9·008$;
       or $B = 141° 11′$, $C = 7° 40′$, $c = 1·279$;

324.   (vii) $A = 50° 10′$, $B = 69° 50′$, $c = 10·15$;
     (viii) $A = 15° 23′$, $C = 11° 19′$, $b = 28·62$;
      (ix) $A = 35° 04′$, $C = 14° 56′$, $c = 5·382$;
       (x) $A = 30° 29′$, $C = 37° 31′$, $b = 19·19$.

Exercise 15d

   1. (i) 5·385 cm, (ii) 21° 48′, (iii) 33° 41′.

   2. 17 m, 28° 04′, 33° 41′.

page
324. **3.** (i) 7 cm, (ii) 51° 03′, (iii) 63° 37′.
 4. (i) 70° 34′, (ii) 19 cm, (iii) 96° 22′.
 5. 17·8 m, 12·8 m. **6.** 70·8 m, 11° 55′.
325. **7.** 11·5 m, 16° 06′, 21° 03′. **8.** 15·59 cm., 31° 18′.
 9. 43·6 m, 10° 02′. **10.** 15° 11′. **11.** 35° 16′.
 12. 26° 28′, 44° 12′. **13.** 23·6 m. **14.** 10° 40′, 5° 23′.
326. **15.** 2° 43′. **16.** 25° 24′. **17.** 19° 44′.

Exercise 16a

329. **1.** (i) 90°, (ii) 45°, (iii) 60°, (iv) 120°,
  (v) 30°, (vi) 270°, (vii) 450°, (viii) 720°,
  (ix) 900°, (x) 240°, (xi) 630°, (xii) 135°.
 2. (i) 2π, (ii) $\frac{1}{2}\pi$, (iii) $\frac{1}{4}\pi$, (iv) $\frac{1}{12}\pi$,
  (v) $\frac{1}{3}\pi$, (vi) $\frac{2}{3}\pi$, (vii) $\frac{5}{3}\pi$, (viii) $\frac{3}{2}\pi$,
  (ix) 3π, (x) $\frac{1}{6}\pi$, (xi) $\frac{5}{6}\pi$, (xii) $\frac{5}{2}\pi$.
 3. 8 cm. **4.** 9·6 cm. **5.** 6 cm. **6.** $\frac{4}{5}$ rad.
 7. 3 cm². **8.** 4 rad. **9.** 12 cm. **10.** 4 cm².

Exercise 16b

 1. (i) $\frac{1}{8}\pi$, (ii) 6π, (iii) $\pi/900$, (iv) $5\pi/24$.
 2. (i) 72°. (ii) 5°, (iii) 105°, (iv) 630°.
330. **3.** 2·705 cm. **4.** $3/\pi$. **5.** 1·2 rad, 68° 45′. **6.** 6·434 cm.
 7. (i) 0·841(5), (ii) 0·296, (iii) − 2·185,
  (iv) − 0·544, (v) − 0·842, (iv) 1·220.
 8. 4·032 cm².
 9. (i) 150·80 cm², (ii) 62·35 cm², (iii) 88·44 cm².
 10. 24·14 cm². **11.** 22·37 cm².
 12. $\frac{1}{2}r^2(2\pi - \theta + \sin\theta)$. **13.** $\frac{1}{2}a^2(2\phi + \sin 2\phi)$.

Exercise 16c

333. **1.** 64° 52′, 1·132 rad. **2.** 42° 21′, 0·739 rad.
 3. 0°, 0 rad; 76° 44′, 1·339 rad.
 4. − 23° 29′, − 0·410 rad; 68° 11′, 1·190 rad.
 5. 0°, 0 rad; 31° 46′, 0·555 rad; 105° 57′, 1·849 rad.
 6. 15° 34′, 111° 18′; 0·272, 1·943.
 7. − 119° 33′, 13° 17′; − 2·087, 0·232.

page
333. **8.** 66° 47′, 1·166 rad.   **9.** 132° 21′, 2·310 rad.
        10. 144° 27′, 2·521 rad.

334. **Qu. 2.** (i) 6 deg/s, (ii) 1 rev/min.
      Qu. 3. (i) 3000 deg/s, (ii) $2\frac{1}{7}$ deg/h.

Exercise 16d

336. **1.** (i) $\frac{1}{60}$ rev/min, (ii) $\frac{1}{10}$ deg/s, (iii) $\pi/1800$ rad/s.
      2. (i) 1200 deg/s, (ii) $20\pi/3$ rad/s.
      3. (i) 1536 rev/min, (ii) 161 rad/s.
      4. 0·263 rad/h.
      5. (i) 100π rad/s, (ii) 17·0 cm/s.
      6. (i) 3·89 rad/s, (ii) 15·6 cm/s.
      7. (i) 40π rad/s, (ii) 1·57 m/s.

337. **8.** (i) 35·2 rad/s, (ii) 336 rev/min.
      9. 1600 rev/min. **10.** 128 rad/s.
      11. 1·23 m/s, 0·524 m/s.
      12. $1·99 \times 10^{-7}$ rad/s, 30 km/s. **13.** 4·8 km/h.

340. **Qu. 4.** (i) $1\frac{1}{2}$, (ii) 2, (iii) $\frac{1}{2}$, (iv) $\frac{1}{2}$, (v) sin a, (vi) cos a,
      (vii) $\frac{2}{9}$, (viii) 2, (ix) sec^2 a.

341. **Qu. 6.** $2 \cos \frac{1}{2}(A + B) \sin \frac{1}{2}(A - B)$.

343. **Qu. 9.** (i) $- 3 \sin 3x$, (ii) $2 \sin x \cos x = \sin 2x$,
      (iii) $4 \cos 2x$, (iv) $- 3 \cos^2 x \sin x$.

Exercise 16e

344. **1.** (i) $- 2 \sin 2x$, (ii) $6 \cos 6x$, (iii) $- 3 \sin (3x - 1)$,
      (iv) $2 \cos (2x - 3)$, (v) $15 \sin 5x$, (vi) $8 \cos 4x$,
      (vii) $- 6 \cos \frac{3}{2}x$, (viii) $\cos \frac{1}{2} (x + 1)$, (ix) $2x \cos x^2$.
      2. (i) $- \frac{1}{3} \cos 3x + c$,   (ii) $\frac{1}{3} \sin 3x + c$,
      (iii) $- \frac{1}{2} \cos 4x + c$,   (iv) $\sin 2x + c$,
      (v) $\frac{1}{12} \cos 6x + c$,   (vi) $\frac{9}{2} \sin 4x + c$,
      (vii) $- \frac{1}{3} \cos (2x + 1) + c$,
      (viii) $\frac{3}{2} \sin (2x - 1) + c$, (ix) $- \frac{4}{3} \cos \frac{1}{2}x + c$.
      3. (i) $2 \sin x \cos x = \sin 2x$,
      (ii) $- 8 \cos x \sin x = - 4 \sin 2x$,
      (iii) $- 3 \cos^2 x \sin x$,   (iv) $6 \sin^2 x \cos x$,

page

344.   (v) $-12 \cos^3 x \sin x$,   (vi) $\dfrac{\cos x}{2\sqrt{(\sin x)}}$,

     (vii) $\dfrac{-\sin x}{2\sqrt{(\cos x)}}$,

    (viii) $-6 \cos 3x \sin 3x = -3 \sin 6x$,

    (ix) $4 \sin 2x \cos 2x = 2 \sin 4x$,

    (x) $-18 \sin^2 3x \cos 3x$,

    (xi) $24 \sin^3 2x \cos 2x$,   (xii) $\dfrac{\cos 2x}{\sqrt{(\sin 2x)}}$.

   4. (i) $\cos x - x \sin x$,   (ii) $\sin 2x + 2x \cos 2x$,

    (iii) $x(2 \sin x + x \cos x)$,

    (iv) $\cos^2 x - \sin^2 x = \cos 2x$,

    (v) $(x \cos x - \sin x)/x^2$,

    (vi) $-(2x \sin 2x + \cos 2x)/x^2$,

    (vii) $(\sin x - x \cos x)/\sin^2 x$,

   (viii) $x(2 \cos x + x \sin x)/\cos^2 x$,

    (ix) $\sec^2 x$,   (x) $-\operatorname{cosec}^2 x$,

    (xi) $\sec x \tan x$,   (xii) $-\operatorname{cosec} x \cot x$.

   5. (i) 1 m, (ii) 2 m/s², (iii) 0·983 s.

   6. (i) $\frac{1}{3}\pi$ s, (ii) $-\frac{3}{2}\sqrt{3}$ cm/s, (iii) -5, $3\frac{3}{4}$, -3 cm/s².

345.   **7.** (i) 5, (ii) -20.   **8.** (i) 0·841, (ii) $\frac{5}{3}\sqrt{5}$, (iii) $-\frac{17}{3}\sqrt{5}$.

   9. $\frac{2}{3}$.   **10.** 2π.   **11.** $\frac{1}{3}\pi + \frac{1}{2}\sqrt{3}$, $\frac{1}{3}\pi^2 + 1$.

   12. $\sqrt{3} - \frac{1}{3}\pi$, $2 - \frac{1}{8}\pi^2$.   **14.** $\frac{1}{2}(1 + \cos 2x)$.

346. **17.** $\frac{1}{2}$.

Exercise 16f

347.   **1.** (i) $2 \sec^2 2x$,   (ii) $-3 \operatorname{cosec}^2 3x$,

    (iii) $6 \sec 2x \tan 2x$,   (iv) $-\operatorname{cosec} \frac{1}{2}x \cot \frac{1}{2}x$,

    (v) $-2 \sec^2 (2x + 1)$,

    (vi) $\sec (3x - 2) \tan (3x - 2)$,

    (vii) $6 \operatorname{cosec}^2 (3x + 2)$,   (viii) $-2x \operatorname{cosec}^2 x^2$,

    (ix) $(\sec^2 \sqrt{x})/2\sqrt{x}$.

   2. (i) $2 \tan x \sec^2 x$,   (ii) $2 \sec^2 x \tan x$,

    (iii) $-6 \cot^2 x \operatorname{cosec}^2 x$,   (iv) $-6 \operatorname{cosec}^2 x \cot x$,

    (v) $-4 \sec^2 2x \tan 2x$,

    (vi) $-3 \operatorname{cosec}^2 3x \cot 3x$,

    (vii) $\sec^3 2x \tan 2x$,   (viii) $8 \operatorname{cosec}^4 x \cot x$,

    (ix) $(\sec^2 x)/2\sqrt{(\tan x)}$.

page
347. **3.**   (i) $\tan x + x \sec^2 x$,   (ii) $\sec x (\sec^2 x + \tan^2 x)$,

(iii) $x(2 \cot x - x \csc^2 x)$,

(iv) $3 \csc x(1 - x \cot x)$,

(v) $- \csc x (\csc^2 x + \cot^2 x)$,

(vi) $(x \sec^2 x - \tan x)/x^2$,

(vii) $\sec x (x \tan x - 2)/x^3$,

(viii) $x \sin x$,   (ix) $2x \sec^2 x \tan x$.

348. **4.**   (i) $\tfrac{1}{2} \tan 2x + c$,   (ii) $3 \sec x + c$,

(iii) $2 \cot \tfrac{1}{2}x + c$,   (iv) $- \tfrac{1}{9} \csc 3x + c$,

(v) $\sec^2 x + c$, *or* $\tan^2 x + c$, (vi) $\tan x + c$,

(vii) $\sec x + c$,   (viii) $- \tfrac{1}{2} \cot 2x + c$,

(ix) $- \tfrac{1}{2} \csc 2x + c$.

5. $1 - \tfrac{1}{4}\pi$.   **6.** 2π.   **7.** (i) $2\sqrt{3}$, (ii) $5\sqrt{5}$, (iii) $5\sqrt{3}$.

8. (i) $1\cdot767$, (ii) $2\cdot007$, (iii) $0\cdot9884$, (iv) $0\cdot5735$.

9. $0\cdot9$ m.   **10.** $0\cdot02$ m.

12. $\cot^2 x = \csc^2 x - 1$, $- \cot x - x + c$.

Exercise 16g

349. **1.** (i) 72°, (ii) 150°, (iii) $67\tfrac{1}{2}^\circ$, (iv) 105°.

2. (i) $11\pi/6$, (ii) $5\pi/18$, (iii) $5\pi/12$, (iv) $2\pi/15$.

3. (i) $0\cdot909$, (ii) $1\cdot139$, (iii) $3\cdot903$, (iv) $- 0\cdot987$.

4. $10\cdot5$ cm.   **5.** $1\tfrac{3}{5}$ cm., 29° $50'$.   **7.** 171 cm^2.

8. 110° $51'$, $1\cdot9345$ rad.   **9.** 29° $30'$, $0\cdot515$ rad.

10. 60°, 72°, 144°, $\tfrac{1}{3}\pi$, $\tfrac{2}{5}\pi$, $\tfrac{4}{5}\pi$ rad.   **11.** π rad/s.

12. (i) $\tfrac{1}{720}$ rev/min, (ii) $\pi/21\,600$ rad/s.

13. 44 rad/s, 238 km/h.

350. **14.** $6\cdot9$ rad/s.   **15.** $42\cdot4$ rev/min.

16. (i) 1, (ii) $\tfrac{2}{3}$, (iii) $- \sin a$.

18.   (i) $3 \cos 3x$,   (ii) $\tfrac{1}{2} \sec^2 \tfrac{1}{2}x$,

(iii) $- 2x \sin x^2$,   (iv) $- (\sin x)/2\sqrt{(\cos x)}$,

(v) $- 6 \csc^3 x \cot x$,   (vi) $2 \sin x$,

(vii) $- 18 \sec^3 2x \tan 2x$,

(viii) $(\cos 2x)/\sqrt{(\sin 2x)}$,   (ix) $12 \tan 2x \sec^2 2x$.

19.   (i) $\tfrac{1}{2} \sin 2x + c$,   (ii) $- \tfrac{1}{2} \cos (2x - 1) + c$,

(iii) $6 \sin \tfrac{1}{2}x + c$,   (iv) $2 \tan \tfrac{1}{2}x + c$,

(v) $- \csc x + c$,   (vi) $\tfrac{1}{2} \sec 2x + c$,

(vii) $- \csc x + c$,   (viii) $\tfrac{1}{2} \tan 2x + c$,

(ix) $- \tfrac{1}{2} \cos x^2 + c$.

page

350. 20.   (i) $\sin x + x \cos x$,

      (ii) $\cos x \cos 2x - 2 \sin x \sin 2x$,

      (iii) $2x \tan x (\tan x + x \sec^2 x)$,

      (iv) $\sec x(x \tan x - 1)/x^2$,

      (v) $- (2 \sin 3x \sin 2x + 3 \cos 3x \cos 2x)/\sin^2 3x$,

      (vi) $\cos x \tan 2x + 2 \sin x \sec^2 2x$,

      (vii) $(x \cos x - 2 \sin x)/x^3$,

      (viii) $x^2 \sin x$.

351. 23. (i) 5, (ii) $- 20$, (iii) 10.   **24.** (i) $\frac{15}{8}\sqrt{15}$ cm/s,

    (ii) 14 cm.

    25. (i) 1, (ii) $\frac{4}{3}\sqrt{3}$, (iii) $\frac{1}{2}\pi$, (iv) $\frac{1}{4}$.

    26. (i) 0·4962; (ii) 0·7092.   **28.** $- \frac{3}{2}\sqrt{3}$.

    29. $- 5\sqrt{3}$, max.; $5\sqrt{3}$, min.

Exercise 17a

354. **1.** $x^2 + y^2 = 25$.

355. **2.** $x^2 + y^2 - 6x - 2y + 6 = 0$.

    3. $4x - 10y + 29 = 0$.   **4.** $5x - 3y - 4 = 0$.

    5. $x + 1, y^2 = 2x + 1$.   **6.** $x^2 = 4y$.

    7. $2x^2 + 2y^2 - x - 1 = 0$.

    8. $3x^2 + 3y^2 + 36x - 38y + 159 = 0$.

    9. $3x^2 - y^2 = 48$.   **10.** $3x^2 + 4y^2 = 48$.

    11. $x^2 + y^2 = 9$.   **12.** $y^2 = 4ax$.

    13. $3x^2 + 4y^2 = 12$.   **14.** $y = 0$.

    15. $2x + 3y - 13 = 0$.   **16.** $x^2 + y^2 = 1$.

Exercise 17b

358. **1.**   (i) $(y - 2)(y + 5) + (x + 3)(x - 4) = 0$;

      (ii) $(y - 1)(y - 4) + (x - \frac{1}{2})(x + \frac{3}{2}) = 0$;

      (iii) $y(y - a) + x(x - a) = 0$;

      (iv) $(y - y_1)(y - y_2) + (x - x_1)(x - x_2) = 0$.

    2. (i) $x^2 + y^2 = 36$, (ii) $4x^2 + y^2 = 64$.

    3. $x^2 + y^2 = 16$.   **4.** $xy = 3y + 4x$.

    5. $xy = 3$.   **6.** $y = 6x^2 + 1$.

    7. $y^2 = 8x + 4$.   **9.** $x^2 + y^2 = 4$.

    10. $4x^2 + 4y^2 - 8x + 3 = 0$.

    11. $2y = x^2 + x + 2$.   **12.** $2x + 3y - 13 = 0$.

page
359. **13.** $xy = 2x + 3y$.   **14.** $x^2 - 4xy + 5y^2 = 4$.

     15. $y^2 - xy - y + 2x = 0$. **16.** $2x + 3y - 13 = 0$.

     17. $x^2 + y^2 = 2$.   **18.** $x^2 + y^2 + 4ax = 0$.

     19. (i) $y^2 = 81 - 18x$, (ii) $y^2 = 1 - 2x$.

362. **Qu. 1.** (i) $y = x \pm \sqrt{7}$, (ii) $y = \sqrt{3}x \pm \sqrt{13}$.

     Qu. 2. (i) $(0, 0)$; (ii) $(0, 0)$, $(3, 6)$.

Exercise 17c

363. **1.**   (i) $4x - y - 4 = 0$, $x + 4y - 18 = 0$;

       (ii) $4x - y - 2 = 0$, $x + 4y - 9 = 0$;

       (iii) $y + 2 = 0$, $x + 1 = 0$;

       (iv) $x + y + 1 = 0$, $x - y - 3 = 0$;

       (v) $6x + y + 4 = 0$, $x - 6y + 50 = 0$;

       (vi) $x + y - 4 = 0$, $x - y = 0$;

       (vii) $2x - 3y + 1 = 0$, $3x + 2y - 5 = 0$.

     2. (i) $(\frac{1}{2}, 2)$; (ii) $(2, -2)$; (iii) $(6, \frac{2}{3})$; (iv) $(-\frac{5}{2}, -\frac{9}{4})$.

     3. $(1, 0)$, $(3, 0)$; $2x + y - 2 = 0$, $x - 2y - 1 = 0$;

       $2x - y - 6 = 0$, $x + 2y - 3 = 0$.

     4. $5x - y - 11 = 0$, $3x + y + 3 = 0$.

     5. $x + 2y - 1 = 0$, $x - 2y + 1 = 0$, $(0, \frac{1}{2})$.

     6. $(0, 0)$, $(1, 1)$; $x = 0$, $2y - x - 1 = 0$; $y = 0$,

       $y - 2x + 1 = 0$.

     7. $4y - x + 48 = 0$, $(48, 0)$.

     8. $9x - y - 27 = 0$, $9x - y + 5 = 0$.

     9. $x + y \pm 4 = 0$.

364. **11.** $(-\frac{1}{4}, -\frac{7}{4})$.   **12.** 0, 2; $y = 0$, $y - 4x + 4 = 0$.

     13. $x + 4y - 4c = 0$.

     14. (i) $(0, 0)$; (ii) $(1, 1)$, $(-1, -1)$; (iii) $(\frac{3}{5}, \frac{3}{5})$.

     16. $3x - 8y \pm 10 = 0$.   **17.** $x - y \pm 4 = 0$.

     18. $n^2 = a^2l^2 + b^2m^2$.

Exercise 17d

     1. $2x - 16y + 41 = 0$.   **2.** $y^2 = 8(x - 2)$.

365. **3.** $3x^2 + 4y^2 - 24x + 36 = 0$.

     4. $8x^2 - y^2 - 18ax + 9a^2 = 0$.

     5. $x^2 + y^2 + 4x = 0$.   **6.** $3x^2 + 3y^2 + 8x = 0$.

     7. $(x - a)(x - c) + (y - b)(y - d) = 0$.

page

365. 8. (i) $x^2 + y^2 = 144$, (ii) $x^2 + 9y^2 = 324$.

9. $x^2 + y^2 = 36$. 10. $xy = 4$.

11. $bx + ay = xy$. 12. $4x^2 + y^2 - 4x - 8 = 0$.

13. $7, 7x - y + 1 = 0, (-\frac{1}{6}, -\frac{1}{6})$.

14. $x + 2y = 12, (-6, 9)$.

366. 15. $x + y - 4 = 0, x + 9y - 12 = 0$.

16. $\pm 1, x - y + 1 = 0, 4x + 4y - 11 = 0$.

17. $9y - 27x = 19, (-\frac{2}{3}, \frac{1}{9})$.

18. $8x - 28y + 49 = 0$.

19. $y = x + 1 + 2/x, (-2, -2); x - 2y + 6 = 0,$
$x - 2y - 2 = 0$.

20. $2\frac{4}{5}, x + 3y - 7 = 0$.

21. $y = x^3 - x^2 - x + 2, 7x - y - 10 = 0,$
$x + 7y - 30 = 0$.

22. $6x + 12y \pm 5 = 0$.

CHAPTER 18

369. Qu. 1. $g = -a, f = -b, c = (a^2 + b^2 - r^2)$.

Exercise 18a

370. 1. (i) $x^2 + y^2 - 4x - 6y + 12 = 0$;

(ii) $x^2 + y^2 + 6x - 8y = 0$;

(iii) $9x^2 + 9y^2 - 12x + 6y + 1 = 0$;

371. (iv) $x^2 + y^2 + 10y = 0$;

(v) $x^2 + y^2 - 6x + 7 = 0$;

(vi) $144x^2 + 144y^2 + 72x - 96y - 47 = 0$.

2. (i) $1, (-2, 3)$; (ii) $2, (1, 2)$; (iii) $\frac{3}{2}, (\frac{3}{2}, 0)$;

(iv) $\frac{7}{2}, (-\frac{3}{2}, 2)$; (v) $\frac{1}{4}\sqrt{2}, (-\frac{1}{4}, -\frac{1}{4})$; (vi) $1, (\frac{1}{3}, \frac{1}{2})$;

(vii) $\sqrt{(a^2 + b^2)}, (a, b)$; (viii) $\sqrt{(g^2 + f^2 - c)},$
$(-g, -f)$.

3. (i), (iv) if $a > 0$, (vi) if $b = 0$, (vii) if $c < 0$.

4. $x^2 + y^2 - 4x - 2y - 15 = 0$.

5. $x^2 + y^2 + 2x - 4y - 20 = 0$.

6. $(5, 3), \sqrt{10}; x^2 + y^2 - 10x - 6y + 24 = 0$.

7. $x^2 + y^2 - 8x - 10y + 21 = 0$.

8. $x^2 + y^2 - 8x - 8y + 12 = 0$.

9. $x^2 + y^2 - 4x + 6y + 4 = 0$.

page

372. **10.** $x^2 + y^2 + \cdot 2x + 2y - 8 = 0$.   **11.** 4, 6.

12. The y-axis is a tangent.

13. $x^2 + y^2 \pm 8x - 10y + 16 = 0$.

14. $x^2 + y^2 - 4x - 4y + 4 = 0$.

15. $(2, 1)$, $x^2 + y^2 - 4x - 2y - 45 = 0$.

16. $x^2 + y^2 - 16x + 8y - 5 = 0$.

17. (i) $x^2 + y^2 + 4x - 2y = 0$,

(ii) $x^2 + y^2 - 10x - 8y + 28 = 0$,

(iii) $x^2 + y^2 - 2x - 49 = 0$.

18. $(4, 0)$, 2.   **19.** $(4, 1)$, 3.

20. $x^2 + y^2 - 4x - 6y + 9 = 0$.

375. **Qu. 2.** (i) 0, (ii) No real length.

Exercise 18b

1. (i) $3x - y = 0$;   (ii) $x - 4y + 17 = 0$;

(iii) $4w + y - 11 = 0$;   (iv) $3x + y - 8 = 0$;

(v) $4x + 9y + 5 = 0$.

2. (i) $\sqrt{10}$, (ii) $\sqrt{15}$, (iii) $\sqrt{29}$, (iv) $2\sqrt{7}$,

(v) $\sqrt{(x_1^2 + y_1^2 - a^2)}$, (vi) $\sqrt{c}$.

3. 5.   **4.** $x - y - 1 = 0$, $x + y - 5 = 0$.

5. $(23, 0)$, $(0, 7\frac{2}{3})$, $88\frac{1}{3}$.   **6.** $\sqrt{13}$.

376. **7.** 4.   **8.** $\sqrt{(X^2 + Y^2 - 4)}$, $2x - 5 = 0$.

9. $2x + 3y - 6 = 0$.

11. $(3, \frac{9}{4})$, $(-3, -\frac{9}{4})$;

$16x^2 + 16y^2 - 96x - 72y + 81 = 0$,

$16x^2 + 16y^2 + 96x + 72y + 81 = 0$.

12. $(7, 4)$.

377. **Qu. 3.** $X^2 + Y^2 - 1$, $X^2 + Y^2 - 6X - 8Y + 21$;

$3x + 4y - 11 = 0$.

Qu. 4. $x + y = 0$.

Exercise 18c

378. **2.** $(0, 0)$.   **4.** $(1, 2)$.   **5.** $(-2, 5)$.   **6.** $y = 2x$.

7. $x^2 + y^2 - 10x - 8y + 33 = 0$.

379. **8.** $4x - 3y - 18 = 0$, $13\frac{1}{2}$.   **9.** $2\sqrt{3}$.

10. $x^2 + y^2 = a^2$.   **11.** $(0, 0)$, $3x + y = 0$.

12. $x + y - 1 = 0$.   **13.** $2x - 5 = 0$.   **14.** 5.

512 ANSWERS

page

379. 15. $x^2 + y^2 - 10x - 4y + 4 = 0.$

16. $(3, 1), (7\frac{1}{2}, 2\frac{1}{2}); (5, 0).$ **17.** $x^2 + y^2 - 5x - y = 0.$

18. $(2, -3), (-11, -3); x^2 + y^2 - 4x + 6y = 0,$
$x^2 + y^2 + 22x + 6y + 117 = 0.$

380. 20. $3x^2 - y^2 - 12x + 9 = 0.$ **22.** $h^2 + k^2 = a^2.$

23. $y^2 = 4(5 - 2x).$

24. $x^2 + y^2 + 12x = 0; (-6, 0), 6.$

25. $a.$ **26.** $x^2 + y^2 + 4x - 9 = 0.$

CHAPTER 19

381. **Qu. 1.**   (i) $y - x = -1,$ (ii) $y + 2x = -1,$
(iii) $2y - x = -12,$ (iv) $3y + x = 13,$
(v) $5y + 7x = -9,$ (vi) $4y - 3x = 7,$

382.   (vii) $6y + 5x = -39,$
(viii) $3y - 4x = 23,$ (ix) $yt - x = at^2,$
(x) $y + tx = at^3 + 2at,$
(xi) $y \sin \theta + x \cos \theta = a,$
(xii) $t^2y + x = 2ct.$

Qu. 2.   (i) $2x - 3y = -2,$ (ii) $3x + 4y = 0,$
(iii) $6x - 5y = -43,$
(iv) $2x + 3y = 7,$
(v) $y + tx = k + th,$
(vi) $bx - ay = bx_1 - ay_1,$
(vii) $y - t^2x = c/t - ct^3,$

Qu. 3.   (i) $x/3 + y/2 = 1,$ (ii) $y/2 - x = 1,$
(iii) $2x + 5y = 1,$ (iv) $4y - 3x = 1.$

383. **Qu. 4.**   (i) $x/3 + y/2 = 1,$ (ii) $y/5 - x = 1,$
(iii) $3y/2 - 2x = 1.$

Qu. 5. $p \sec \alpha, p \csc \alpha, x \cos \alpha + y \sin \alpha = p.$

384. **Qu. 6.** (i) $\tan^{-1} \frac{1}{2}$, (ii) $\tan^{-1} \frac{1}{3}$, (iii) $\tan^{-1} \frac{17}{6}$, (iv) $\tan^{-1} \frac{1}{7}$,

(v) $\tan^{-1} \frac{19}{9}$, (vi) $\tan^{-1} \left| \dfrac{\sin \alpha - \cos \alpha}{\sin \alpha + \cos \alpha} \right|.$

385. **Qu. 7.** (i) 4, (ii) -4, (iii) 5, (iv) -2, (v) -9, (vi) 9,
(vii) -7, (viii) 2, (ix) $-11.$

Exercise 19a

388. 1. (i) $(5\frac{1}{2}, 6\frac{2}{3})$, (ii) $(1\frac{1}{5}, 2\frac{3}{5})$, (iii) $(1, -3\frac{1}{4})$, (iv) $(1, -3)$,
(v) $(-10, -15)$, (vi) $(29, -14)$,

(vii) $\left(\dfrac{5a + 6b}{8}, \dfrac{10a + 3b}{8}\right)$, (viii) $\left(\dfrac{p}{p+q}, \dfrac{3q + 5p}{p+q}\right)$,

(ix) $\left(\dfrac{m_2 x_2 + m_1 x_1}{m_1 + m_2}, \dfrac{m_2 y_2 + m_1 y_1}{m_1 + m_2}\right)$.

2. (i) $(4\frac{1}{3}, 3)$, (ii) $(3, -2)$, (iii) $(-1\frac{2}{3}, -1\frac{1}{3})$,
(iv) $\{\frac{1}{2}(a + c), \frac{1}{2}(b + d)\}$,
(v) $\{\frac{1}{3}(x_1 + x_2 + x_3), \frac{1}{3}(y_1 + y_2 + y_3)\}$.

3. (i) $5:3$, $3\frac{3}{4}$; (ii) $6, -19$; (iii) $3, 2:3$; (iv) $-7, 1$;
(v) $0, -6:1$; (vi) $2, 6$; (vii) $-4:5, -7$.

389. 4.   (i) $y - 3x + 9 = 0$;   (ii) $2y + x = 0$;
(iii) $5y - 2x - 3 = 0$; (iv) $4y + 3x + 12 = 0$;
 (v) $y - 6x + 16 = 0$; (vi) $4y - 9x - 3 = 0$;
(vii) $y = 2$;   (viii) $2y + x - 4 = 0$;
(ix) $3y - 4x - 13 = 0$;   (x) $6y + x - 19 = 0$;
(xi) $y + x - 1 = 0$;   (xii) $y - t^2 x = k - t^2 h$.

5. (i) $\tan^{-1}\frac{1}{3}$, (ii) $\tan^{-1} 2$, (iii) $\tan^{-1}\frac{7}{24}$, (iv) $\tan^{-1}\frac{12}{5}$,

(v) $\tan^{-1}\left|\dfrac{m_1 - m_2}{1 + m_1 m_2}\right|$,   (vi) $\tan^{-1}\left|\dfrac{a_1 b_2 - a_2 b_1}{a_1 a_2 + b_1 b_2}\right|$.

6. $m_1 m_2 = -1$.   7. $5y^2 = 8(x - 3)$.
8. $x^2 + y^2 - 16x + 60 = 0$.
9. $3x^2 - y - 6x + 5 = 0$.   10. $x^2 + y^2 + 6x = 0$.
11. $2, -\frac{1}{2}$.

390. 12. $-(\sqrt{2} + 1)$.
13. (i) $\tan^{-1} 2$, (ii) $\tan^{-1}\frac{7}{24}$, (iii) $\tan^{-1}\frac{8}{11}$,
(iv) $\tan^{-1} 2$, $\tan^{-1}\frac{3}{19}$, $\tan^{-1}\frac{6}{43}$.
16. $x^2 + y^2 - 6x \pm 2y + 8 = 0$.

Exercise 19b

395. 2. (i) $r = a$, (ii) $\theta = \alpha$, (iii) $r = a \sec \theta$,
(iv) $r = a \operatorname{cosec} \theta$, (v) $r = a \cos \theta$, (vi) $r = 2a \sin \theta$,
(vii) $a^2 = r^2 + c^2 - 2cr \cos \theta$,
(viii) $r = 2a/(1 + \cos \theta)$.

page
396. **5.** (i) $r = a$, (ii) $r^2 = a^2 \sec 2\theta$, (iii) $\theta = 0$,
(iv) $r = 2a/(1 + \cos \theta)$, (v) $r = 2 \sin \theta$,
(vi) $r^2 = 2c^2 \operatorname{cosec} 2\theta$.

6. (i) $x^2 + y^2 = 4$, (ii) $(x^2 + y^2 - ax)^2 = a^2(x^2 + y^2)$,
(iii) $x^2 + y^2 - ax = 0$, (iv) $x^4 + x^2y^2 = a^2y^2$,
(v) $(x^2 + y^2)^3 = 4a^2(x + y)^4$, (vi) $4xy = c^2$,
(vii) $x^2 + y^2 = (l - ex)^2$, (viii) $y^2 = 4ax$.

7. (i) 1, $60°$; (ii) $2\sqrt{2}$, $-45°$; (iii) 2, $\tan^{-1}\frac{4}{3}$;
(iv) 2, $\tan^{-1}(-\frac{12}{5})$; (v) $\frac{1}{5}\sqrt{10}$, $\tan^{-1}3$;
(vi) $c/\sqrt{(a^2 + b^2)}$, $\tan^{-1}(b/a)$.

399. **Qu. 8.** (i) 1, (ii) $\frac{9}{13}$, (iii) $\frac{1}{2}\sqrt{26}$, (iv) $3\sqrt{2}$, (v) $\frac{16}{17}\sqrt{34}$,
(vi) $\frac{2}{13}\sqrt{13}$, (vii) $\frac{3}{5}a$, (viii) $\frac{4}{5}q$,
(ix) $\frac{1}{13}(12X - 5Y + 7)$, (x) $\frac{1}{17}(8x_1 - 15y_1)$.

Exercise 19c

401. **1.** (i) $4\frac{1}{5}$, (ii) $2\frac{1}{13}$, (iii) $\frac{10}{17}\sqrt{17}$, (iv) 0, (v) $\frac{38}{29}\sqrt{29}$,
(vi) $1\frac{1}{5}$, (vii) $\frac{6}{41}\sqrt{41}$, (viii) $\frac{5}{13}\sqrt{13}$, (ix) p,
(x) $\frac{1}{13}(5X - 12Y + 1)$, (xi) $\frac{2}{17}c$,
(xii) $\frac{1}{5}(4y_1 - 3x_1 + 2)$.

2. (i) $3x - y - 2 = 0$, $x + 3y - 4 = 0$;
(ii) $7x - 7y + 4 = 0$, $x + y - 2 = 0$;
(iii) $17x + 17y - 4 = 0$, $7x - 7y - 4 = 0$;
(iv) $x + (1 \pm \sqrt{2})y - 1 = 0$.

3. (i) $8x - 4y + 17 = 0$, (ii) $8y + 1 = 0$,
(iii) $4x + 12y + 5 = 0$.

402. **4.** $4x^2 - 4xy + y^2 - 20x + 30y + 65 = 0$.

5. $4x + 3y - 24 = 0$. **6.** $x^2 - 2xy + 3y^2 = 2$.

7. $7x^2 - 2xy + 7y^2 - 40x - 40y + 48 = 0$.

8. $y - 2x + 6 = 0$, $y - 2x - 4 = 0$.

9. $4y - 3x - 15 = 0$, $4y - 3x + 35 = 0$.

10. $8x + 6y - 43 = 0$. **11.** $2x - 3y \pm 13 = 0$.

12. $x^2 + y^2 - 4x - 14y + 49 = 0$.

13. $x^2 + y^2 - 60x - 40y + 1040 = 0$.

14. $2x - y - 18 = 0$. **15.** $n^2 = a^2(l^2 + m^2)$.

403. **16.** $x^2 + y^2 - 12x - 6y + 36 = 0$.

Exercise 19d

page
407. 2. (i) $\frac{4}{3}$, $\frac{4}{3}$; (ii) $\pm \frac{3}{2}$, $\pm 3a$.

408.   (iii) $- 2$, $- \frac{1}{3}$; (iv) $60°$, $(\sqrt{3}/2)b$.

3. (i) $(y - 2)^2 = x - 1$,   (ii) $x^3 = y^2$,
   (iii) $xy = 1$,   (iv) $2x + y - 5 = 0$,
   (v) $y^2 = 4ax$,   (vi) $xy^2 = 1$,
   (vii) $5x + y - 13 = 0$,   (viii) $4x^2 - 9y^2 = 144$,
   (ix) $4x^2 + 9y^2 = 36$,   (x) $y = 1 - 2x^2$,
   (xi) $9x^2 - 16y^2 = 144$,   (xii) $x = 4y^3 - 3y$.

4. (i) $x = t^4$, $y = t^5$;
   (ii) $x = t - 2$, $y = t^2 - 2t$;

   (iii) $x = \dfrac{2}{t^2 - 1}$, $y = \dfrac{2t}{t^2 - 1}$;

   (iv) $x = \dfrac{1}{1 - t^3}$, $y = \dfrac{t}{1 - t^3}$;

   (v) $x = \dfrac{3t}{1 + t^3}$, $y = \dfrac{3t^2}{1 + t^3}$;

   (vi) $x = \dfrac{1}{t - 2}$, $y = \dfrac{t}{t - 2}$;

   (vii) $x = \dfrac{1}{1 + t^2}$, $y = \dfrac{t}{1 + t^2}$;

   (viii) $x = \dfrac{1 - t}{t}$, $y = 1 - t$.

5. $3x - 2y + 1 = 0$.   6. $- \frac{4}{13}$, $- \frac{4}{3}$.

7. $(\frac{1}{2}t^2, \frac{3}{2}t)$, $2y^2 = 9x$.   8. $y^2 = 8ax$.

9. $x = y(2x - 1)^2$.   10. $(1, 1)$, $(- 1, - 1)$, $\sqrt{2}$.

11. $(a, 2a)$, inflexion; $(4a, -4a)$, minimum.

409. 12. (i) $(p + q)y - 2x = 2pq$,
   (ii) $pqy - x + (p + q) = 0$,
   (iii) $(p^2 + pq + q^2)y - x = pq(p + q)$,
   (iv) $(pq - 1)y - 2pqx + 2(p + q) = 0$.

13. $(a, 2a)$, $\frac{3}{2}\sqrt{2}a$.   14. $- 1, 2$; $(1, - 2)$, $(4, 4)$.

15. $(1, \pm 2)$, $(4, \pm 4)$.   16. $c^2t^4 - a^2t^2 + c^2 = 0$, $\sqrt{2}a$.

Exercise 19e

410. **1.**　(i) $2y + 3x - 1 = 0,\ 2x - 3y - 5 = 0$;

(ii) $16y + 4x - 33 = 0,\ 4x - y + 1 = 0$;

(iii) $x + y + a = 0,\ x - y - 3a = 0$;

(iv) $y + x + 2c = 0,\ y - x = 0$;

(v) $2y + x + 9 = 0,\ 2x - y + 3 = 0$;

(vi) $3\sqrt{3}y + 2x - 12 = 0,\ 6\sqrt{3}x - 4y - 5\sqrt{3} = 0$.

2.　(i) $ty - 2x - t^3 = 0,\ 2y + tx = 6t^2 + t^4$;

(ii) $ty - x - at^2 = 0,\ y + tx = 2at + at^3$;

(iii) $y - tx + t^4 = 0,\ ty + x = 3t^5 + 4t^3$;

(iv) $t^2y + x - 2ct = 0,\ y - t^2x = c/t - ct^3$;

(v) $bx \cos t + ay \sin t = ab$,

$ax \sin t - by \cos t = \frac{1}{2}(a^2 - b^2) \sin 2t$;

(vi) $bx \sec t - ay \tan t = ab$,

$ax \sin t + by = (a^2 + b^2) \tan t$.

3.　(i) $(p + q)y - 2x = 2pq,\ py - x = p^2$;

(ii) $y + pq(p + q)x = p^3 + pq + q^3$,

$y + 2p^3x = 3p^2$;

(iii) $pqy + x = c(p + q),\ p^2y + x = 2cp$;

(iv) $bx \cos \frac{1}{2}(p + q) + ay \sin \frac{1}{2}(p + q)$

$= ab \cos \frac{1}{2}(p - q),\ bx \cos p + ay \sin p = ab$.

4. $2x + y - 12a = 0,\ (9a, -6a)$.　**5.** $(-\frac{1}{8}c, -8c)$.

6. $(-\frac{1}{2}, 4)$.

7. $yt - x = at^2$; $2, 4$; $2y - x = 4a,\ 4y - x = 16a$.

8. $y + x = 2c,\ 9y + x = 6c$.

9. $y + 2x = 12a,\ y - 4x + 72a = 0$.

10. $(-c/t^3, -ct^3)$.

Exercise 19f

412.　**4.** $yt - x = at^2,\ y + tx = 2at + at^3$.

414.　**17.** $y^2 = a(x - 3a)$.

19. $2, -1$; $y + 2x = 12a,\ y - x + 3a = 0$.

20. $y - x - a = 0,\ 4y - x - 16a = 0$.

Exercise 19g

1. $(3\frac{1}{3}, 1)$.

415.　**2.** $(4\frac{2}{3}, 3\frac{1}{12})$.　**3.** $25xy - 10y = 9$.　**4.** $\tan^{-1} \frac{2 \cdot 0}{2}$.

5. $0,\ \tan^{-1} \frac{8}{6 \cdot 3}$.　**9.** $r = 2a/(1 - \cos \theta)$.　**10.** $6\frac{1}{4}$.

page
415. 11. $r = a(1 - \cos\theta)$, $x^2 + 2ay = a^2$.

12. $\frac{7}{25}x - \frac{24}{25}y = \frac{2}{5}, \frac{2}{5}$.

13. (i) $2x - 8y + 7 = 0$, $4x + y + 4 = 0$;
    (ii) $12x + 4y - 13 = 0$, $2x - 6y + 7 = 0$.

14. $x^2 - 2xy + y^2 + 8x + 8y = 0$.

15. $x^2 + y^2 - 2x - 2y - 6 = 0$.

16. $x - 3y + 5 = 0$, $x - 3y + 25 = 0$.

18. (i) $x^4 = ay^3$; (ii) $y^2 = 4x^2(1 - x^2)$.

19. $2xy = x + 1$.

21. $x - 2y + 2 = 0$, $2x + y - 11 = 0$.

22. $2t^3y + x = 3t^2$, $2xy^2 = 1$.

24. $x - y + a = 0$, $x - 5y + 25a = 0$.

25. $(\frac{121}{9}a, -\frac{22}{3}a)$.

CHAPTER 20

Note: Approximate answers have generally been rounded to 2 or 3 significant figures. The reader should not assume from the form of an answer that the result is exact.

page
420. Qu. 1. (i) s varies as the square of t,
    (ii) V varies as the cube of r,
    (iii) y varies inversely as the square of x,
    (iv) T varies as the square root of l,
    (v) p varies inversely as v,
    (vi) the square of T varies as the cube of d.

421. Qu. 2. W is increased by a factor of (i) 8, (ii) 27.

423. Qu. 3.

(a) (i) $p = kq$, (ii) $p = \dfrac{k}{v}$, (iii) $v = kx^2$,

(iv) $U = k\sqrt{l}$, (v) $F = kc^2$, (vi) $H = \dfrac{l_0}{d^2}$,

(vii) $T = \dfrac{l_0}{\sqrt{g}}$, (viii) $A = ks^n$, (ix) $A^3 = kv^2$.

(b) (i) $\dfrac{p_1}{p_2} = \dfrac{q_1}{q_2}$, (ii) $\dfrac{p_1}{p_2} = \dfrac{v_2}{v_1}$, (iii) $\dfrac{v_1}{v_2} = \dfrac{x_1^2}{x_2^3}$.

423.   (iv) $\dfrac{U_1}{U_2} = \dfrac{\sqrt{l_1}}{\sqrt{l_2}}$,   (v) $\dfrac{F_1}{F_2} = \dfrac{c_1{}^2}{c_2{}^2}$,   (vi) $\dfrac{H_1}{H_2} = \dfrac{d_2{}^2}{d_1{}^2}$,

     (vii) $\dfrac{T_1}{T_2} = \dfrac{\sqrt{g_2}}{\sqrt{g_1}}$,   (viii) $\dfrac{A_1}{A_2} = \dfrac{s_1{}^n}{s_2{}^n}$,   (ix) $\dfrac{A_1{}^3}{A_2{}^3} = \dfrac{v_1{}^2}{v_2{}^2}$.

425.   **Qu. 4.** l is increased by a factor of (i) 4, (ii) 9.
        T is increased by a factor of $\sqrt{2}$.

     Qu. 5. 1·1 s.

426.   **Qu. 6.** $w = \dfrac{3 \cdot 21 \times 10^{10}}{d^2}$.

     Qu. 7. w is multiplied by (1) $\frac{1}{4}$, (ii) $\frac{1}{9}$.

     Qu. 8. (i) c varies as p,
            (ii) C varies as a^2 over a limited range,
            (iii) w varies as r^3,
            (iv) l varies inversely as b,
            (v) S varies as l^2,
            (vi) A varies as a^2,
            (vii) a varies as $\sqrt{A}$,
            (viii) V varies as a^3,
            (ix) a varies as $\sqrt[3]{V}$.

Exercise 20a

427.   **1.** 5·70 cm².   **2.** 29·9 km, $d \simeq 3 \cdot 57 \sqrt{h}$.
     3. 155 cm, $l \simeq 24 \cdot 8(5) \, T^2$.
     4. 25 m.   **5.** 0·242 kg, 11·0 cm, $m = \frac{1}{810} d^2$.

428.   **6.** (i) $C = 200\pi r$, yes, (ii) $C = 5 \cdot 08\pi r$, yes.
     7. $2 \cdot 8 \times 10^5$ N/m².   **8.** 3584.
     9. 15·6 cm³, 11·2 cm.   **10.** 10·4 k, $v = \frac{8}{5} \sqrt{(5l)}$.
     11. 3168, 3·3 mm.

429.   **12.** (i) y varies as t^6, (ii) p varies inversely as r^2.
     13. $F = \frac{175}{54} v^2$.   (i) 2360 N, (ii) 43·6 km/h .
     14. $H \simeq 0 \cdot 000\,182 v^3$, 4·92 kW.
     15. Increases approx. 7% in speed and 15% in acceleration.
     16. 1·59.   **17.** Increase $\simeq 0 \cdot 05\%$.

page

430. 19. 1 h 37 min, $T \simeq 2\cdot87 \times 10^{-6}d^{3/2}$.
    20. 275:1472.

432. **Qu. 9.** $y = kxz^3$.   **Qu. 10.** $W \propto \dfrac{hr^2}{t}$.

    Qu. 11. (i) $T = kmr^2$, (ii) $T \propto mr^2$.

    Qu. 12. $F = k\dfrac{mv^2}{r}$.

433. **Qu. 13.** (i) $\dfrac{z_1}{z_2} = \dfrac{x_1 y_1{}^2}{x_2 y_2{}^2}$,   (ii) $\dfrac{z_1}{z_2} = \dfrac{y_1 x_2{}^2}{y_2 x_1{}^2}$,

     (iii) $\dfrac{z_1}{z_2} = \dfrac{x_1{}^3 y_1{}^2}{x_2{}^3 y_2{}^2}$,   (iv) $\dfrac{z_1}{z_2} = \dfrac{x_1 y_1}{x_2 y_2}$,

     (v) $\dfrac{z_1}{z_2} = \dfrac{x_1{}^2 y_1{}^2}{x_2{}^2 y_2{}^2}$,   (vi) $\dfrac{z_1}{z_2} = \dfrac{y_2 \sqrt{x_1}}{y_1 \sqrt{x_2}}$.

436. **Qu. 14.** (i) $C = K + kx^3$, k, K constants;
            (ii) £11·95.

Exercise 20b

    1. 15·7 cm².

437. **2.** 454·5 rev/min.   **3.** 18 litres/s.   **4.** 675 kJ.

    5. 185 cm³, $V \simeq 51\cdot6\dfrac{T}{p}$.   **6.** 0·70 kW, $0\cdot001\dfrac{V^2}{R}$.

    7. 570 s⁻¹, $f \simeq 14\cdot2\dfrac{\sqrt{F}}{l}$.

438. **8.** 76 s⁻¹.   **9.** 4·91 s.
    10. The former. Ratio 16:15.
    11. The latter. Ratio 27:32.

439. **12.** The former. Ratio 16:15.
    13. £230. **14.** £1·95, $C = 1\cdot35 + 0\cdot003n$.
    15. 88 m, $s = 10t + 3t^2$.
    16. 54·4 m, $s = 0\cdot2v + 0\cdot006v^2$, 50 km/h.

page

439. 17. $C = 0.05 + 0.04$m.

440. 18. 70p, 125.   **19.** $28\frac{1}{2}$ cm², $S = 2x^2 + \dfrac{40}{x}$.

20. $V = \pi rh^2 - \frac{1}{3}\pi h^3$.   **21.** $\frac{1}{4}n^4 + \frac{1}{2}n^3 + \frac{1}{4}n^2$.

442. Qu. **15.** $v = 4.2t$.

444. Qu. **16.** The cost in labour and materials before any copies are run off is £2·10.

445. Qu. **17.** (i) $y = -3x + 121.8$,
(ii) $y = 13.3x - 68.2$,
(iii) $y = -11.9x + 211.5$.
Qu. **18.** $y = 12x - 72$; 6 cm.

446. Qu. **19.** $R = 0.0005v^3$.

447. Qu. **20.** $k = 0.49$.

453. Qu. **22.** $k = 2070$, $a = 1.05$.   3700.

Exercise 20c

454. 1. $A = 1.84D$. Yes.
2. $m = 1.02t$; relative density = gradient $\times \frac{100}{9}$.
3. $h_2 = 0.34h_1$.

455. 4. $d = 7 + 6.8n$, taking n to be 0, $7\frac{1}{2}$, $19\frac{1}{2}$.
5. (i) £390, (ii) £87.   **6.** 11 km/litre, about 17 litres.

456. 7. $\theta = 62 - \frac{1}{2}t$. No: cooler bodies lose heat more slowly.

8. Yes. About 219 litres. **9.** Yes. $l = \dfrac{15}{r}$.

10. $m = 0.338d^2$.

457. 11. $y = 3.50 - 0.025x^2$.

12. $p = 10.6 - \dfrac{9.6}{i}$; $i = \dfrac{9.6}{10.6 - p}$.

$$\left(\text{Theoretically, } i = \dfrac{10}{11 - p}.\right)$$

13. $P = 0.199s^{1.50}$.

page

458. 14. $p \propto \dfrac{1}{\sqrt{n}}$.

15. $I = 1 \cdot 6 r^2$.

16. $w = 3 \cdot 94(0 \cdot 846)^n$.

17. $0 \cdot 0071$.

459. 18. $1 \cdot 28$.

19. $0 \cdot 005$,

460. 20. $T = 100(1 \cdot 008)^{-x}$. No.

21. $0 \cdot 84$, 142.

461. 22. $P = 15900(1 \cdot 13)^x$ thousands; $1 \cdot 012$, or $1 \cdot 2\%$ per annum.